Artificial Intelligence-based Smart Power Systems

Artificial Intelligence-based Smart Power Systems

Edited by

Sanjeevikumar Padmanaban
Department of Electrical Engineering, Information Technology, and Cybernetics,
University of South-Eastern Norway, Porsgrunn, Norway

Sivaraman Palanisamy
World Resources Institute (WRI) India, Bengaluru, India

Sharmeela Chenniappan
Department of Electrical and Electronics Engineering, Anna University, Chennai, India

Jens Bo Holm-Nielsen
Department of Energy Technology, Aalborg University, Aalborg, Denmark

IEEE PRESS
WILEY

Published by John Wiley & Sons, Inc., Hoboken, New Jersey.
Published simultaneously in Canada.

For general information on our other products and services or for technical support, please contact our Customer Care Department within the United States at (800) 762-2974, outside the United States at (317) 572-3993 or fax (317) 572-4002.

Wiley also publishes its books in a variety of electronic formats. Some content that appears in print may not be available in electronic formats. For more information about Wiley products, visit our web site at www.wiley.com.

Library of Congress Cataloging-in-Publication Data Applied for:

Hardback ISBN: 9781119893967

Cover Design: Wiley
Cover Image: © Ekaterina Goncharova/Getty Images

Set in 9.5/12.5pt STIXTwoText by Straive, Chennai, India

Contents

Editor Biography *xv*
List of Contributors *xvii*

1   Introduction to Smart Power Systems *1*
Sivaraman Palanisamy, Zahira Rahiman, and Sharmeela Chenniappan
1.1   Problems in Conventional Power Systems *1*
1.2   Distributed Generation (DG) *1*
1.3   Wide Area Monitoring and Control *2*
1.4   Automatic Metering Infrastructure *4*
1.5   Phasor Measurement Unit *6*
1.6   Power Quality Conditioners *8*
1.7   Energy Storage Systems *8*
1.8   Smart Distribution Systems *9*
1.9   Electric Vehicle Charging Infrastructure *10*
1.10   Cyber Security *11*
1.11   Conclusion *11*
       References *11*

2   Modeling and Analysis of Smart Power System *15*
Madhu Palati, Sagar Singh Prathap, and Nagesh Halasahalli Nagaraju
2.1   Introduction *15*
2.2   Modeling of Equipment's for Steady-State Analysis *16*
2.2.1   Load Flow Analysis *16*
2.2.1.1   Gauss Seidel Method *18*
2.2.1.2   Newton Raphson Method *18*
2.2.1.3   Decoupled Load Flow Method *18*
2.2.2   Short Circuit Analysis *19*
2.2.2.1   Symmetrical Faults *19*
2.2.2.2   Unsymmetrical Faults *20*
2.2.3   Harmonic Analysis *20*
2.3   Modeling of Equipments for Dynamic and Stability Analysis *22*
2.4   Dynamic Analysis *24*
2.4.1   Frequency Control *24*
2.4.2   Fault Ride Through *26*
2.5   Voltage Stability *26*
2.6   Case Studies *27*
2.6.1   Case Study 1 *27*
2.6.2   Case Study 2 *28*
2.6.2.1   Existing and Proposed Generation Details in the Vicinity of Wind Farm *29*

2.6.2.2   Power Evacuation Study for 50 MW Generation   *30*
2.6.2.3   Without Interconnection of the Proposed 50 MW Generation from Wind Farm on 66 kV Level of 220/66 kV Substation   *31*
2.6.2.4   Observations Made from Table 2.6   *31*
2.6.2.5   With the Interconnection of Proposed 50 MW Generation from Wind Farm on 66 kV level of 220/66 kV Substation   *31*
2.6.2.6   Normal Condition without Considering Contingency   *32*
2.6.2.7   Contingency Analysis   *32*
2.6.2.8   With the Interconnection of Proposed 60 MW Generation from Wind Farm on 66 kV Level of 220/66 kV Substation   *33*
2.7   Conclusion   *34*
         References   *34*

3   Multilevel Cascaded Boost Converter Fed Multilevel Inverter for Renewable Energy Applications   *37*
         Marimuthu Marikannu, Vijayalakshmi Subramanian, Paranthagan Balasubramanian, Jayakumar Narayanasamy, Nisha C. Rani, and Devi Vigneshwari Balasubramanian
3.1   Introduction   *37*
3.2   Multilevel Cascaded Boost Converter   *40*
3.3   Modes of Operation of MCBC   *42*
3.3.1   Mode-1 Switch S_A Is ON   *42*
3.3.2   Mode-2 Switch S_A Is ON   *42*
3.3.3   Mode-3-Operation – Switch S_A Is ON   *42*
3.3.4   Mode-4-Operation – Switch S_A Is ON   *42*
3.3.5   Mode-5-Operation – Switch S_A Is ON   *42*
3.3.6   Mode-6-Operation – Switch S_A Is OFF   *42*
3.3.7   Mode-7-Operation – Switch S_A Is OFF   *42*
3.3.8   Mode-8-Operation – Switch S_A Is OFF   *43*
3.3.9   Mode-9-Operation – Switch S_A Is OFF   *44*
3.3.10   Mode 10-Operation – Switch S_A is OFF   *45*
3.4   Simulation and Hardware Results   *45*
3.5   Prominent Structures of Estimated DC–DC Converter with Prevailing Converter   *49*
3.5.1   Voltage Gain and Power Handling Capability   *49*
3.5.2   Voltage Stress   *49*
3.5.3   Switch Count and Geometric Structure   *49*
3.5.4   Current Stress   *52*
3.5.5   Duty Cycle Versus Voltage Gain   *52*
3.5.6   Number of Levels in the Planned Converter   *52*
3.6   Power Electronic Converters for Renewable Energy Sources (Applications of MLCB)   *54*
3.6.1   MCBC Connected with PV Panel   *54*
3.6.2   Output Response of PV Fed MCBC   *54*
3.6.3   H-Bridge Inverter   *54*
3.7   Modes of Operation   *55*
3.7.1   Mode 1   *55*
3.7.2   Mode 2   *55*
3.7.3   Mode 3   *56*
3.7.4   Mode 4   *56*
3.7.5   Mode 5   *56*
3.7.6   Mode 6   *56*
3.7.7   Mode 7   *58*
3.7.8   Mode 8   *58*
3.7.9   Mode 9   *59*
3.7.10   Mode 10   *59*

3.8   Simulation Results of MCBC Fed Inverter   *60*
3.9   Power Electronic Converter for E-Vehicles   *61*
3.10   Power Electronic Converter for HVDC/Facts   *62*
3.11   Conclusion   *63*
       References   *63*

4   **Recent Advancements in Power Electronics for Modern Power Systems-Comprehensive Review**
       on DC-Link Capacitors Concerning Power Density Maximization in Power Converters   *65*
       Naveenkumar Marati, Shariq Ahammed, Kathirvel Karuppazaghi, Balraj Vaithilingam, Gyan R. Biswal,
       Phaneendra B. Bobba, Sanjeevikumar Padmanaban, and Sharmeela Chenniappan
4.1   Introduction   *65*
4.2   Applications of Power Electronic Converters   *66*
4.2.1 Power Electronic Converters in Electric Vehicle Ecosystem   *66*
4.2.2 Power Electronic Converters in Renewable Energy Resources   *67*
4.3   Classification of DC-Link Topologies   *68*
4.4   Briefing on DC-Link Topologies   *69*
4.4.1 Passive Capacitive DC Link   *69*
4.4.1.1 Filter Type Passive Capacitive DC Links   *70*
4.4.1.2 Filter Type Passive Capacitive DC Links with Control   *72*
4.4.1.3 Interleaved Type Passive Capacitive DC Links   *74*
4.4.2 Active Balancing in Capacitive DC Link   *75*
4.4.2.1 Separate Auxiliary Active Capacitive DC Links   *76*
4.4.2.2 Integrated Auxiliary Active Capacitive DC Links   *78*
4.5   Comparison on DC-Link Topologies   *82*
4.5.1 Comparison of Passive Capacitive DC Links   *82*
4.5.2 Comparison of Active Capacitive DC Links   *83*
4.5.3 Comparison of DC Link Based on Power Density, Efficiency, and Ripple Attenuation   *86*
4.6   Future and Research Gaps in DC-Link Topologies with Balancing Techniques   *94*
4.7   Conclusion   *95*
       References   *95*

5   **Energy Storage Systems for Smart Power Systems**   *99*
       Sivaraman Palanisamy, Logeshkumar Shanmugasundaram, and Sharmeela Chenniappan
5.1   Introduction   *99*
5.2   Energy Storage System for Low Voltage Distribution System   *100*
5.3   Energy Storage System Connected to Medium and High Voltage   *101*
5.4   Energy Storage System for Renewable Power Plants   *104*
5.4.1 Renewable Power Evacuation Curtailment   *106*
5.5   Types of Energy Storage Systems   *109*
5.5.1 Battery Energy Storage System   *109*
5.5.2 Thermal Energy Storage System   *110*
5.5.3 Mechanical Energy Storage System   *110*
5.5.4 Pumped Hydro   *110*
5.5.5 Hydrogen Storage   *110*
5.6   Energy Storage Systems for Other Applications   *111*
5.6.1 Shift in Energy Time   *111*
5.6.2 Voltage Support   *111*
5.6.3 Frequency Regulation (Primary, Secondary, and Tertiary)   *112*
5.6.4 Congestion Management   *112*
5.6.5 Black Start   *112*
5.7   Conclusion   *112*
       References   *113*

6   **Real-Time Implementation and Performance Analysis of Supercapacitor for Energy Storage**   *115*
Thamatapu Eswararao, Sundaram Elango, Umashankar Subramanian, Krishnamohan Tatikonda, Garika Gantaiahswamy, and Sharmeela Chenniappan
6.1   Introduction   *115*
6.2   Structure of Supercapacitor   *117*
6.2.1   Mathematical Modeling of Supercapacitor   *117*
6.3   Bidirectional Buck–Boost Converter   *118*
6.3.1   FPGA Controller   *119*
6.4   Experimental Results   *120*
6.5   Conclusion   *123*
      References   *125*

7   **Adaptive Fuzzy Logic Controller for MPPT Control in PMSG Wind Turbine Generator**   *129*
Rania Moutchou, Ahmed Abbou, Bouazza Jabri, Salah E. Rhaili, and Khalid Chigane
7.1   Introduction   *129*
7.2   Proposed MPPT Control Algorithm   *130*
7.3   Wind Energy Conversion System   *131*
7.3.1   Wind Turbine Characteristics   *131*
7.3.2   Model of PMSG   *132*
7.4   Fuzzy Logic Command for the MPPT of the PMSG   *133*
7.4.1   Fuzzification   *134*
7.4.2   Fuzzy Logic Rules   *134*
7.4.3   Defuzzification   *134*
7.5   Results and Discussions   *135*
7.6   Conclusion   *139*
      References   *139*

8   **A Novel Nearest Neighbor Searching-Based Fault Distance Location Method for HVDC Transmission Lines**   *141*
Aleena Swetapadma, Shobha Agarwal, Satarupa Chakrabarti, and Soham Chakrabarti
8.1   Introduction   *141*
8.2   Nearest Neighbor Searching   *142*
8.3   Proposed Method   *144*
8.3.1   Power System Network Under Study   *144*
8.3.2   Proposed Fault Location Method   *145*
8.4   Results   *146*
8.4.1   Performance Varying Nearest Neighbor   *147*
8.4.2   Performance Varying Distance Matrices   *147*
8.4.3   Near Boundary Faults   *148*
8.4.4   Far Boundary Faults   *149*
8.4.5   Performance During High Resistance Faults   *149*
8.4.6   Single Pole to Ground Faults   *150*
8.4.7   Performance During Double Pole to Ground Faults   *151*
8.4.8   Performance During Pole to Pole Faults   *151*
8.4.9   Error Analysis   *152*
8.4.10   Comparison with Other Schemes   *153*
8.4.11   Advantages of the Scheme   *154*
8.5   Conclusion   *154*
      Acknowledgment   *154*
      References   *154*

9 **Comparative Analysis of Machine Learning Approaches in Enhancing Power System Stability** *157*
Md. I. H. Pathan, Mohammad S. Shahriar, Mohammad M. Rahman, Md. Sanwar Hossain, Nadia Awatif, and Md. Shafiullah
9.1 Introduction *157*
9.2 Power System Models *159*
9.2.1 PSS Integrated Single Machine Infinite Bus Power Network *159*
9.2.2 PSS-UPFC Integrated Single Machine Infinite Bus Power Network *160*
9.3 Methods *161*
9.3.1 Group Method Data Handling Model *161*
9.3.2 Extreme Learning Machine Model *162*
9.3.3 Neurogenetic Model *162*
9.3.4 Multigene Genetic Programming Model *163*
9.4 Data Preparation and Model Development *165*
9.4.1 Data Production and Processing *165*
9.4.2 Machine Learning Model Development *165*
9.5 Results and Discussions *166*
9.5.1 Eigenvalues and Minimum Damping Ratio Comparison *166*
9.5.2 Time-Domain Simulation Results Comparison *170*
9.5.2.1 Rotor Angle Variation Under Disturbance *170*
9.5.2.2 Rotor Angular Frequency Variation Under Disturbance *171*
9.5.2.3 DC-Link Voltage Variation Under Disturbance *173*
9.6 Conclusions *173*
References *174*

10 **Augmentation of PV-Wind Hybrid Technology with Adroit Neural Network, ANFIS, and PI Controllers Indeed Precocious DVR System** *179*
Jyoti Shukla, Basanta K. Panigrahi, and Monika Vardia
10.1 Introduction *179*
10.2 PV-Wind Hybrid Power Generation Configuration *180*
10.3 Proposed Systems Topologies *181*
10.3.1 Structure of PV System *181*
10.3.2 The MPPTs Technique *183*
10.3.3 NN Predictive Controller Technique *183*
10.3.4 ANFIS Technique *184*
10.3.5 Training Data *186*
10.4 Wind Power Generation Plant *187*
10.5 Pitch Angle Control Techniques *189*
10.5.1 PI Controller *189*
10.5.2 NARMA-L2 Controller *190*
10.5.3 Fuzzy Logic Controller Technique *192*
10.6 Proposed DVRs Topology *192*
10.7 Proposed Controlling Technique of DVR *193*
10.7.1 ANFIS and PI Controlling Technique *193*
10.8 Results of the Proposed Topologies *196*
10.8.1 PV System Outputs (MPPT Techniques Results) *196*
10.8.2 Main PV System outputs *196*
10.8.3 Wind Turbine System Outputs (Pitch Angle Control Technique Result) *198*
10.8.4 Proposed PMSG Wind Turbine System Output *199*
10.8.5 Performance of DVR (Controlling Technique Results) *203*
10.8.6 DVRs Performance *203*
10.9 Conclusion *204*
References *204*

11   **Deep Reinforcement Learning and Energy Price Prediction** *207*
        Deepak Yadav, Saad Mekhilef, Brijesh Singh, and Muhyaddin Rawa
        Abbreviations *207*
11.1   Introduction *208*
11.2   Deep and Reinforcement Learning for Decision-Making Problems in Smart Power Systems *210*
11.2.1   Reinforcement Learning *210*
11.2.1.1   Markov Decision Process (MDP) *210*
11.2.1.2   Value Function and Optimal Policy *211*
11.2.2   Reinforcement Learnings to Deep Reinforcement Learnings *212*
11.2.3   Deep Reinforcement Learning Algorithms *212*
11.3   Applications in Power Systems *213*
11.3.1   Energy Management *213*
11.3.2   Power Systems' Demand Response (DR) *215*
11.3.3   Electricity Market *216*
11.3.4   Operations and Controls *217*
11.4   Mathematical Formulation of Objective Function *218*
11.4.1   Locational Marginal Prices (LMPs) Representation *219*
11.4.2   Relative Strength Index (RSI) *219*
11.4.2.1   Autoregressive Integrated Moving Average (ARIMA) *219*
11.5   Interior-point Technique & KKT Condition *220*
11.5.1   Explanation of Karush–Kuhn–Tucker Conditions *220*
11.5.2   Algorithm for Finding a Solution *221*
11.6   Test Results and Discussion *221*
11.6.1   Illustrative Example *221*
11.7   Comparative Analysis with Other Methods *223*
11.8   Conclusion *224*
11.9   Assignment *224*
        Acknowledgment *225*
        References *225*

12   **Power Quality Conditioners in Smart Power System** *233*
        Zahira Rahiman, Lakshmi Dhandapani, Ravi Chengalvarayan Natarajan, Pramila Vallikannan,
        Sivaraman Palanisamy, and Sharmeela Chenniappan
12.1   Introduction *233*
12.1.1   Voltage Sag *234*
12.1.2   Voltage Swell *234*
12.1.3   Interruption *234*
12.1.4   Under Voltage *234*
12.1.5   Overvoltage *234*
12.1.6   Voltage Fluctuations *234*
12.1.7   Transients *235*
12.1.8   Impulsive Transients *235*
12.1.9   Oscillatory Transients *235*
12.1.10   Harmonics *235*
12.2   Power Quality Conditioners *235*
12.2.1   STATCOM *235*
12.2.2   SVC *235*
12.2.3   Harmonic Filters *236*
12.2.3.1   Active Filter *236*
12.2.4   UPS Systems *236*
12.2.5   Dynamic Voltage Restorer (DVR) *236*

12.2.6   Enhancement of Voltage Sag *236*
12.2.7   Interruption Mitigation *237*
12.2.8   Mitigation of Harmonics *241*
12.3   Standards of Power Quality *244*
12.4   Solution for Power Quality Issues *244*
12.5   Sustainable Energy Solutions *245*
12.6   Need for Smart Grid *245*
12.7   What Is a Smart Grid? *245*
12.8   Smart Grid: The "Energy Internet" *245*
12.9   Standardization *246*
12.10   Smart Grid Network *247*
12.10.1   Distributed Energy Resources (DERs) *247*
12.10.2   Optimization Techniques in Power Quality Management *247*
12.10.3   Conventional Algorithm *248*
12.10.4   Intelligent Algorithm *248*
12.10.4.1 Firefly Algorithm (FA) *248*
12.10.4.2 Spider Monkey Optimization (SMO) *250*
12.11   Simulation Results and Discussion *254*
12.12   Conclusion *257*
          References *257*

13   **The Role of Internet of Things in Smart Homes** *259*
          Sanjeevikumar Padmanaban, Mostafa Azimi Nasab, Mohammad Ebrahim Shiri, Hamid Haj Seyyed Javadi,
          Morteza Azimi Nasab, Mohammad Zand, and Tina Samavat
13.1   Introduction *259*
13.2   Internet of Things Technology *260*
13.2.1   Smart House *261*
13.3   Different Parts of Smart Home *262*
13.4   Proposed Architecture *264*
13.5   Controller Components *265*
13.6   Proposed Architectural Layers *266*
13.6.1   Infrastructure Layer *266*
13.6.1.1 Information Technology *266*
13.6.1.2 Information and Communication Technology *266*
13.6.1.3 Electronics *266*
13.6.2   Collecting Data *267*
13.6.3   Data Management and Processing *267*
13.6.3.1 Service Quality Management *267*
13.6.3.2 Resource Management *267*
13.6.3.3 Device Management *267*
13.6.3.4 Security *267*
13.7   Services *267*
13.8   Applications *268*
13.9   Conclusion *269*
          References *269*

14   **Electric Vehicles and IoT in Smart Cities** *273*
          Sanjeevikumar Padmanaban, Tina Samavat, Mostafa Azimi Nasab, Morteza Azimi Nasab, Mohammad Zand, and
          Fatemeh Nikokar
14.1   Introduction *273*
14.2   Smart City *275*

14.2.1   Internet of Things and Smart City *275*
14.3   The Concept of Smart Electric Networks *275*
14.4   IoT Outlook *276*
14.4.1   IoT Three-layer Architecture *276*
14.4.2   View Layer *276*
14.4.3   Network Layer *277*
14.4.4   Application Layer *278*
14.5   Intelligent Transportation and Transportation *278*
14.6   Information Management *278*
14.6.1   Artificial Intelligence *278*
14.6.2   Machine Learning *279*
14.6.3   Artificial Neural Network *279*
14.6.4   Deep Learning *280*
14.7   Electric Vehicles *281*
14.7.1   Definition of Vehicle-to-Network System *281*
14.7.2   Electric Cars and the Electricity Market *281*
14.7.3   The Role of Electric Vehicles in the Network *282*
14.7.4   V2G Applications in Power System *282*
14.7.5   Provide Baseload Power *283*
14.7.6   Courier Supply *283*
14.7.7   Extra Service *283*
14.7.8   Power Adjustment *283*
14.7.9   Rotating Reservation *284*
14.7.10   The Connection between the Electric Vehicle and the Power Grid *284*
14.8   Proposed Model of Electric Vehicle *284*
14.9   Prediction Using LSTM Time Series *285*
14.9.1   LSTM Time Series *286*
14.9.2   Predicting the Behavior of Electric Vehicles Using the LSTM Method *287*
14.10   Conclusion *287*
            Exercise *288*
            References *288*

15   Modeling and Simulation of Smart Power Systems Using HIL *291*
            Gunapriya Devarajan, Puspalatha Naveen Kumar, Muniraj Chinnusamy, Sabareeshwaran Kanagaraj,
            and Sharmeela Chenniappan
15.1   Introduction *291*
15.1.1   Classification of Hardware in the Loop *291*
15.1.1.1 Signal HIL Model *291*
15.1.1.2 Power HIL Model *292*
15.1.1.3 Reduced-Scaled HIL Model *292*
15.1.2   Points to Be Considered While Performing HIL Simulation *293*
15.1.3   Applications of HIL *293*
15.2   Why HIL Is Important? *293*
15.2.1   Hardware-In-The-Loop Simulation *294*
15.2.2   Simulation Verification and Validation *295*
15.2.3   Simulation Computer Hardware *295*
15.2.4   Benefits of Using Hardware-In-The-Loop Simulation *296*
15.3   HIL for Renewable Energy Systems (RES) *296*
15.3.1   Introduction *296*
15.3.2   Hardware in the Loop *297*
15.3.2.1 Electrical Hardware in the Loop *297*

15.3.2.2   Mechanical Hardware in the Loop   *297*
15.4   HIL for HVDC and FACTS   *299*
15.4.1   Introduction   *299*
15.4.2   Modular Multi Level Converter   *300*
15.5   HIL for Electric Vehicles   *301*
15.5.1   Introduction   *301*
15.5.2   EV Simulation Using MATLAB, Simulink   *302*
15.5.2.1   Model-Based System Engineering (MBSE)   *302*
15.5.2.2   Model Batteries and Develop BMS   *302*
15.5.2.3   Model Fuel Cell Systems (FCS) and Develop Fuel Cell Control Systems (FCCS)   *303*
15.5.2.4   Model Inverters, Traction Motors, and Develop Motor Control Software   *304*
15.5.2.5   Deploy, Integrate, and Test Control Algorithms   *304*
15.5.2.6   Data-Driven Workflows and AI in EV Development   *305*
15.6   HIL for Other Applications   *306*
15.6.1   Electrical Motor Faults   *306*
15.7   Conclusion   *307*
   References   *308*

16   **Distribution Phasor Measurement Units (PMUs) in Smart Power Systems**   *311*
   Geethanjali Muthiah, Meenakshi Devi Manivannan, Hemavathi Ramadoss, and Sharmeela Chenniappan
16.1   Introduction   *311*
16.2   Comparison of PMUs and SCADA   *312*
16.3   Basic Structure of Phasor Measurement Units   *313*
16.4   PMU Deployment in Distribution Networks   *314*
16.5   PMU Placement Algorithms   *315*
16.6   Need/Significance of PMUs in Distribution System   *315*
16.6.1   Significance of PMUs – Concerning Power System Stability   *316*
16.6.2   Significance of PMUs in Terms of Expenditure   *316*
16.6.3   Significance of PMUs in Wide Area Monitoring Applications   *316*
16.7   Applications of PMUs in Distribution Systems   *317*
16.7.1   System Reconfiguration Automation to Manage Power Restoration   *317*
16.7.1.1   Case Study   *317*
16.7.2   Planning for High DER Interconnection (Penetration)   *319*
16.7.2.1   Case Study   *319*
16.7.3   Voltage Fluctuations and Voltage Ride-Through Related to DER   *320*
16.7.4   Operation of Islanded Distribution Systems   *320*
16.7.5   Fault-Induced Delayed Voltage Recovery (FIDVR) Detection   *322*
16.8   Conclusion   *322*
   References   *323*

17   **Blockchain Technologies for Smart Power Systems**   *327*
   A. Gayathri, S. Saravanan, P. Pandiyan, and V. Rukkumani
17.1   Introduction   *327*
17.2   Fundamentals of Blockchain Technologies   *328*
17.2.1   Terminology   *328*
17.2.2   Process of Operation   *329*
17.2.2.1   Proof of Work (PoW)   *329*
17.2.2.2   Proof of Stake (PoS)   *329*
17.2.2.3   Proof of Authority (PoA)   *330*
17.2.2.4   Practical Byzantine Fault Tolerance (PBFT)   *330*
17.2.3   Unique Features of Blockchain   *330*

17.2.4   Energy with Blockchain Projects *330*

17.2.4.1   Bitcoin Cryptocurrency *331*

17.2.4.2   Dubai: Blockchain Strategy *331*

17.2.4.3   Humanitarian Aid Utilization of Blockchain *331*

17.3   Blockchain Technologies for Smart Power Systems *331*

17.3.1   Blockchain as a Cyber Layer *331*

17.3.2   Agent/Aggregator Based Microgrid Architecture *332*

17.3.3   Limitations and Drawbacks *332*

17.3.4   Peer to Peer Energy Trading *333*

17.3.5   Blockchain for Transactive Energy *335*

17.4   Blockchain for Smart Contracts *336*

17.4.1   The Platform for Smart Contracts *337*

17.4.2   The Architecture of Smart Contracting for Energy Applications *338*

17.4.3   Smart Contract Applications *339*

17.5   Digitize and Decentralization Using Blockchain *340*

17.6   Challenges in Implementing Blockchain Techniques *340*

17.6.1   Network Management *341*

17.6.2   Data Management *341*

17.6.3   Consensus Management *341*

17.6.4   Identity Management *341*

17.6.5   Automation Management *342*

17.6.6   Lack of Suitable Implementation Platforms *342*

17.7   Solutions and Future Scope *342*

17.8   Application of Blockchain for Flexible Services *343*

17.9   Conclusion *343*

         References *344*

18   **Power and Energy Management in Smart Power Systems** *349*

         Subrat Sahoo

18.1   Introduction *349*

18.1.1   Geopolitical Situation *349*

18.1.2   Covid-19 Impacts *350*

18.1.3   Climate Challenges *350*

18.2   Definition and Constituents of Smart Power Systems *351*

18.2.1   Applicable Industries *352*

18.2.2   Evolution of Power Electronics-Based Solutions *353*

18.2.3   Operation of the Power System *355*

18.3   Challenges Faced by Utilities and Their Way Towards Becoming Smart *356*

18.3.1   Digitalization of Power Industry *359*

18.3.2   Storage Possibilities and Integration into Grid *360*

18.3.3   Addressing Power Quality Concerns and Their Mitigation *362*

18.3.4   A Path Forward Towards Holistic Condition Monitoring *363*

18.4   Ways towards Smart Transition of the Energy Sector *366*

18.4.1   Creating an All-Inclusive Ecosystem *366*

18.4.1.1   Example of Sensor-Based Ecosystem *367*

18.4.1.2   Utilizing the Sensor Data for Effective Analytics *368*

18.4.2   Modular Energy System Architecture *370*

18.5   Conclusion *371*

         References *373*

Index *377*

Editor Biography

Sanjeevikumar Padmanaban received his PhD degree in electrical engineering from the University of Bologna, Bologna, Italy, 2012. He was an Associate Professor at VIT University from 2012 to 2013. In 2013, he joined the National Institute of Technology, India, as a faculty member. Then, he served as an Associate Professor with the Department of Electrical and Electronics Engineering, University of Johannesburg, South Africa, from 2016 to 2018. From March 2018 to February 2021, he was an Assistant Professor with the Department of Energy Technology, Aalborg University, Esbjerg, Denmark. He continued his activities from March 2021 as an Associate Professor with the CTIF Global Capsule (CGC) Laboratory, Aarhus University, Herning, Denmark. Presently, he is a Full Professor in Electrical Power Engineering with the Department of Electrical Engineering, Information Technology, and Cybernetics, University of South-Eastern Norway, Norway. He is a fellow of the Institution of Engineers, India, the Institution of Electronics and Telecommunication Engineers, India, and the Institution of Engineering and Technology, UK. He is listed among the world's top two scientists (from 2019) by Stanford University, USA.

Sivaraman Palanisamy was born in Vellalur, Madurai District, Tamil Nadu, India. He completed schooling in Govt. Higher Secondary School, Vellalur, received his BE in Electrical and Electronics Engineering and ME in Power Systems Engineering from Anna University, Chennai, India, in 2012 and 2014, respectively. He has more than eight years of industrial experience in the field of power system analysis, grid code compliance studies, electric vehicle charging infrastructure, solar PV system, wind power plant, power quality studies, and harmonic assessments. Presently he is working as a Program Manager – EV charging infrastructure at WRI India (major contribution done before joining this position). He is an expert in power system simulation software such as ETAP, PSCAD, DIGSILENT POWER FACTORY, PSSE, and MATLAB. He is a working group member of various IEEE standards such as P2800.2, P2418.5, P1854, and P3001.9. He is an IEEE Senior Member, a Member of CIGRE, Life Member of the Institution of Engineers (India), and The European Energy Center (EEC). He is also a speaker who is well versed on both national and international standards.

Sharmeela Chenniappan holds a BE in Electrical and Electronics Engineering, ME in Power Systems Engineering from Annamalai University, Chidambaram, India, and a PhD in Electrical Engineering from Anna University, Chennai, India, in 1999, 2000, and 2009, respectively. She received her PG Diploma in Electrical Energy Management and Energy Audit from Annamalai University, Chidambaram in 2010. At present, she holds the post of Professor in the Department of EEE, CEG campus, Anna University, Chennai, India. She has more than 21 years of teaching/research experience and has taught various subjects to undergraduate and postgraduate students. She did a number of research projects and consultancy work in renewable energy, Electric Vehicle Charging Infrastructure, power quality, and design of PQ compensators for various industries. She is an IEEE Senior Member, a Fellow of the Institution of Engineers (India), and a Life Member of CBIP, ISTE, and SSI.

Jens Bo Holm-Nielsen currently works as Head of the Esbjerg Energy Section at the Department of Energy Technology, Aalborg University. Through his research, he helped establish the Center for Bioenergy and Green Engineering in 2009 and served as the head of the research group. He has vast experience in the field of biorefinery concepts and biogas production, in particular anaerobic digestion. He has implemented bio-energy systems projects in various provinces in Denmark and European states. He served as the technical advisor for many industries in this field. He has executed many large-scale European Union and United Nation projects in research aspects of bioenergy, biorefinery processes, the full biogas chain, and green engineering. He was a member on invitation with various capacities in committees for over 250 various international conferences and organizer of international conferences, workshops, and training programs in Europe, Central Asia, and China. His focus areas are renewable energy, sustainability, and green jobs for all.

List of Contributors

Ahmed Abbou
Department of Electrical Engineering
Mohammed V University in Rabat
Mohammadia School of Engineers
Rabat
Morocco

Shobha Agarwal
Department of Higher Technical Education and Skill
Development
Jharkhand University
Ranchi
India

Shariq Ahammed
PES
Valeo India Private Limited
Chennai
India

Nadia Awatif
Department of Electrical, Electronic and
Communication Engineering
Military Institute of Science and Technology
Dhaka
Bangladesh

Gyan R. Biswal
Department of Electrical and Electronics Engineering
Veer Surendra Sai University of Technology (VSSUT)
Burla
Odisha
India

Phaneendra B. Bobba
Department of Electrical and Electronics Engineering
Gokaraju Rangaraju Institute of Engineering and
Technology (GRIET)
Hyderabad
India

Satarupa Chakrabarti
School of Computer Engineering
KIIT University
Bhubaneswar
India

Soham Chakrabarti
School of Computer Engineering
KIIT University
Bhubaneswar
India

Ravi Chengalvarayan Natarajan
Department of Electrical and Electronics Engineering
Vidya Jyothi Institute of Technology
Hyderabad
Telangana
India

Sharmeela Chenniappan
Department of Electrical and Electronics Engineering
Anna University
Chennai
India

Khalid Chigane
Department of Electrical Engineering
Mohammed V University in Rabat
Mohammadia School of Engineers
Rabat
Morocco

Muniraj Chinnusamy
Department of EEE
Knowledge Institute of Technology
Salem
India

Gunapriya Devarajan
Department of EEE
Sri Eshwar College of Engineering
Coimbatore
India

Lakshmi Dhandapani
Department of Electrical and Electronics Engineering
Academy of Maritime Education and Training
(AMET)
Chennai
India

Sundaram Elango
Department of Electrical and Electronics Engineering
Coimbatore Institute of Technology
Coimbatore
India

Thamatapu Eswararao
Department of Electrical and Electronics Engineering
Coimbatore Institute of Technology
Coimbatore
India

Garika Gantaiahswamy
Department of Electrical and Electronics Engineering
JNTU Kakinada
Andhra Loyola Institute of Engineering and
Technology
Vijayawada
Andhra Pradesh
India

A. Gayathri
Department of EEE
Sri Krishna College of Technology
Coimbatore
Tamil Nadu
India

Md. Sanwar Hossain
Department of Electrical and Electronic Engineering
Bangladesh University of Business and Technology
Dhaka
Bangladesh

Bouazza Jabri
Department of Physical
LCS Laboratory
Faculty of Sciences
Mohammed V University in Rabat
Rabat
Morocco

Hamid Haj Seyyed Javadi
Department of Mathematics and Computer Science
Shahed University
Tehran
Iran

Jayakumar Narayanasamy
Department of EEE
The Oxford College of Engineering
Bommanahalli
Bangalore
India

Sabareeshwaran Kanagaraj
Department of EEE
Karpagam Institute of Technology
Coimbatore
India

Kathirvel Karuppazaghi
PES
Valeo India Private Limited
Chennai
India

Meenakshi Devi Manivannan
Department of Electrical and Electronics Engineering
Thiagarajar College of Engineering
Madurai
India

Saad Mekhilef
Power Electronics and Renewable Energy Research
Laboratory
Department of Electrical Engineering
University of Malaya
Kuala Lumpur
Malaysia

and

School of Science, Computing and Engineering
Technologies
Swinburne University of Technology
Hawthorn
Vic
Australia

and

Smart Grids Research Group
Center of Research Excellence in Renewable Energy
and Power Systems
King Abdulaziz University
Jeddah
Saudi Arabia

Naveenkumar Marati
PES
Valeo India Private Limited
Chennai
India

Marimuthu Marikannu
Department of EEE
Saranathan College of Engineering
Trichy
India

Rania Moutchou
Department of Electrical Engineering
Mohammed V University in Rabat
Mohammadia School of Engineers
Rabat
Morocco

Geethanjali Muthiah
Department of Electrical and Electronics Engineering
Thiagarajar College of Engineering
Madurai
India

Nagesh Halasahalli Nagaraju
Power System Studies
Power Research & Development Consultants Pvt Ltd
Bengaluru
India

Morteza Azimi Nasab
CTIF Global Capsule
Department of Business Development and Technology
Aarhus University
Herning
Denmark

Mostafa Azimi Nasab
CTIF Global Capsule
Department of Business Development and Technology
Aarhus University
Herning
Denmark

and

Department of Electrical and Computer Engineering
Boroujerd Branch
Islamic Azad University
Boroujerd
Iran

Puspalatha Naveen Kumar
Department of EEE
Sri Eshwar College of Engineering
Coimbatore
India

Fatemeh Nikokar
Department of Business Development and Technology
CTIF Global Capsule
Aarhus University
Herning
Denmark

Sanjeevikumar Padmanaban
Department of Electrical Engineering, Information Technology, and Cybernetics
University of South-Eastern Norway
Porsgrunn
Norway

Sivaraman Palanisamy
World Resources Institute (WRI) India
Bengaluru
India

Madhu Palati
Department of Electrical and Electronics Engineering
BMS Institute of Technology and Management
Affiliated to Visvesvaraya Technological University
Doddaballapur Main Road, Avalahalli
Yelahanka
Bengaluru
India

P. Pandiyan
Department of EEE
KPR Institute of Engineering and Technology
Coimbatore
Tamil Nadu
India

Basanta K. Panigrahi
Department of Electrical Engineering
Institute of Technical Education & Research
SOA University
Bhubaneswar
India

Paranthagan Balasubramanian
Department of EEE
Saranathan College of Engineering
Trichy
India

Md. I. H. Pathan
Department of Electrical and Electronic Engineering
Hajee Mohammad Danesh Science and Technology University
Dinajpur
Bangladesh

Sagar Singh Prathap
Energy and Power Sector
Center for Study of Science Technology and Policy
Bengaluru
India

Zahira Rahiman
Department of Electrical and Electronics Engineering
B.S. Abdur Rahman Crescent Institute of Science & Technology
Chennai
India

Mohammad M. Rahman
Information and Computing Technology Division
Hamad Bin Khalifa University
College of Science and Engineering
Doha
Qatar

Hemavathi Ramadoss
Department of Electrical and Electronics Engineering
Thiagarajar College of Engineering
Madurai
India

Nisha C. Rani
Department of EEE
The Oxford College of Engineering
Bommanahalli
Bangalore
India

Muhyaddin Rawa
Smart Grids Research Group
Center of Research Excellence in Renewable Energy
and Power Systems
King Abdulaziz University
Jeddah
Saudi Arabia

and

Department of Electrical and Computer Engineering
Faculty of Engineering
K.A. CARE Energy Research and Innovation Center
King Abdulaziz University
Jeddah
Saudi Arabia

Salah E. Rhaili
Department of Electrical Engineering
Mohammed V University in Rabat
Mohammadia School of Engineers
Rabat
Morocco

V. Rukkumani
Department of EIE
Sri Ramakrishna Engineering College
Coimbatore
Tamil Nadu
India

Subrat Sahoo
Hitachi Energy Research
Vasteras
Sweden

Tina Samavat
CTIF Global Capsule
Department of Business Development and Technology
Aarhus University
Herning
Denmark

S. Saravanan
Department of EEE
Sri Krishna College of Technology
Coimbatore
Tamil Nadu
India

Md. Shafiullah
King Fahd University of Petroleum & Minerals
Interdisciplinary Research Center for Renewable
Energy and Power Systems
Dhahran
Saudi Arabia

Mohammad S. Shahriar
Department of Electrical Engineering
University of Hafr Al-Batin
Hafr Al Batin
Saudi Arabia

Logeshkumar Shanmugasundaram
Department of Electronics and Communication
Engineering
Christ the King Engineering College
Coimbatore
India

Mohammad Ebrahim Shiri
Mathematics and Computer Science Department
Amirkabir University of Technology
Tehran
Iran

Jyoti Shukla
Department of Electrical Engineering
Poornima College of Engineering
RTU
Jaipur
India

Brijesh Singh
Department of Electrical and Electronics Engineering
KIET Group of Institutions
Ghaziabad
India

Umashankar Subramanian
Renewable Energy Laboratory
Department of Communications and Networks
Prince Sultan University
College of Engineering
Riyadh
Saudi Arabia

Aleena Swetapadma
School of Computer Engineering
KIIT University
Bhubaneswar
India

and

Renewable Energy Laboratory
Department of Communications and Networks
Prince Sultan University
College of Engineering
Riyadh
Saudi Arabia

Krishnamohan Tatikonda
Department of Electrical and Electronics Engineering
JNTU Kakinada
Andhra Loyola Institute of Engineering and
Technology
Vijayawada
Andhra Pradesh
India

Balraj Vaithilingam
PES
Valeo India Private Limited
Chennai
India

Pramila Vallikannan
Department of Electrical and Electronics Engineering
B.S. Abdur Rahman Crescent Institute of Science &
Technology
Chennai
India

Monika Vardia
Department of Electrical Engineering
Poornima College of Engineering
RTU
Jaipur
India

Devi Vigneshwari Balasubramanian
Department of EEE
The Oxford College of Engineering
Bommanahalli
Bangalore
India

Vijayalakshmi Subramanian
Department of EEE
Saranathan College of Engineering
Trichy
India

Deepak Yadav
Power Electronics and Renewable Energy Research
Laboratory
Department of Electrical Engineering
University of Malaya
Kuala Lumpur
Malaysia

Mohammad Zand
CTIF Global Capsule
Department of Business Development and Technology
Aarhus University
Herning
Denmark

1

Introduction to Smart Power Systems

Sivaraman Palanisamy[1], Zahira Rahiman[2], and Sharmeela Chenniappan[3]

[1] *World Resources Institute (WRI) India, Bengaluru, India*
[2] *Department of Electrical and Electronics Engineering, B.S. Abdur Rahman Crescent Institute of Science & Technology, Chennai, India*
[3] *Department of Electrical and Electronics Engineering, Anna University, Chennai, India*

1.1 Problems in Conventional Power Systems

The conventional power system is generally classified as power generation, power transmission, and power distribution systems. The power is generated from thermal plants, nuclear plants, or hydroplants at remote locations and this is transmitted to the load center through a power transmission system [1]. The distribution system is used to distribute the electric power to various end-users. It has limited control and visibility of power flows from generation to the end user's load. Some of the problems associated with conventional systems are limited visibility in power flows, limited control, delay in measurement and control, higher energy losses in transmission and distribution systems, poor power quality, etc. [2].

1.2 Distributed Generation (DG)

The distributed generation (DG) is used to produce the electric power closer to the load center or end-user loads to reduce the energy loss in the transmission as well as distribution system and improve the voltage profile. The sources of DG can be both renewable energy sources (like solar, wind, and fuel cells), and nonrenewable energy sources (like diesel generators). These sources as simply called distributed energy resources (DERs) [3]. Generally, these DGs are interconnected with the primary or secondary distribution systems based on their rating. Figure 1.1 shows the single-line diagram of a 100 kW rooftop solar PV system as DG connected to the 415 V, 50 Hz secondary distribution system.

Figure 1.2 shows the single-line diagram of a 1 MW rooftop solar PV system as DG connected to the 11 kV, 50 Hz primary distribution system.

The intermittency is one of the major challenges of using renewable energy sources such as solar PV and wind energy conversion systems as DG. Due to intermittence, the output power from the solar PV system and wind energy conversion system also varies throughout the operation resulting in power balance and stability issues [4]. The impact of intermittency can be reduced to a certain extent by using a complex software program/tool to predict the energy output based on various historical data.

Artificial Intelligence-based Smart Power Systems, First Edition.
Edited by Sanjeevikumar Padmanaban, Sivaraman Palanisamy, Sharmeela Chenniappan, and Jens Bo Holm-Nielsen.
© 2023 The Institute of Electrical and Electronics Engineers, Inc. Published 2023 by John Wiley & Sons, Inc.

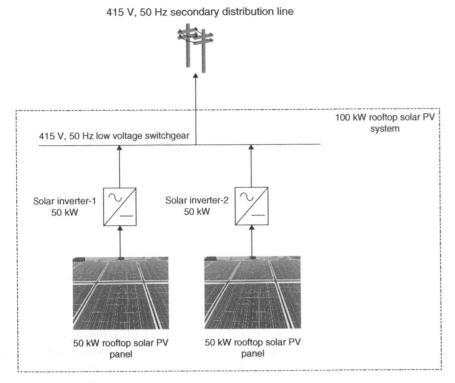

415 V, 50 Hz secondary distribution line

415 V, 50 Hz low voltage switchgear

100 kW rooftop solar PV system

Solar inverter-1
50 kW

Solar inverter-2
50 kW

50 kW rooftop solar PV panel

50 kW rooftop solar PV panel

Figure 1.1 Single line diagram of a rooftop solar PV system connected to the secondary distribution system.

1.3 Wide Area Monitoring and Control

Power grids are the most complicated and essential systems in today's life. The risk of experiencing a wide variety of faults and failures is increasing [5]. The unpredictable and cascaded events of faults lead to a blackout, and they have an impact on a large range of consumers. Many grid codes allow the frequency within the specified tolerance limits. Hence, flexibility in frequency leads to under drawl or over drawl of real power, as well as under generation or over a generation by the utilities. This results in the overloading of transmission lines and under voltage or over voltage of the grid. Also, unpredictability, intermittency, and variability of renewable energy integration pose challenges in grid operation. Conventional Supervisory Control and Data Acquisition (SCADA) systems are limited to steady-state measurements and cannot be used for observing the system dynamics behavior. To overcome the drawbacks of a conventional system, one of the most recent advancements in modern power grids is wide-area monitoring (WAM). With the developments of WAM, power system dynamic behavior is monitored closely in real-time. So that the faults in the power grid can be identified and protected in a wider range [6].

The overall goal of using WAM is to improve protection and to develop new protection concepts that will make blackouts less probable and much less severe even if they do occur. The following are the key areas where WAM can help to protect power systems.

1. Dealing with large-scale interruptions
2. Taking the appropriate precautions to mitigate the impact of failed systems
3. Ignoring relay settings that are incompatible with the current system configuration
4. Achieving a reasonable balance between security and dependability

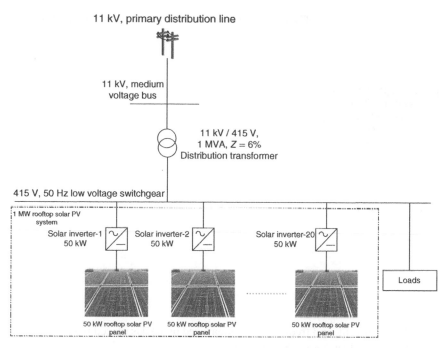

Figure 1.2 Single line diagram of a rooftop solar PV system connected to the primary distribution system.

The purpose of protection is to safeguard specific elements of the power system as well as the security of the power system as a whole.

In the case of main equipment protection, WAM plays a significant role. This is due to the fact that primary protection must consistently offer a very fast response to any failure on the element that it safeguards. WAM, on the other hand, can be a beneficial tool for increasing system performance due to the slower response time necessary for backup protection and the fact that it protects a zone of the system. Wide-area measurements have the potential to enable the development of supervisory methods for backup protection, more complex types of system protection, and altogether new protection concepts. Examples of these protection functions are

1. Dynamic relays adjust their parameters in response to changes in the system condition.
2. Multiterminal line protection has been improved.
3. Predictive end-of-line protection, which monitors the distant location breaker and replaces the under-reaching Zone 1 with an instantaneous characteristic if it is open.
4. Modify relay settings temporarily to prevent malfunction during cold load pickup.
5. Employ the capability of modern relays to self-monitor to find hidden faults and use the IEC 61850 hot-swap capabilities to eliminate them.
6. Artificial controlled microgrids provide an adaptive controlled divergence to prevent an uncontrolled system separation.

WAM gathers data from remote places throughout the power grid and integrates them in real-time into a single snapshot of the power system for a given time. Synchronized measurement technology (SMT) is a crucial component of WAM because it allows measurements to be correctly timestamped, typically using global positioning system (GPS) timing signals. The data may be simply merged with these timestamps, and phase angle measurements can be made with a common reference [7]. Figure 1.3 shows the generic WAMS architecture based on phasor measurement units (PMUs). PMUs, phasor data concentrators (PDCs), communication networks, data storage, and application software are the primary components of WAM. The number of substation PDCs is determined

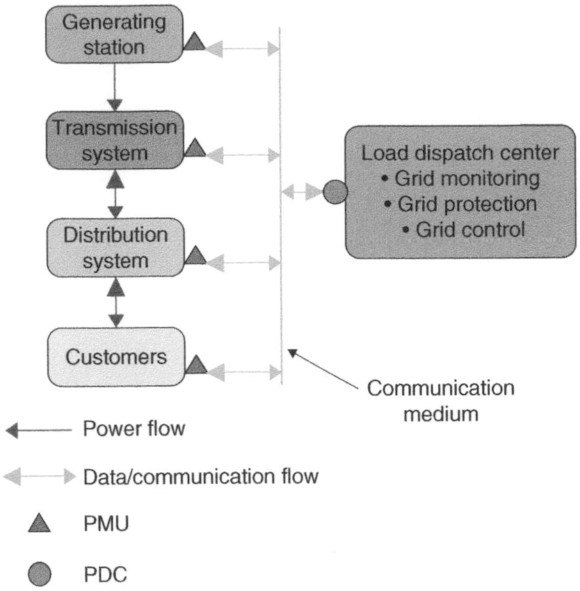

Figure 1.3 Block diagram of wide-area monitoring and control.

by the power system requirements. Voltage, current, and frequency are measured by PMUs placed in substations. These readings are routed straight to the central PDC or a substation PDC.

The following functions are available at the PDC substation:

✓ Synchronization of date and time
✓ Gathers info from PMUs
✓ Analyzes collected data
✓ Data is sent to the central PDC
✓ Communicates data with the regional SCADA
✓ Data is archived locally
✓ Carries out local data analysis and security actions

1.4   Automatic Metering Infrastructure

The name Advanced Metering Infrastructure or simply AMI refers to the entire infrastructure, which includes everything from smart meters to two-way communication networks to control center equipment, as well as all the applications that allow for the gathering and transfer of energy usage data in real-time. The backbone of the smart grid [8] is AMI, which enables two-way connectivity with customers. Error-free meter reading from remote, network problem and its diagnosis, load profile/patterns, energy audits/consumptions, and partial load curtailment in place of load shedding are all potential objectives of AMI. The typical building blocks of AMI are shown in Figure 1.4.

AMI is made up of several hardware and software components that all work together to measure energy consumption and send data about it to utility companies and customers [8]. The key technological components of AMI are,

☐ **Smart Meters**: Advanced meter devices that could gather data of electrical parameters at various intervals and transfer the data to the utility via fixed communication networks, as well as receiving information from the utility such as pricing signals and relaying it to the consumer [9].

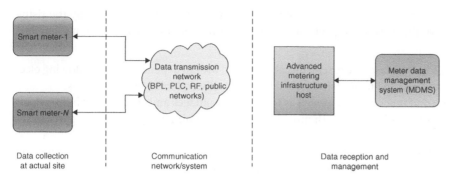

Figure 1.4 Basic building blocks of AMI.

- **Communication Network**: Smart meters can provide data to utility companies and vice versa. The advanced communication networks allow two-way communication between smart meters and utility companies. For these applications, networks like Broadband over Powerline (BPL), Power Line Communications (PLC), Fiber Optic Communication, Fixed Radio Frequency (RF), or public networks (e.g. landline, cellular, paging) are used [10].
- **Meter Data Acquisition System**: Data is collected from smart meters over a communication network and sent to the meter data management system (MDMS) using software applications on the Control Centre hardware and DCUs (Data Concentrator Units).

  MDMS Metering: receives the information, stores it, and analyzed it by the host system.
- **Home Area Network (HAN)**: It can be a consumer-side extension of AMI, allowing for easier communication between household appliances and AMI, and thus better load control by both the utility and the consumer [11].

The benefits of AMI are multifold and can be generally categorized as follows:

Operational Benefits: The entire system benefits from AMI since it improves meter reading accuracy, detects energy theft, and responds to power outages while removing the need for an on site meter reading.

Financial Benefits: Utility companies financially benefit from AMI because it lowers equipment and maintenance costs, enables faster restoration of electric service during outages, and streamlines the billing process.

Customer Benefits: Electric customers benefit from AMI because it detects meter faults early, allows for speedier service restoration, and improves billing accuracy and flexibility. AMI also offers time-based tariff choices, which can help consumers save money and better manage their energy usage.

Security Benefits: AMI technology allows for better monitoring of system resources, reducing the risk of cyber-terrorist networks posing a threat to the grid.

In spite of various advantages, AMI deployment faces three significant challenges: higher capital costs or investments, connection or interoperability with other grid systems, and standardization.

High Capital Costs: A full-scale implementation of AMI necessitates investments in all hardware and software components, including smart meters, network infrastructures, and network management software, as well as costs associated with meter installation and maintenance.

Integration: Customer Information Systems (CISs), Geographical Information Systems (GISs), Outage Management Systems (OMSs), Work Management System (WMS), Mobile Workforce Management (MWM), SCADA/DMS, Distribution Automation System (DAS), and other utilities' information technology systems essentially integrated with AMI.

Standardization: Compatibility standards must be created, as they are the keys to properly connecting and sustaining an AMI-based grid system. They set universal requirements for AMI technology, deployment, and general operations.

Investing in AMI to modernize the power grid system will alleviate several grid stresses caused by the rising power demands. AMI will improve three critical aspects of power grid infrastructure such as system reliability, energy cost, and electricity theft.

System Reliability: AMI technology increases electricity distribution and overall dependability by allowing electricity distributors to identify and respond to electric demand automatically, reducing power outages.

Energy Costs: Increased stability and functionality, as well as fewer power outages and streamlined billing operations, will greatly reduce the expenses involved with providing and maintaining the grid, resulting in significantly cheaper electricity bills.

Electricity Theft: Electricity theft is a prevalent problem in Society. AMI systems that track energy usage will aid in monitoring power in real-time, resulting in enhanced system transparency.

1.5 Phasor Measurement Unit

A phasor measurement unit or simply PMU is a crucial measurement tool that is used on electric power systems to improve grid operators' visibility on the huge power grid network/system [12]. It measures the parameter called a phasor and it provides the information/data of magnitude and phase angle of voltage or current at a particular location [13]. This information/data shall be used to find the operating frequency at a particular time instant and examine the condition of the system as shown in Figure 1.5.

A PMU may provide up to 60 measurements per second. As compared with a typical SCADA-based system, the measurements per second are higher in PMU. A typical SCADA-based system will provide the data (one measurement data in two to four seconds time interval) [14]. The main advantage of using PMU over conventional SCADA system is PMU can collect the data of all PMU at a particular time through GPS. This means, that collected data across the power grid are time-synchronized. Because of this reason, PMUs are also called synchro phasors [15].

The information collected from the PMU conveys to the system operator whether the main electrical parameters such as voltage, current, and frequency are within the specified limit with tolerance or not. The capability of the PMU is as follows,

- Line congestion: prediction, analysis, and manage
- Analyzing the event after the disturbance or fault (post fault analysis)
- Instability and stress detection
- Inefficiencies detection

In this decade, several thousands of PMUs are successfully installed and commissioned in transmission and/or distribution grids across the globe. A PMU can be integrated with smart controllers, and this will reduce the manual operations required by the SCADA system in decision making and control. Due to this feature, the grid becomes robust and efficient, it allows the more integration of renewable powers, DERs, and microgrids.

The report on Unified Real-Time Dynamic State Measurement (URTDSM) by Power Grid Corporation of India Ltd. (PGCIL) shows the importance of PMU data (data from various lines at time-stamped) is useful for prediction and post fault event analysis. PGCIL followed the philosophy stated below for installing the PMUs across India, installation of PMUs on substations at 400 kV level above, all generating stations at 220 kV level and above, HVDC terminals, important inter-regional connection points, inter-national connection points, etc. Also, the provision of PDC at all State Load Dispatch Centers (SLDCs), Regional Load Dispatch Centers (RLDCs), and National Load Dispatch Center (NLDC) [7].

The PMU is used to measure the magnitude and phase angle of bus voltage and line current phasor. PMU takes the bus PT input for voltage and line CT input for current at the substation as well as GPS time signal. The PMU presently available in the market can measure one set of bus voltage (three-phase) and two sets of line current (three-phase). The typical arrangement of PMU in substation and Main Phasor Data Concentrator (MPDC)/Sub Phasor Data Concentrator (SPDC) in load dispatch center is shown in Figure 1.6 [7].

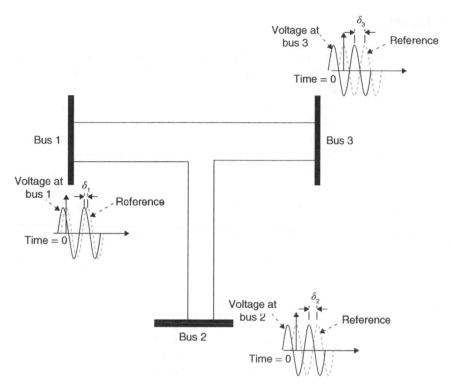

Figure 1.5 Transmission line data.

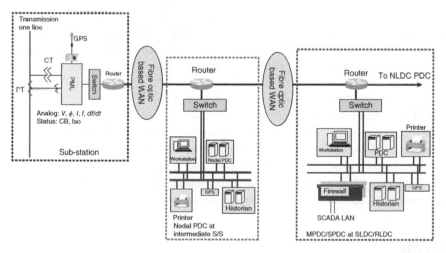

Figure 1.6 Typical arrangement of PMU in substation and PDC in the load dispatch center.

The PMU output from the substation is communicated to PDC through a Local Area Network (LAN) switch and router. A PDC at the load dispatch center is used to receive the data from the multiple PMUs of different substations. Also, PDC communicates with other PDCs and transfers the data. The PDCs align/store the PMU data by the time-stamped label and create the time-synchronized dataset. All the PDCs are connected locally with computers/host workstations, printers, and operators' cabinets via Ethernet.

1.6 Power Quality Conditioners

The IEEE Std 1100-2005 [16] defines the power quality as the concept of powering and grounding electronic equipment in a manner that is suitable to the operation of that equipment and compatible with the premise wiring system and other connected equipment. The IEEE Std 1159-2019 [17] categorizes the various power quality phenomena which are transients, short-duration RMS variation, long-duration RMS variation, imbalance, waveform distortion, voltage fluctuations, and power frequency variations [17, 18].

The power quality conditioners like Distribution STATCOM (DSTATCOM), Dynamic Voltage Restorer (DVR), and Unified Power Quality Conditioners (UPQCs) are widely used for dynamic reactive power compensation, voltage profile (voltage sag and swell, under-voltage and over-voltage) enhancement during the fault or other events. These power quality conditioners inject dynamically controlled reactive power to enhance the voltage profile.

The harmonics are one of the important power quality parameters and they are generally defined as the deviation of voltage and/or current from ideal sinusoidal waveshape due to non-linear characteristics of the connected loads and power electronics (inverter) based generations. IEEE Std 519-2014 IEEE recommended practice and requirements for harmonic control in electric power systems and provided the recommended harmonic limits for both voltage and current at the point of common coupling. If the harmonic injection from the plant exceeds the recommended limits specified in IEEE 519, suitable mitigation measures (e.g. harmonic filters) shall be proposed and tuned to mitigate the harmonics propagation into the power systems [19].

Various types of harmonic filters are available in the market for harmonic mitigation. They are passive filter, active filter, and hybrid filter. Passive filters are one of the cheaper and cost-effective solutions for harmonic mitigation [16]. It contains passive elements such as resistors, capacitors, and inductors. Based on the characteristics, passive filters are further classified into the low-pass filter, high-pass filter, band-pass filter, and tuned filters. The low pass filter is used to cancel the higher-order harmonics, while the high-pass filter is used to cancel the lower-order harmonics. A band-pass filter is used to cancel the order or frequency of the harmonic outside the range of band frequency. The tuned filters are used to cancel one specific harmonic order or frequency.

Shunt Active Power Filter is used to mitigate the current harmonics propagation into the system, reactive power compensation as well as balance the unbalanced currents. It injects the equal compensating currents with a 180° opposite phase shift for harmonic cancelation. This can also be used for voltage profile enhancement and reactive power compensation. It employs the micro-controller-based control circuits to estimate the magnitude of harmonic contents of each order [19–22]. The active filter is expensive as compared with passive harmonic filters and it is suitable for the place where plant loading pattern varies with respect to time [23].

A hybrid filter is a combination of both passive and active harmonic filters. It has the combined characteristics of both passive and active harmonic filters. The passive part of the filter is used for mitigation of harmonic current injected by the loads which are operating constantly throughout the day and the active harmonic filter is used to mitigate the harmonic currents injected by the loads whose operations vary with respect to time [24]. The hybrid filter is economic as compared with the active harmonic filter of the same rating. This filter is suitable for the locations where partial loads are constant throughout their operation and partial loads vary with respect to time.

1.7 Energy Storage Systems

The energy storage systems (ESSs) are widely used to store the energy whenever the surplus power is available/the cost of electricity is low and discharge the stored energy during the scarcity/cost of electricity is high. Concerns about the environment, such as global warming, have become global issues. As a result, the use of renewable energy sources like solar and wind power, as well as Smart Grids that effectively use all sorts of power sources, is seen as a very promising technology [25]. Electricity grids achieve reliable power supply by balancing supply and demand to the best of their abilities. However, when the use of solar power and other renewable energy sources, which have variable production, grows, the grid's power supply may become unstable [26]. This poses several difficulties. To overcome such difficulties, an ESS has been required in the power system [27]. In an emergency, energy storage devices can also be used as a backup power supply [28]. ESSs are extremely adaptable, and they may be used to satisfy the needs of a wide range of customers and in a variety of industries [29, 30]. These include renewable energy power producers, grid equipment such as transmission and distribution equipment, as well as commercial buildings, factories, and residences [31].

The grid-connected ESSs are used for both energy management applications and power quality applications such as load balancing or leveling, reducing the intermittency and smoothing the renewable energy integration, peak demand shifting or shaving, and providing the uninterrupted power supply to end-user loads, voltage and frequency regulation, voltage sag mitigation, reactive power control, and management, black start and islanded mode of operation, Volt/var control, etc. in the power systems [32]. The amount of renewable energy penetration keeps increasing every year which demands the requirements of grid-connected ESSs. The ESS is used to smooth the power output from renewable energy sources such as solar PV and wind energy conversion systems and level the load pattern.

The ESS can be classified in many ways, but one of the most useful is one based on the duration and frequency of power delivery. They are (i) short-term (seconds to minutes), (ii) medium-term (day storage), and (iii) long-term ESS (weekly to monthly) [33].

The short-term ESS (less than 25 minutes) [33] shall be used for spinning reserve, peak shaving, UPS, primary and secondary frequency control, electric vehicles (EVs), etc.

The medium-term ESS (1–10 hours) [33] shall be used for load leveling, tertiary frequency control, UPS, tertiary frequency control, etc.

Finally, the long-term ESS (from 50 hours to less than three weeks) [33] shall be used for long-term services during periods whenever power output from renewable power plants (solar and wind) is limited (known as "dark-calm periods").

The short-term services possibly will be provided via flywheels, superconductive magnetic coils, and super-capacitors. The medium-term services possibly will be provided by pumped hydropower, compressed air ESS, thermoelectric storage, and electrochemical ESS, such as lithium-ion, lead-acid, high temperature, and flow batteries [34, 35]. Hydrogen or natural gas storage systems can provide long-term services.

1.8 Smart Distribution Systems

The electric power distribution systems are used to distribute the electric power to end customers designed to deliver power from substations to customers. In order to operate the end user's equipment to an efficient and satisfactory level, the incoming power supply from the distribution company shall have high reliability and quality. Nowadays, end-users are using sophisticated equipment for easy control of their applications. This sophisticated equipment requires a high-quality and reliable input power supply for their trouble-free day-to-day operations [36]. In order to provide a high-quality and reliable power supply, the conventional distribution systems are changed as smart distribution systems with additional devices or equiment's and technology in place. The faults are one of the major problems in the distribution system and affect the incoming power supply to end-users until the fault is rectified. The majority of the faults are transients in nature. Hence, the auto recloser is used to close the circuit after the fault incident with the intended time delay. This auto recloser function reduces the human intervention to switch ON the circuit breaker manually after the fault. Also, fault identification sensors are used to easily identify the faults which are permanent in nature. The use of fault identification sensors reduces the time and effort required to examine the fault location through physical inspection.

Another important problem concerning the conventional distribution system is the lack of remote monitoring and control, and grid automation capability [2]. In smart distribution, these features are included and the distribution operator at the control center or central monitoring and control location can easily monitor what is happening in the grid and promptly take the necessary actions in case of any extreme faults/events. Modern intelligent devices, smart grid applications like WAMS, and distribution PMUs are improved the distribution system monitoring and controllability functions.

The DG is used to produce the energy locally closer to the end-user loads. It reduces the energy losses in the distribution lines. The DG uses DERs which are available closer to end-users like solar PV systems, wind energy conversion systems, and fuel cells for power generation. The main problems of solar PV systems and wind energy conversion systems are uncertainties and variable output power [37]. The ESSs are used to smooth the output power from the renewable energy systems. Nowadays, smart distribution system employs the distributed ESSs connected across the distribution system to smooth the output power from DERs like solar PV systems.

The recent development and advancement in multi-disciplinary engineering are EVs. EVs are emerging in this decade due to the rapid depletion of fossil fuels and environmental pollution. These EVs are charged by means of an electric power supply from the distribution system. The charging characteristics of EVs are not uniform and vary with respect to the state of charge of the battery [38]. The large penetration of these EVs on the distribution system shall change the loading pattern of the distribution system.

1.9 Electric Vehicle Charging Infrastructure

The EVs are an alternative to conventional internal combustion engine-based vehicles. Nowadays, the usage of EVs is rapidly increasing and will be connected and recharged across the distribution systems. The EV charging infrastructure or electric vehicle supply equipment (EVSE) is used to recharge the batteries of EVs. In the smart distribution system, EVSE is the main component of the EVs ecosystem. It is essential to revamp the existing electric distribution infrastructure with modern technologies and equipment to cater to the large penetration of EV loads [39]. Hence, the necessary planning action to be taken care of and the upgradation of existing electrical infrastructure or new electrical infrastructure is required at different levels.

The power rating of EV chargers is very based on vehicle type (two-wheelers, three-wheelers, and four-wheelers) and type of charging (slow charging and fast charging). The typical rating of an EV charger is in the range of 500 watts (W) to 500 kW.

Based on the location of the charging unit (EVSE), EV charging is classified into an Onboard charger and OFF-board charger. If EVSE is placed inside the vehicle, then it is called an Onboard charger, i.e. charging units are kept within the vehicle. If the EVSE is placed outside the vehicle, then it is called an Off-board charger, i.e. charging units are kept outside the vehicle.

Based on the physical location of charging, EV charging is further classified into residential charging and public charging. The EV owners or users charging their vehicles in their residential homes is called residential charging. The advantage of residential charging is the lesser cost per energy usage in kWh and the disadvantage is taking more time (in the range of five to eight hours) for recharging. On the other hand, public charging stations are used for commercial vehicles. Most public charging station employs fast chargers for recharging commercial vehicles in a lesser time duration [40].

Due to technological advancement, nowadays Vehicle to Grid (V2G) concept becomes more popular. In this V2G concept, the stored energy available in the vehicle is again fed back to the grid during peak hours as a distributed ESS.

1.10 Cyber Security

Cyber security is one of the key features of smart power systems. It is used to improve service reliability. Hence, it is brought into the power system in the planning and development stage itself. It ensures the legacy of security and protection of data. The highly secured data can be achieved from a properly planned strategy, and it provides flexibility in operation and efficiency. The utilities should implement a comprehensive, integrated, well-monitored, and regularly updated cyber security program [41]. Cyber security is essential for the various elements of smart power systems such as smart meters, SCADA systems, substations, and control systems, and cyber security actions are applied to all the participants of power systems, i.e. generating station, transmission system, and distribution systems.

1.11 Conclusion

The conventional power system has limited power-generation plants or sources connected to the bulk transmission grid and it powers the millions of end-users across it. It has limited control and visibility of power flows from generating plants to the end-users. This chapter discussed the various technological developments and recent advancements such as DG, WAM, and control using automatic metering infrastructure, PMU, power quality conditioners, EV charging infrastructure, smart distribution systems, and ESSs, and cyber security in smart power systems.

References

1 Sivaraman, P., Sharmeela, C., Mahendran, R., and Thaiyal Nayagi, A. (2020). *Basic Electrical and Instrumentation Engineering*. Wiley.

2 Sivaraman, P. and Sharmeela, C. (2020). Existing issues associated with electric distribution system. In: *Handbook of Research on New Solutions and Technologies in Electrical Distribution Networks* (ed. B. Kahn, H.H. Alhelou and G. Hayek), 1–31. Hershey, PA: IGI Global.

3 Pandey, S.K., Mohanty, S.R., and Kishor, N. (2013). A literature survey on load–frequency control for conventional and distribution generation power systems. *Renewable and Sustainable Energy Reviews* 25: 318–334.

4 Sharmeela, C., Sivaraman, P., Sanjeevikumar, P., and Holm-Nielsen, J.B. (2021). *Microgrid Technologies*. Wiley.

5 Leger, A.S., Spruce, J., Banwell, T. et al. (2016). Smart grid testbed for wide-area monitoring and control systems. In: *2016 IEEE/PES Transmission and Distribution Conference and Exposition (T&D)*, 1–5. IEEE.

6 Malik, H., Ahmad, M.W., Alotaibi, M.A., and Almutairi, A. (2022). Development of wide area monitoring system for smart grid application. *Journal of Intelligent & Fuzzy Systems* 42 (2): 827–839.

7 Unified Real-Time Dynamic State Measurement (URTDSM) Report (2012). Power Grid Corporation of India Ltd Gurgaon Feb'12. PGCIL, https://cea.nic.in/wp-content/uploads/2020/03/1st-1.pdf (accessed 14 September 2022).

8 Jaiswal, D. and Thakare, M. (2022), Overview of an Advanced Metering Infrastructure Based on Smart Meters (Feb 26, 2022). *Proceedings of the 3rd International Conference on Contents, Computing & Communication (ICCCC-2022)*, Available at SSRN: https://ssrn.com/abstract=4056050 or http://dx.doi.org/10.2139/ssrn.4056050.

9 Martins, J.F., Pronto, A.G., Delgado-Gomes, V., and Sanduleac, M. (2019). Smart meters and advanced metering infrastructure. In: *Pathways to a Smarter Power System* (ed. A. Tascikaraogu and O. Erdinc), 89–114. Academic Press.

10 Ghasempour, A. (2016). Optimum number of aggregators based on power consumption, cost, and network lifetime in advanced metering infrastructure architecture for smart grid internet of things. In: *2016 13th IEEE Annual Consumer Communications & Networking Conference (CCNC)*, 295–296. IEEE.

11 Rahmann, C. and Alvarez, R. (2022). The role of smart grids in the low carbon emission problem. In: *Planning and Operation of Active Distribution Networks*, 455–485. Cham: Springer Lecture Notes in Electrical Engineering book series (LNEE, volume 826).

12 Hampannavar, S., Teja, C.B., Swapna, M. et al. (2020). Performance improvement of M-class phasor measurement unit (PMU) using hamming and blackman windows. In: *2020 IEEE International Conference on Power Electronics, Smart Grid and Renewable Energy (PESGRE2020)*, 1–5. IEEE.

13 Sodhi, R., Srivastava, S.C., and Singh, S.N. (2011). Multi-criteria decision-making approach for multistage optimal placement of phasor measurement units. *IET Generation, Transmission and Distribution* 5 (2): 181–190.

14 Narendra, K., Gurusinghe, D.R., and Rajapakse, A.D. (2012). Dynamic performance evaluation and testing of phasor measurement unit (PMU) as per IEEE C37. 118.1 Standard. In: *Doble Client Committee Meetings International Protection Testing Users Group,* (PTUG), Chicago, IL, USA, vol. 10.

15 Roscoe, A.J. (2013). Exploring the relative performance of frequency-tracking and fixed-filter phasor measurement unit algorithms under C37. 118 test procedures, the effects of interharmonics, and initial attempts at merging P-class response with M-class filtering. *IEEE Transactions on Instrumentation and Measurement* 62 (8): 2140–2153.

16 IEEE Standards Association. (2005). IEEE Std1100-2005-IEEE recommended practice for powering and grounding electronic equipment. In *Institute of Electrical and Electronics Engineers*.

17 "IEEE Recommended Practice for Monitoring Electric Power Quality," in *IEEE Std 1159-2019 (Revision of IEEE Std 1159-2009)*, 1–98, 13 August 2019, https://doi.org/10.1109/IEEESTD.2019.8796486.

18 Sivaraman, P. and Sharmeela, C. (2021). Power quality and its characteristics. In: *Power Quality in Modern Power Systems* (ed. P. Sanjeevikumar, C. Sharmeela and J.B. Holm-Nielsen), 1–60. Academic Press.

19 Sivaraman, P. and Sharmeela, C. (2021). Power system harmonics. In: *Power Quality in Modern Power Systems* (ed. P. Sanjeevikumar, C. Sharmeela and J.B. Holm-Nielsen), 77–85. Academic Press.

20 Mahendran, R., Sivaraman, P., and Sharmeela, C. (2015). Three phase grid interfaced renewable energy source using active power filter. In: *5th International Exhibition & Conference, GRIDTECH 2015*, 77–85. New Delhi, India.

21 Singh, B., Al-Haddad, K., and Chandra, A. (1999). A review of active filters for power quality improvement. *IEEE Transactions on Industrial Electronics* 46 (5): 960–971.

22 Patel, M.A., Patel, A.R., Vyas, D.R., and Patel, K.M. (2009). Use of PWM techniques for power quality improvement. *International Journal of Recent Trends in Engineering* 1 (4): 99–102.

23 Kale, M. and Ozdemir, E. (2005). Harmonic and reactive power compensation with shunt active power filter under non-ideal mains voltage. *Electric Power Systems Research* 74 (3): 363–370.

24 Akagi, H. (1996). New trends in active filters for power conditioning. *IEEE Transactions on Industry Applications* 32 (6): 1312–1322.

25 Baker, J.N. and Collinson, A. (1999). Electrical energy storage at the turn of the millennium. *Power Engineering Journal* 13 (3): 107–112.

26 Sivaraman, P., Sharmeela, C., and Kothari, D.P. (2017). Enhancing the voltage profile in distribution system with 40 GW of solar PV rooftop in Indian grid by 2022: a review. In: *1st International Conference on Large Scale Grid Integration Renewable Energy*, September 2017 in New Delhi, India.

27 Lawder, M.T., Suthar, B., Northrop, P.W.C. et al. (2014). Battery energy storage system (BESS) and battery management system (BMS) for grid-scale applications. *Proceedings of the IEEE* 102 (6): 1014–1030.

28 Chen, H., Cong, T.N., Yang, W. et al. (2009). Progress in electrical energy storage system: a critical review. *Progress in Natural Science* 19 (3): 291–312.

29 Feehally, T., Forsyth, A.J., Todd R. et al. (2016). Battery energy storage systems for the electicity grid: UK research facilities. In: *The 8th IET International Conference on Power Electronics, Machines and Drives (PEMD) 2016*. IET Power Electronics, Machines and Drives 2016, 19–21 April 2016, Glasgow, UK. Institution of Engineering and Technology. ISBN 978-1-78561-188-9, https://doi.org/10.1049/cp.2016.0257.

30 Gallardo-Lozano, J., Milanés-Montero, M.I., Guerrero-Martínez, M.A., and Romero-Cadaval, E. (2012). Electric vehicle battery charger for smart grids. *Electric Power Systems Research* 90: 18–29.

31 Xu, X., Bishop, M., Oikarinen, D.G., and Hao, C. (2016). Application and modeling of battery energy storage in power systems. *Journal of Power Energy System* 2 (3): 82–90.

32 Sivaraman, P. and Sharmeela, C. (2017). Battery energy storage system addressing the power quality issue in grid connected wind energy conversion system. In: *1st International Conference on Large-Scale Grid Integration Renewable Energy in India*, 1–3. New Delhi, India.

33 EmilioGhiani, G.P. (2018). Impact of renewable energy sources and energy storage technologies on the operation and planning of smart distribution networks. In: *Operation of Distributed Energy Resources in Smart Distribution Networks* (ed. K. Zare and S. Nojavan), 25–48.

34 Wang, G., Konstantinou, G., Townsend, C.D. et al. (2016). A review of power electronics for grid connection of utility scale battery energy storage systems. *IEEE Transactions on Sustainable Energy* 7 (4): 1778–1790.

35 Xavier, L.S., Amorim, W., Cupertino, A.F. et al. (2019). Power converters for battery energy storage systems connected to medium voltage systems: a comprehensive review. *BMC Energy* 1 (1): 1–15.

36 Sivaraman, P. and Sharmeela, C. (2020). Introduction to electric distribution system. In: *Handbook of Research on New Solutions and Technologies in Electrical Distribution Networks* (ed. B. Kahn, H.H. Alhelou and Ghassan), 1–31. Hershey, PA: IGI Global.

37 Eltawil, M.A. and Zhao, Z. (2010). Grid-connected photovoltaic power systems: technical and potential problems – a review. *Renewable and Sustainable Energy Reviews* 14 (1): 112–129.

38 Khalid, M.R., Alam, M.S., Sarwar, A., and Asghar, M.J. (2019). A comprehensive review on electric vehicles charging infrastructures and their impacts on power-quality of the utility grid. *eTransportation* 1: 100006.

39 Sivaraman, P., Sharmeela, C., and Logeshkumar, S. (2021). Charging infrastructure layout and planning for plug-in-electric vehicles. In: *Cable Based and Wireless Charging Systems for Electric Vehicles: Technology, Control, Management and Grid Integration* (ed. P. Sanjeevikumar), 1–24. UK: IET.

40 Sivaraman, P. and Sharmeela, C. (2021). Power quality problems associated with electric vehicle charging infrastructure. In: *Power Quality in Modern Power Systems* (ed. I.S. Padmanaban, C. Sharmeela and J.B. Holm-Nielsen), 151–161. Academic Press.

41 Teymouri, A., Mehrizi-Sani, A., and Liu, C.C. (2018). Cyber security risk assessment of solar PV units with reactive power capability. In: *IECON 2018-44th Annual Conference of the IEEE Industrial Electronics Society*, 2872–2877. IEEE.

2

Modeling and Analysis of Smart Power System

Madhu Palati[1], Sagar Singh Prathap[2], and Nagesh Halasahalli Nagaraju[3]

[1]*Department of Electrical and Electronics Engineering, BMS Institute of Technology and Management, Affiliated to Visvesvaraya Technological University, Doddaballapur Main Road, Avalahalli, Yelahanka, Bengaluru, India*
[2]*Energy and Power Sector, Center for Study of Science Technology and Policy, Bengaluru, India*
[3]*Power System Studies, Power Research & Development Consultants Pvt Ltd, Bengaluru, India*

2.1 Introduction

In recent days, the demand for electrical energy is increasing due to industrial growth, standard of living, etc. The generation capacity has to be increased to meet the increase in load demand [1]. One of the methods is the setting up new power plants, which increases the investment cost. Due to the environmental pollution and depletion of fossil fuels, nowadays renewable energy-based power plants are becoming more popular in power and energy sector [2]. The energy generated from renewable energy sources is irregular/not uniform in nature which may cause instability of the grid. One of the methods to minimize the grid instability is by employing energy storage technologies [3]. Another method is to reduce the losses by proper analysis and implementation of the distribution system [3]. Due to the advancement in technology, the conventional grid along with new emerging technology, communication devices, electrical equipment, and renewable energy sources leads to the smooth control of electrical power system networks. Proper planning and control of the power system are required to make it effective and act as a smart grid. The implementation proper modeling and analysis of a smart grid are very much necessary for the smooth operation of the smart grid.

Smart grid consists of different layers, power system network, communication system, sensor system, and software system. To meet the increase in demand, integrating the number of renewable energy sources, connected devices, and deployment of smart meters is increasing. Proper control and integration between each layer are necessary to make the system effective by maintaining proper security. To protect the modern grid from malicious software's, cyber-attacks, high-level security has to be provided. Maintaining a smart grid also requires measurement techniques to capture the data at each end, transmit the data, analyze the data, and take proper control action accordingly. In a smart grid, information exchange takes place in both ways from power utilities to consumers and vice versa.

Due to the increase in demand for power, many types of energy sources are integrated into the power system network to make the system reliable. The government is showing interest in the generation of power through renewable energy sources; one of the main renewable sources is power generation through the wind. Generation through wind is cost-effective and does not produce any CO_2 emissions. The generation is strongly dependent on the wind. If forecasting is done with a certain degree of uncertainness, it leads to a mismatch in the generation schedule that has been submitted transmission network operator. To compensate for this loss in generation, another plant generation has to be altered. Power evacuation from this wind farm is a major concern as it requires reactive power for the excitation of induction generators that are used in wind generation [4]. Feasibility studies have to be done to properly evaluate and evacuate the wind power from the wind farm. As of February 2022 total operating capacity of wind power is 40 130 MW, the Government of India through National Wind Energy has installed more than 800 wind monitoring stations all over the country (https://www.indianwindpower.com/wind-energy-in-india.php). Apart from this, India is blessed with offshore wind generation and the target set by the government is 30 MW alone through offshore wind generation by 2030 (https://mnre.gov.in/wind/current-status/). Currently, most of the installations happened in seven states: Andhra Pradesh, Gujarat, Karnataka, Madhya Pradesh, Maharashtra, Rajasthan, and Tamil Nadu.

Proper planning and forecasting are required to evacuate the wind power. As the generation of wind power is dynamic in nature, the injection of wind power into the grid pose challenges to the grid. To main the voltage profile during the light load conditions, as a practice, some of the transmission lines are opened. This results in the tripping of wind turbine generators (WTGs). Therefore, load flow analysis has to be done to estimate the necessary parameters that are impacting the grid stability. Power flow analysis gives information about reactive power consumption, overloaded lines, and weak buses with low volt profiles [5, 6]. The power that is generated from the plants has to be evacuated and distributed to the load centers. Therefore, there is a need to study power evacuation; in this chapter feasibility of various aspects is studied in evacuating the power and case studies are studied.

In this chapter, the need for modeling a power system, advantages of single line diagram, representation of quantities in per unit system, and components used for construction of a single line diagram are presented. Types of disturbance in power system and their consequences on the system behavior were studied. Steady state analysis, load flow analysis, and methods to perform load flow studies are discussed. Need for short circuit analysis, types of faults, and their impact on the power system under different operating conditions are presented. Causes of harmonics in the power system, influence of harmonics, and mitigation of harmonics are discussed. Effect of increase or decrease in load demand on frequency and the methods to regain frequency stability under such conditions are discussed. Effect of terminal voltage at different buses due to sudden closing, or opening of lines, transmission lines subjected to overload, and methods to maintain voltage stability are presented. Two case studies are studied to analyze the power system when subjected to disturbance.

2.2 Modeling of Equipment's for Steady-State Analysis

The interconnected system is complex; the whole system can be represented as a single-line diagram which is a pictorial representation of various components used in the power system. It is preferred to express voltage, current, power, and impedance in per unit rather than actual values. Per unit (pu) system is more advantageous since most of the manufacturers specify impedance in pu and don't differ much from the impedance values of machines with different ratings, the pu value of transformer on both sides of a transformer is the same. Pu values of each component used in the single line diagram will be represented using standard symbols published by IEEE with respective pu values as shown in Figure 2.1.

To study the behavior of power system under different operating conditions, the whole power system has to be modeled in a form of simulation model. The model has to be constructed with different block and each block represents the component of the power system network. The main components of the power system network are overhead lines, transformers, underground cables, shunt reactors, capacitor banks, etc., and all these components are modeled as a two-port model [6]. The disturbances that occur on power system are sequenced or occurrence of combination of different faults simultaneously will be implemented on developed model and simulation is performed.

2.2.1 Load Flow Analysis

Load flow analysis is also called as power flow analysis; this is applied to a power system to find the bus voltages, real and reactive power flow flowing in different lines, power factor at each bus or node, etc. This information helps

Figure 2.1 Single-line diagram of power system network.

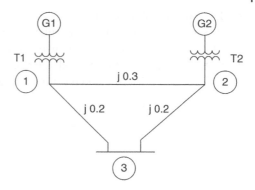

Table 2.1 Known and unknown quantities of various buses.

S. no	Bus type	Quantities specified	Quantities to be determined
1	Load bus or *PQ* bus	*P*, *Q*	|*E*|, δ
2	Generator bus or *PV* bus	*P*, |*E*|	*Q*, δ
3	Slack bus or reference bus	|*E*|, δ	*P*, *Q*

an electrical engineer to investigate the lines which are overloaded; information related to generators overloading, bus voltage violations, buses with low power factor, losses in the system, and the system can be designed in a better way to overcome the above problems. Also, for designing a new system this load flow analysis has to be computed. Also, for expanding the existing network it is very much required. Computing load flow analysis manually is a tedious process and involves a lot of time. Many iterations have to be done to converge the solution, which becomes difficult for any engineer to solve the system by numerical methods. Therefore, the required data is fed to the computer, and load flow analysis is run by choosing the optimal method and the results are obtained [7].

Load flow analysis is important for designing and planning of a new power system network and also when changes are made to the existing network, i.e. adding or removing the transmission lines based on future demand. Load flow analysis gives steady-state voltages at all buses for a given load. This has to be carried out for different loading conditions and the whole system has to be analyzed. By performing load flow analysis, information such as voltage violation at the substation, overloaded lines, overloaded transformers and power flow in the transmission line as well as in the transformers can be known. This study also helps in deciding the economic operation of a power system.

There are four main quantities which are required for performing load flows; they are a magnitude of voltage (|*E*|), phase angle (δ), real power (*P*), and reactive power (*Q*). Depending on the prior values of these quantities available, the power system buses are classified into

1. Load bus
2. Generator bus
3. Slack bus

The known and unknown parameters of these buses are listed in Table 2.1.
The methods for solving the load flow problems are as follows:

1. Gauss–Seidel method
2. Newton Raphson method
3. Decoupled Load flow method

2.2.1.1 Gauss Seidel Method

Gauss Seidel method is an iterative method used to solve linear algebraic equations; at each iteration the solution is updated with new voltages till convergence is reached or till the magnitudes of voltages do not differ much with the tolerance value. In this method, the rate of convergence is slow and cannot be applied for large power system. The voltage at bus i is calculated using the expression

$$V_i = \frac{1}{Y_{ii}} \left[\frac{P_i - jQ_i}{V_i^*} - \sum_{\substack{k=1 \\ K \neq i}}^{n} Y_{ik} V_k \right] \quad i = 2, 3, \ldots, n \tag{2.1}$$

2.2.1.2 Newton Raphson Method

The Newton Raphson method is used to solve nonlinear algebraic equations, and this method is superior compared to Gauss Seidel method, convergence happens at fast rate, and can be applied to complex power system also. The equation used in this method can be expressed in Matrix form

$$[\Delta P \Delta Q] = [J_1 J_2 J_3 J_4][\Delta \delta \Delta |E|] \tag{2.2}$$

where $J_1, J_2, J_3,$ and J_4 are Jacobian elements and are partial derivatives of ΔP and ΔQ with respect to δ and $|E|$, respectively. Real power and reactive power are calculated using the expression

$$P_{i,\text{cal}} = G_{ii}|V_i|^2 + \sum_{\substack{k=1 \\ K \neq i}}^{n} |V_i||V_k| (G_{ik} \cos \delta_{ik} + B_{ik} \sin \delta_{ik}) \tag{2.3}$$

$$Q_{i,\text{cal}} = -B_{ii}|V_i|^2 + \sum_{\substack{k=1 \\ K \neq i}}^{n} |V_i||V_k| (G_{ik} \sin \delta_{ik} + B_{ik} \cos \delta_{ik}) \tag{2.4}$$

Power residues are calculated using

$$\Delta P_i^k = P_{i,\text{Sch}} - P_{i,\text{cal}} \tag{2.5}$$

$$\Delta Q_i^k = Q_{i,\text{Sch}} - Q_{i,\text{cal}} \tag{2.6}$$

At the end of each iteration bus voltages, phase angle is updated and the iteration has to be continued till the difference in the real power and reactive power is less than the tolerance value

$$\delta_i^{k+1} = \delta_i^k + \Delta \delta_i^k \tag{2.7}$$

$$\left| E_i^{k+1} \right| = \left| E_i^k \right| + \Delta \left| E_i^k \right| \tag{2.8}$$

$$\left| \Delta p_i^k \right| \leq \varepsilon \tag{2.9}$$

$$\left| \Delta Q_i^k \right| \leq \varepsilon \tag{2.10}$$

2.2.1.3 Decoupled Load Flow Method

In Newton Raphson method, in each and every iteration inverse matrix of Jacobian matrix has to be computed, for a large interconnected power system, the size of Jacobian matrix increases and computing the inverse matrix takes time for every iteration. To overcome this problem, decoupled load flow method is used. The change in voltage magnitude at any bus will primarily affect reactive power and does not affect real power. Therefore, the terms with derivative of real power with respect to voltage are made zero. The change in voltage phase angle at any bus will primarily affect real power and does not affect reactive power. Therefore, the terms with a derivative of reactive power with respect to voltage phase are made zero. With the above considerations, the equations expressed in matrix form are shown using an equation

$$[\Delta P \Delta Q] = [H\, 0\, 0\, L] \left[\Delta\delta \; \frac{\Delta|V|}{|V|} \right] \tag{2.11}$$

$$\frac{\Delta P}{|V|} = [B'][\Delta\delta] \tag{2.12}$$

$$\frac{\Delta Q}{|V|} = [B''] \left[\frac{\Delta|V|}{|V|} \right] \tag{2.13}$$

where B' is the imaginary part of Y_{BUS} except slack bus and B'' is the imaginary part of Y_{BUS} for all load buses.

2.2.2 Short Circuit Analysis

A normal operating system sometimes may undergo disruption due to disturbance in the system like insulation breakdown, shorting of lines, overheating of equipment due to loose contacts, overloading of lines, cables, etc., which are termed as faults in the power system. Therefore, there is a need to study the current magnitudes in each line when there is an occurrence of fault in the system. The fault can occur at generator terminals or in transmission lines or at load terminals. Magnitude of current has to be determined for different fault locations and accordingly the protection system and scheme have to be designed to clear the fault before the equipment's get damaged [8, 9]. Handono et al. [10] developed a model for protecting a building against the short circuits and developed protective system was designed based on short circuit studies.

In power system, the faults can be broadly classified into two types

1. Symmetrical faults
2. Unsymmetrical faults

2.2.2.1 Symmetrical Faults

Symmetrical faults are all three phases coming in contact with another or all the three phases coming in contact with the ground. Here, the system remains in balanced condition; therefore, finding the fault current in any one phase is enough and that will be same fault current flowing through the other two phases. The network is represented in Thevenin's equivalent circuit and the fault current can be estimated by using the equation

$$I_f = \frac{V\text{th}}{Z\text{th} + Z_f} \tag{2.14}$$

where,

I_f is the fault current,
Vth is the open circuit voltage across the fault incident terminals,
Zth is the equivalent impedance seen from the fault incident terminals,
Z_f is the fault impedance.

Also, the fault current can be estimated using Z_{BUS}

1. Form Y_{BUS} and compute $Z_{BUS,}$
2. Get the value of Z_f
3. Find the fault current at any node k and current in the line i to j is given by

$$I_k = \frac{E_k}{Z_{kk} + Z_f} \tag{2.15}$$

$$I_{ij} = \frac{E_{ij}}{Z_{ij} + Z_f} \tag{2.16}$$

2.2.2.2 Unsymmetrical Faults

These faults cause unbalance in the system and these can be analyzed using symmetrical components [8]. Unsymmetrical faults are categorized into two types

1. Shunt type
2. Series type

Shunt type of faults is classified as

(i) Single line to ground fault (SLG)
(ii) Line to line fault (LL)
(iii) Double line to ground fault (LLG)

Series type of faults is classified as

(i) One conductor open
(ii) Two conductors open

The fault current in SLG is given by

$$I_f = \frac{3E_a}{Z_1 + Z_2 + Z_0} \tag{2.17}$$

The fault current in LL is given by

$$I_f = \frac{-j\sqrt{3}E_a}{Z_1 + Z_2} \tag{2.18}$$

The fault current in LLG is given by

$$I_{a1} = \frac{3E_a}{Z_1 + \frac{Z_2 Z_0}{Z_2 + Z_0}} \tag{2.19}$$

$$I_{a0} = \frac{-E_a + I_{a1} Z_1}{Z_0} \tag{2.20}$$

$$I_f = 3I_{a0} \tag{2.21}$$

where

I_{a1}, I_{a2}, and I_{a0} are positive; negative; and zero sequence currents.
Z_1, Z_2, and Z_0 are positive; negative; and zero sequence impedances.
E_a is the prefault voltage.

2.2.3 Harmonic Analysis

The power system network is operated mostly at 50 or 60 Hz. The converters, inverters, and power electronic circuits, switching power supplies produce distortion in voltage and current waveforms in terms of multiples of the fundamental frequency. This, results in distortion in waveform which is different from sinusoidal waveform often termed as harmonics. Most of the harmonics are generated due to nonlinear loads connected to the power system network. Harmonics cause overheating of equipment, power loss, heating of neutral conductor and cause damage to the equipment. Also, harmonics cause interference in the nearby communication lines, cause resonance and often leading to abnormal operation of protective equipment. Therefore, there is a need to identify and mitigate the harmonics. Figure 2.2 shows the sinusoidal and distorted waveform.

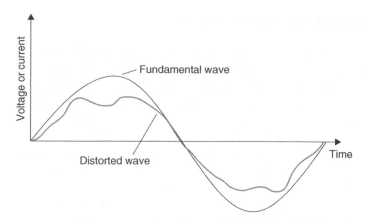

Figure 2.2 Fundamental sinusoidal waveform and distorted waveform.

Total Harmonic distortion (THD) of current and voltage waveforms is expressed as

$$\text{THD}_i = \sqrt{\frac{\Sigma_{n=2}\, i_n^2}{i_1^2}} \tag{2.22}$$

$$\text{THD}_v = \sqrt{\frac{\Sigma_{n=2}\, v_n^2}{v_1^2}} \tag{2.23}$$

where i_n and v_n are the rms values of nth Fourier components of current and voltage waveforms $i(t)$ and $v(t)$, respectively [11].

$$\text{THD}_{\text{even}} = \sqrt{\frac{i_2^2 + i_4^2 + i_6^2 + \ldots + i_{2n+2}^2}{i_1^2}} \tag{2.24}$$

$$\text{THD}_{\text{odd}} = \sqrt{\frac{i_3^2 + i_5^2 + i_7^2 + \ldots + i_{2n+1}^2}{i_1^2}} \tag{2.25}$$

$$\text{THD} = \sqrt{\frac{i_2^2 + i_3^2 + i_4^2 + \ldots + i_n^2}{i_1^2}} \tag{2.26}$$

Electrical utility companies must monitor the harmonic levels in the distribution system and analysis of harmonics is very important for utility companies [12]. If attention is not paid to these harmonics, level of power quality gets reduced. Many researchers have worked and developed different techniques to estimate, measure the harmonics present in the system [13, 14]. Some of the methods include Fast Fourier transform, wavelet method, statistical approach method, least square method, neural networks, phasor measuring units, dynamic approach-based algorithm [15–20] were proposed by the researchers to study the harmonic distortion in power system network. By using line reactors harmonic, current distortion can be reduced and these need to be designed to withstand full load current [21]. Passive filters are economical and simple, are used to compensate the harmonics, and does not have a significant effect on voltage and power factor [22]. Active filters are costly and complex in size, but these filters can effectively mitigate harmonics in converter and inverter circuits [23]. In distribution networks, distribution Static synchronous compensator (D-STATCOM) is used to improve power factor and mitigate the harmonics at the source end, as the network size increases, components required increases and installation cost increases rapidly [24].

2.3 Modeling of Equipments for Dynamic and Stability Analysis

The main aim of power system is to generate the power and distribute it to the customers by maintaining the voltage, current, and frequency values under the prescribed limits even during disturbances in the power system. It has to get sustained in case of small disturbances without losing its stability. Even during the large disturbance, the system has to be designed to withstand these disturbances without causing any damage to the system. To maintain the system in stable condition, Proper coordination between the relays associated with circuit breakers and the dynamics governed by generators, load and control devices should be maintained. Typically, the fault is cleared by the circuit breaker between 3 and 10 cycles. Due to advancement in the technology, modern circuit breakers are capable of clearing the fault in 1.5 cycles itself. Certain instances there is delay in operation time of relay leading to variation of voltage and current magnitudes caused by system dynamics which cannot be ignored. Due to increased load demand and during the heavy load condition the power flow in the lines has exceeded the limits, leading to cascading tripping of lines and causing instability. Conventional stability methods used are not able to handle this situation, this led to many researchers to work on new methods of determining the stability. For simulating the transient studies, many computer simulation packages [25] are available in the market like ETAP, EMTP-RV, MIPOWER, PSCAD, PSSE, and MATLAB. All the dynamic elements, generators, load, control devices are represented using differential equations. Each component used in the power system network will have block created in the software package, based on the requirement the specification, rating of each component can be changed accordingly and a developed model in ETAP software is shown in Figure 2.3. All the components are integrated as per the single line diagram and different analysis can be performed, switching operations can be selected, i.e. which circuit breaker to be opened or closed, type of system radial system or ring main system can be decided [26].

The Steady state Stability problem is classified as

1. Steady State Stability

   (i) Static Stability
       The stability which can be maintained without the help of automatic control devices
   (ii) Dynamic Stability
       The stability which can be maintained with the help of automatic control devices

2. Transient Stability

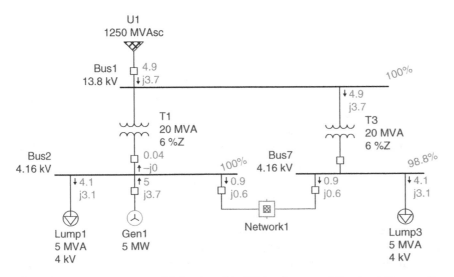

Figure 2.3 Power system model developed in ETAP software and the simulation results.

The ability of the power system to maintain synchronism under large disturbances like short circuits, generation loss, load loss and line outages is called transient stability. This depends on initial operating condition and severity of disturbance.

By performing transient stability analysis, we can obtain the information about system, whether it can maintain stability during the disturbance. Helpful in determining the power transmission capability between two interconnected systems. Also, these studies will help the design engineer in the planning stage to develop a new system or modification to the existing system.

Power system consists of many interconnected synchronous machines. During normal operating conditions, the relative position is fixed between the rotor axis and resultant magnetic field, this is often termed as power angle or torque angle. During the disturbance, the rotor accelerates or decelerates. This relative motion is non-linear and is expressed as a second order differential equation and is called as swing equation, this equation describes the swing of the rotor during the disturbance. This causes changes in the network structure, therefore power angle curve prior to fault, during fault and post fault will be different. The time taken to clear the fault is known as fault clearing time. If the fault exists for a long time, system remaining in unstable state increases. Therefore, the fault has to be cleared at the earliest and the maximum time taken to clear the fault before losing stability is called critical fault clearing time. Most of the faults occur in the power system are SLGs and often gets cleared automatically. Therefore, to maintain continuity of power supply, modern circuit breakers are equipped with auto-reclosure facility. If the fault still persists the breakers open permanently. Transient stability of a power system can be determined using a Swing equation. Swing equation is given by

$$M \frac{\mathrm{d}^2 \delta_m}{\mathrm{d}t^2} = P_a = P_m - P_e \qquad (2.27)$$

where,

M is the Inertia constant.
δ_m is the angular displacement in mechanical radians.
P_m is the shaft power.
P_e is the electrical power output.
P_a is the acceleration power.

One of the important methods to determine the transient stability is using Equal area criteria method. Figure 2.4 shows power angle versus fault clearing time.

Even after the fault is cleared, If the power angle increases indefinitely with time, then the system is unstable till the machine loses synchronism. If the power angle increases, reaches maximum and starts decreasing as shown in Figure 2.4 the system becomes stable. In Multi machine system the swing equation has to be solved for each machine. Therefore, numerical methods are used to determine the fault clearing time. Common methods used for solving the equation are

Figure 2.4 Power angle versus time for stable and unstable system.

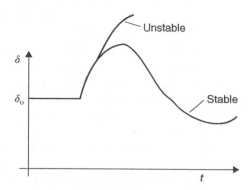

1. Point by point method
2. Modified Euler's method
3. Runge Kutta method
4. Milnes Predictor method

Classification of power system stability:

1. Rotor angle stability
   (i) Small signal stability
   (ii) Transient stability
2. Frequency stability
   (i) Short term stability
   (ii) Long term stability
3. Voltage stability
   (i) Large disturbance voltage stability
   (ii) Small disturbance voltage stability

Rotor angle stability is the ability of power system in which the interconnected synchronous machines remain in synchronism after getting subjected to disturbance. During the disturbance the acceleration or deceleration of rotor happens, resulting one generator to run faster compared to another generator. Rotor angular difference increases, leading to oscillations and instability occur.

2.4 Dynamic Analysis

When a power system undergoes a disturbance, the property of a system to regain its original state of equilibrium is termed as power system stability.

Often there will be a disturbance on the power system, this can be a large disturbance or small disturbance. When a power system is subjected to these disturbances it may deviate from the operating conditions and there are chances of losing stability. During a disturbance most of the time the power system appears to be quasi steady state and changes to transient state for a brief period. The disturbance can be due to faults on the lines such as LL, SLG, LLG, generator outage, load shedding.

2.4.1 Frequency Control

When a system undergoes a disturbance, ability of the power system to maintain steady frequency is termed as frequency stability. If the disturbance is large, it leads imbalance between the generators and load. Instability occurs in the form of frequency swings which lead to outages of generators or tripping of loads.

Due to increase in load demand, integrating new generation plants to the existing system, the load being dynamic in nature, poor coordination between the control devices, protective devices lead to a mismatch between load and generation. Increase in load demand directly affects the load on the generator and the rotor speed will decrease which is directly related to frequency, leading frequency instability. Frequency variation for long duration affect the power system security, stability, operation and performance of the system. Therefore, the system frequency has to be maintained within the nominal value. Frequency controllers are used to maintain the system frequency and the size of the controller depends on the type of disturbance and the magnitude of imbalance between generation and load. Under normal operating conditions small change in load changes the frequency and this can be compensated using primary loop control of synchronous generator. Since every synchronous machine has drooping characteristic, accordingly speed changes and the frequency changes and settle down at new value. The tie line power flow values of an interconnected area will change and these will be different from the scheduled values [27].

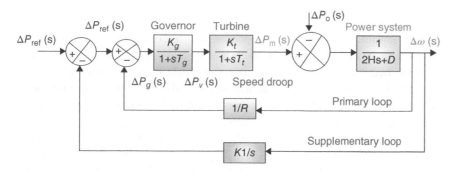

Figure 2.5 Block diagram of automatic generation control.

During the abnormal condition, the variation in frequency increases, secondary loop control or Load frequency control system is activated to bring back the frequency to normal value from a few seconds to a minute's time thereby restoring the frequency and allocating the generation reserve to meet the scheduled power. The Primary loop main function is adjusting the turbine output to match with the load demand. All the generating units which are connected to the system will take up the change in generation and the optimal generation units will be selected. As the load changes there will be changes in the frequency deviation, restoration of frequency is done by supplementary control or secondary loop bringing the frequency deviation to zero. Frequency control which includes a supplementary control is often termed as an Automatic generation control (AGC) and the block diagram with supplementary control is shown in Figure 2.5. In case of severe fault or large disturbance, secondary loop control is not sufficient to handle the situation and will not be able to compensate the frequency which may lead to cascading tripping of line, generator outages. To handle this situation tertiary control is activated using standby power resources. In the worst case if the frequency control is not possible load shedding is done [28].

Power system is interconnected through different areas of generation and load centers as shown in Figure 2.6. The power can be distributed from one area to another through tie lines. Load frequency control regulates the power flow between the areas by maintaining frequency constant.

Area control error is given by

$$\text{ACE} = (P_{\text{tie}} - P_{\text{sch}}) + B_f \Delta_f = \Delta P_{\text{tie}} + B_f \Delta_f \tag{2.28}$$

$P_{\text{tie}}, P_{\text{sch}}, B_f, \Delta_f$ are tie line power, scheduled power, frequency bias constant and frequency deviation, respectively. ACE is negative indicates that net power flow is low and generation has to be increased in that area. Each area will have one ACE, if there are two areas interconnected through tie lines, the ACE of these two areas is given by

$$\text{ACE}_1 = \Delta P_{\text{tie}-1} + B_{f1} \Delta_{f1} \tag{2.29}$$

$$\text{ACE}_2 = \Delta P_{\text{tie}-2} + B_{f2} \Delta_{f2} \tag{2.30}$$

If there is a sudden increase of load in area-2 and it is desirable to absorb the demand without change in frequency. This can be achieved by making the net power flow between these two areas zero. Depending on the condition the power flow between the two areas can be controlled by maintaining frequency constant.

Figure 2.6 Interconnection of areas through tie lines.

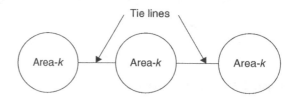

2.4.2 Fault Ride Through

Wind generation is one of the forms of renewable energy source, the increase in load demand is met by these new plants which are commissioned. Wind farms pose lots of faults in the power system and to maintain the steady state operation new grid codes are established. Fault through capability is the capability of the wind turbine and protective devices to remain connected to the power system network during the grid faults. During the grid faults the wind generators have to withstand high over current and power swings, earlier in this type of situation the wind generator was disconnected. Nowadays, disconnection of wind generators is not allowed and it has to stay connected. Authors in [29, 30], have investigated over current relay coordination (OCR) without FRT requirements and with FRT requirements case studies were compared with the new proposed method using a Fault current limiter. The tests were performed on an IEEE 30 and IEEE 33 bus system and the method was able to clear the voltage drop within the specified FRT time. Authors in [31, 32] have studied the impact of FRT capabilities of different generators which vary in terms of size and rating of different manufacturing company components and control strategies were discussed. Voltage swell will appear in the system during the fault recovery time, result in fluctuations in the DC link voltage of renewable energy conversion system (REC), causing reverse power flow and finally trips the entire system [33, 34]. High voltage ride through (HVRT) along with chopper circuit was implemented to reduce the voltage fluctuations. Simulation results were satisfactory and concluded that based on grid code requirement RECs cannot be regulated directly [35].

2.5 Voltage Stability

During the disturbance, the ability of the power system to maintain steady state voltages is termed as voltage stability [36]. Voltage drop occurs when there is large amount of reactive power consumption in the power system network. Large disturbances include sudden loss of load, cascading failure of transmission lines, contingencies. Over excitation of synchronous machine results in rise in capacitive voltage leading to uncontrolled over voltages in the system [37]. Analysis can be done using PV curves and QV curves as shown in Figure 2.7. From PV curves, information related to Variation of real power transfer between one bus to another bus affecting the bus voltages is studied and P_{max} is determined [38]. The Power of a particular area is increased in steps and voltage is observed, PV curve is plotted and real power margin is determined, V_{crit} is the critical voltage corresponding to Maximum Real Power P_{max}. Reactive Power margin is obtained from Q-V curves, from these curve's information related to reactive power absorption or injection for different voltages is studied. These curves are used to assess the voltage stability [5, 39]. The power system operated at low voltage draws more current to produce the power, hence the left side portion is unstable and system cannot be operated in this region. Disadvantage with these methods is at a time only one bus is considered for varying the load, information related to critical buses is not known, power

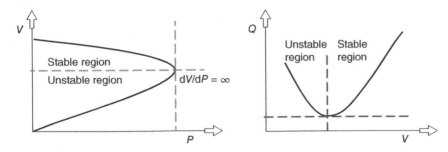

Figure 2.7 *PV* and *QV* characteristics.

flow studies to be done for all buses which is a time-consuming process. Also, the loading point reaches the critical point convergence problem occurs.

Preventive steps that can be taken to avoid voltage collapse are using Load tap changer controller, tap changers are blocked on the current tap which prevents the further deterioration of voltage [40]. The load side voltage is decreased, which simultaneously reduces the load power. By doing load shedding the voltage can be brought down to acceptable limits, but consumers get affected, hence this measure is given the least priority in controlling the voltage stability [41]. Opting for capacitive compensation and turning off the inductive compensation, Increasing the voltage set point of generator increases the generated voltage [42]. Rescheduling the generation in the area where voltage instability exists, but this involves complex procedures [43].

2.6 Case Studies

2.6.1 Case Study 1

In this case study, a power system network in Tirunelveli in India is considered for studying the power flow analysis, short circuit studies. The power system network is interconnected with different types of power generation and largely it consists of a greater number of WTGs. It consists of total 156 buses [44]. Single line diagram of the wind farm network is shown in Figure 2.8. Power flow analysis is conducted, voltage profile and overloading of lines is observed for different conditions of mixed generation of power from different plants. Percentage of wind power generation, Reactive Power and Real Power is shown in Table 2.2 and graph showing Percentage of wind power generation versus Reactive Power and Real Power loss is shown in Figure 2.9. With wind generation as nil, hydro generation meeting the peak demand and thermal generation meeting the base load, only three lines in the range 75–100% got overloaded. With 40% of wind generation and remaining met by another generation, there was no overloading on any of the lines. With 70% of wind penetration, it is observed that nine lines got overloaded ranging from 102% to 115%. With 80% of wind penetration, it is observed that 18 lines got overloaded ranging from 101% to 147% and also at four buses low voltage profile was observed.

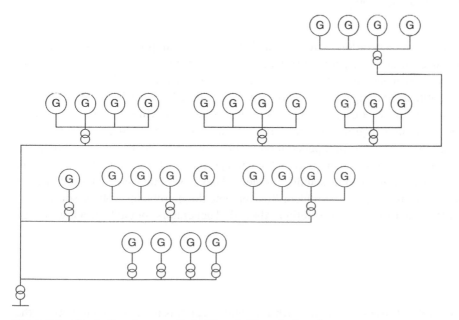

Figure 2.8 Single-line diagram of wind farm in Tirunelveli region.

Table 2.2 Data of wind power generation, reactive power, and power loss.

S. no	% Wind generation	Reactive power (MVAR)	Power loss (MW)
1	10	56.97	19.4
2	20	60.97	21.6
3	30	61.36	22.3
4	40	88.01	29.1
5	50	120.57	41.9
6	60	164.44	60.6
7	70	263.76	84.1
8	80	389.01	114.1

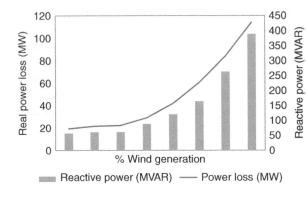

Figure 2.9 Percentage wind generation versus reactive power and real power loss.

It is observed that as the wind power generation increases reactive power consumption increases as well as power loss increases. The transmission lines are getting overloaded and low voltage profile on the number of buses increased. Symmetrical and Unsymmetrical faults were conducted on the network and short circuit studies were analyzed. 28 buses were identified as Strong buses and 7 buses were identified as weak buses. It can be concluded that these weak buses will lead to power evacuation problems if percentage of wind generation increases. Therefore, reactive power compensation is required at wind farms. Dynamic compensation can be planned at substations. Also, it was observed that when the wind penetration was less than 50% the power evacuation was possible with existing lines and there was no overloading, when the wind power penetration into the grid is more than 50% it is causing overloading of lines and leading stability problem. Weak buses that were identified in short circuit analysis are more prone to causing evacuation problem. To overcome these problems, voltage source converter based Multi terminal HVDC system is a feasible solution. Also, wind generators should have a low voltage ride through capability.

2.6.2 Case Study 2

A 50 MW power generated from a Wind firm situated in North Karnataka region is to be interconnected to the 66 kV line of a 220/66 kV substation and is shown in Figure 2.10.

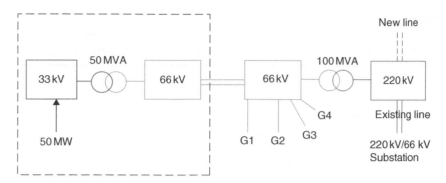

Figure 2.10 Schematic diagram of wind power connected to 220/66 kV substation.

The 50 MW wind generation is pooled at the 33 kV level and stepped up to 66 kV through 50 MVA, 33/66 kV transformer. 66 kV line is connected to 66/220 kV substation through two transformers each of rating 100 MVA. Different scenarios were studied for evacuating the power from the 50 MW wind farm.

2.6.2.1 Existing and Proposed Generation Details in the Vicinity of Wind Farm

Interconnecting Grid station, type of generation, its capacity and status of the line are tabulated in Table 2.3

The 50 MW power has to be evacuated through a double circuit line of 16 km length. The transformer used in the proposed work is 33/66 kV, 50 MVA, with positive sequence impedance of 15% and vector configuration as YNyn0. The transmission line parameters considered for the study are provided in the Table 2.4. The thermal ratings of the conductors have been calculated based on the following assumptions:

Solar radiations = 1045 W/m^2
Wind speed = 2000 m/h
Absorption coefficient = 0.8
Emissivity coefficient = 0.45
Age >1 year

Conductor temperature of 75 °C and ambient temperature of 45 °C.

As per Central Electricity authority (CEA), the voltage levels during normal and Emergency condition are given in Table 2.5.

Table 2.3 Transmission power flow in the vicinity of wind farm.

S. no	Interconnecting grid station	Type of generation	Capacity (MW)	Status
1	66 kV side of substation to G1	Solar	20	Existing
2	66 kV side of substation to G2	Solar	30	Existing
3	66 kV side of substation to G3	Wind + Solar	82	Existing
4	66 kV side of substation to G4	Solar	30	Existing
5	220 kV side of substation to existing line	Wind + Solar	30	Existing
6	220 kV side of substation to new line	Wind	50	New

Table 2.4 Transmission line parameters.

Voltage level (kV)	Type of conductor	R (Ω)	X (Ω)	B (Mhos)	Thermal rating (MVA)
220	Zebra	0.0810	0.3970	1.46E^06	213
220	Drake	0.083	0.428	1.37E^06	220
66	Coyote	0.2598	0.4009	1.44E^06	33.38
66	Cat	0.25984	0.40094	1.443E^006	31.32
66	Rabbit	0.237	0.373	1.5E^006	19.433

Table 2.5 Voltage ratings of different transmission lines under normal and emergency condition.

Nominal voltage (kV)	Normal condition		Emergency condition	
	Maximum (kV)	Minimum (kV)	Maximum (kV)	Minimum (kV)
765	800	728	800	713
400	420	380	420	372
230	245	207	245	202
220	245	198	245	194
132	145	122	145	119
110	123	99	123	97
66	72.5	60	72.5	59

2.6.2.2 Power Evacuation Study for 50 MW Generation

The load flow study constitutes a starting point of steady state analysis of power system and its ability to carry power under various system conditions of generations and loads. The study computes power flows in various circuit elements of the network and complex voltages at all the buses in the system. ETAP is the computational software tool used for this purpose. The study reveals the electrical performance and power flows (real and reactive), voltages for specified conditions when the system is operating under steady state. Load flow studies are of great importance in:

- Reliable evacuation of power from new power plants.
- Planning and designing the future expansion of power systems as well as in determining the best operation of existing systems.
- In computing the magnitude and phase angle of the voltage at each bus.
- In determining real and reactive power loadings of transmission lines and transformers (as well as losses) throughout the system and whether these networks are overloaded or not under normal and specified contingency situations.

In addition to providing a prediction of the loading conditions on power systems, a load flow solution is required as a necessary precursor to:

- System fault study required for designing substation equipment short circuit ratings, earth mat designs, etc.
- Transient stability studies.

Table 2.6 Power flow from the windfarm to grid substations without interconnection.

S. no	Line details	Power flow in MW, +ve: import, −ve: export	
		50 MW loading	60 MW loading
1	66 kV side of substation to G1	−11	−13
2	66 kV side of substation to G3	24	34
3	220 kV side of substation to existing	−73	−85
4	220 kV side of substation to new line	53	50

Following are the considerations made for load flow studies:

- All the loads are assumed as constant *PQ* loads.
- The load projections (bus wise) are as per area transmission utility.
- As per CEA criteria, voltage tolerances considered for 765 and 400 kV voltage levels are ±5% and ±10% for voltage levels 220 kV and below.
- Slack bus specified voltage is set to 1.0 pu and angle at 0°.
- Transmission line length is suitably assumed wherever exact details are not available.

2.6.2.3 Without Interconnection of the Proposed 50 MW Generation from Wind Farm on 66 kV Level of 220/66 kV Substation

Load flow studies are conducted for two cases. Case 1 is with 50 MW and Case 2 is with 60 MW loading under the base case scenario. The observation made from this is tabulated in Table 2.6.

2.6.2.4 Observations Made from Table 2.6

Without interconnection of the proposed 50 MW wind generation from wind farm to 220/66 kV substation (substation), it is observed that the 66 kV lines from substation to G1 and from substation to G3 would be loaded to around 11 and 28 MVA under scenario-1 and around 14 and 41 MVA under scenario-2, respectively. 220 kV lines from substation to Existing line and from substation to new line would be loaded to around 74 MVA/circuit and 53 MVA/circuit under scenario-1 and around 88 MVA/circuit and 51 MVA/circuit under scenario-2, respectively.

It is observed that the 66 kV transmission line corridor from substation – G3 would be loaded beyond the thermal loading limit of 66 kV line as specified by the CEA under both scenarios-1 and 2, respectively. Further the 66 kV line from substation to G3 would be loaded beyond the thermal loading limit of conductor line as specified by the CEA. However, these over loadings are mainly due to the Renewable generations interconnected at 66 kV substation and will not have any impact on the evacuation of proposed 50 MW wind generation. The bus voltages and transformer loadings in the vicinity of wind farms under study are within their acceptable limits in both the scenarios.

2.6.2.5 With the Interconnection of Proposed 50 MW Generation from Wind Farm on 66 kV level of 220/66 kV Substation

Normal condition (i.e. without considering contingency): This case has been analyzed to assess the transmission line loadings in the vicinity of site under study with the interconnection of proposed 50 MW generation from wind farm at 100% capacity factor. Transmission line power flows in the immediate vicinity of wind farm is presented in Table 2.7

- Slack bus specified voltage is set to 1.0 pu and angle at 0°.
- Transmission line length is suitably assumed wherever exact details are not available.

Table 2.7 Power flow from the windfarm to grid substations with interconnection.

S. no	Line details	Power flow in MW, +ve: import, −ve: export	
		50 MW loading	60 MW loading
1	66 kV side of substation to G1	−15	−15
2	66 kV side of substation to G3	20	31
3	220 kV side of substation to existing	−85	−99
4	220 kV side of substation to new line	45	42

Table 2.8 With 50 MW generation line details on 66 and 220 kV side of substation.

S. no	Line details	Power flow in MW, +ve: import, −ve: export	
		Scenario-1	Scenario-2
1	66 kV side of substation to G1	−15	−15
2	66 kV side of substation to G3	20	31
3	220 kV side of substation to existing line	−85	−99
4	220 kV side of substation to new line	45	42

2.6.2.6 Normal Condition without Considering Contingency

Generation line details on 66 and 220 kV side of operation with 50 MW of generation are presented in Table 2.8.

With the interconnection of proposed 50 MW wind generation from wind farm to 220/66 kV substation, it is observed that the 66 kV lines from substation to G1 and from substation to G3 would be loaded to around 17 and 22 MVA under scenario-1 and around 18 and 37 MVA under scenario-2, respectively. 220 kV D/C lines from substation to existing line and from substation to the new proposed line would be loaded to around 87 MVA/circuit and 45 MVA/circuit under scenario-1 and around 103 MVA/circuit and 43 MVA/circuit under scenario-2, respectively. The bus voltages, transmission line, and transformer loadings in the vicinity of wind farm under study are within their acceptable limits in both the scenarios during normal condition except the overloading of the 66 kV corridor from substation – G3.

2.6.2.7 Contingency Analysis

Based on the load flow analysis carried out, it is observed that the critical scenario for evacuating power from proposed 50 MW wind generation at wind farm is scenario-2., i.e. Corresponding demand during combined peak wind and solar generation and local light load along with corresponding conventional generation dispatch. In view of this, following $N-1$ contingency studies have been exercised for the most critical power flow scenario, i.e. scenario-2.

- Outage of 220 kV Single circuit line from substation to existing line.
- Outage of 66 kV line from substation to G1.
- Outage of one unit of 2×100 MVA transformer at 220/66 kV substation.
- Transmission line power flows in the immediate vicinity of wind farm during the above $N-1$ contingencies are presented in Table 2.9.

During the outage of either 220 kV line from substation to existing line or during the outage of 66 kV line from substation to G1, it is observed that all the bus voltages, line, and transformer loadings in the vicinity of wind

Table 2.9 Transmission line power flows in the vicinity of wind farm.

S. no	Contingency details	Power flow in MW, +ve: import, −ve: export			
		66 kV		220 kV	
		SS to G1	SS to G3	SS to existing line	SS to new line
1	Outage of 220 kV line from substation to existing line	−16	30	−175	32
2	Outage of 66 kV line from substation to G1	–	31	103	39
3	Outage of one 100 MVA transformer in substation	20	27	−96	43

farm are within their acceptable limits except the over loading of 66 kV corridor from Substation to G3. During the outage of one unit of 2×100 MVA transformer at 220/66 kV substation, it is observed that the other unit would be loaded to around 113 MVA which is around 113% of the transformer rating. This loading is acceptable for short duration. However, if it persists for long duration, then evacuation would be constrained during this $N-1$ contingency condition. This overloading is observed only under scenario-2. Further chance of occurrence of this contingency and scenario-2 at the same time is a very rare condition.

2.6.2.8 With the Interconnection of Proposed 60 MW Generation from Wind Farm on 66 kV Level of 220/66 kV Substation

Normal condition (i.e. without considering contingency): This case has been analyzed to assess the transmission line loadings in the vicinity of site under study with the interconnection of 60 MW generation from wind farm site at 90% capacity factor. Transmission line power flows in the immediate vicinity of Kudligi wind farm which is presented in Table 2.10.

With the interconnection of proposed 60 MW wind farm generation (90% capacity factor) to 220/66 kV substation, it is observed that the 66 kV lines from substation to G1 and from substation to G3 would be loaded to around 18 MVA and 22 MVA under scenario-1 and around 18 MVA and 37 MVA under scenario-2, respectively. 220 kV D/C lines from substation to existing line and from substation to new line would be loaded to around 89 MVA/circuit and 45 MVA/circuit under scenario-1 and around 104 MVA/circuit and 42 MVA/circuit under scenario-2, respectively.

The bus voltages, transmission line, and transformer loadings in the vicinity of wind farm under study are within their acceptable limits in both the scenarios during normal condition except the over loading of 66 kV corridor from substation to G3. Evacuation of 60 MW wind generation with 90% capacity factor, i.e. 54 MW from wind farm site to 220/66 kV substation is feasible during normal condition. However, if the generation capacity reaches 100% of its installed capacity, then 66 kV D/C coyote conductor line from wind farm pooling substation to 220/66 kV

Table 2.10 With 60 MW generation line details on 66 and 220 kV side of substation.

S. no	Line details	Power flow in MW, +ve: import, −ve: export	
		Scenario-1	Scenario-2
1	66 kV side of substation to G1	−16	−16
2	66 kV side of substation to G3	20	31
3	220 kV side of substation to existing line	−86	−100
4	220 kV side of substation to new line	45	41

substation and the wind farm pooling substation transformer would be loaded beyond the acceptable loading limit resulting in evacuation constraint.

2.7 Conclusion

In this chapter, review work carried out by the researchers in the area of contingency analysis, harmonic analysis, steady-state stability, voltage stability, frequency stability, modeling of power system is studied. The power evacuation from wind farm mainly depends on the network topology. If the capacity of the windfarm is increased, during system events, if the deviation in oscillation of frequency is more than the acceptable limits, then it leads to unstable operation. Also, real and reactive power losses increase and reactive power compensation has to be provided. In short circuit studies, weak buses are identified and the voltage profile of those buses is improved by implementing suitable voltage improvement techniques. The stability problem in wind farm can be minimized by selecting the wind generators with a low fault ride through and most of the researchers concluded that for obtaining FRT solution, the grid codes are taken into account. Also, simulation was carried on two models proposed in the two case studies. Load flow studies, contingency, and short circuit analysis were carried out on the two proposed models. All the parameters related to transformer loading, bus voltages, generator overloading, and transmission line overloading are found to be well within the acceptable limits.

References

1 Strasser, T., Siano, P., and Vyatkin, V. (2015). Guest editorial new trends in intelligent energy systems – an industrial informatics points of view. *IEEE Transactions on Industrial Informatics* 11 (1): 207–209.

2 Brown, M.A. (2014). Enhancing efficiency and renewables with smart grid technologies and policies. *Futures* 58: 21–33.

3 Sparacino, A.R., Reed, G.F., and Kerestes, R.J. et al. (2012). Survey of battery energy storage systems and modelling techniques. In: *Proceedings of IEEE Power and Energy Society General Meeting*, IEEE, San Diego, CA, USA, pp. 1–8.

4 Ranjan, A., Karthikeyan, P., Ahuja, A. et al. (2009). Impact of reactive power in power evacuation from wind turbines. *Journal of Electromagnetic Analysis and Applications* 1: 15–23. https://doi.org/10.4236/jemaa.2009.11004.

5 Kundur, P. (1994). *Power System Stability and Control*. Mc-Grawhill.

6 Ramanujam, R. (2009). *Power System Dynamics Simulation and Analysis*. PHI Learning Pvt Ltd.

7 Martinez, J.A. and Mahseredjian, J. (2011). Load flow calculations in distribution systems with distributed resources. A review. In: *2011 IEEE Power and Energy Society General Meeting*, IEEE, Detroit, MI, USA, pp. 1–8. https://doi.org/10.1109/PES.2011.6039172.

8 Kersting, W.H. and Phillips, W.H. (1990). Distribution system short circuit analysis. In: *Proceedings of the 25th Intersociety Energy Conversion Engineering Conference*, IEEE, Reno, NV, USA, pp. 310–315. https://doi.org/10.1109/IECEC.1990.716901.

9 Sun, P., Jiao, Z., and Hanwen, G. (2021). Calculation of short-circuit current in DC distribution system based on MMC linearization. *Frontiers in Energy Research* 9: 1–9. https://doi.org/10.3389/fenrg.2021.634232.

10 Handono, K., Tukiman, Putra, I.M. et al. (2019). Short circuit analysis on electrical power supply building # 71 BATAN for case reliability study of nuclear power plant electrical protection system. *AIP Conference Proceedings* 2180: 020036. https://doi.org/10.1063/1.5135545.

11 Steinmetz, C.P. (1916). *Theory and Calculation of Alternating Current Phenomena*, vol. 4. McGraw-Hill Book Company, Incorporated.

12 Association, I.S. et al. (2014). IEEE recommended practice and requirements for harmonic control in electric power systems. In: *IEEE Std 519-2014 (Revision of IEEE Std 519-1992)*, https://doi.org/10.1109/IEEESTD.2014.6826459. New York: IEEE.

13 Sekar, T.C. and Rabi, B.J. (2012). A review and study of harmonic mitigation techniques. In: *2012 International Conference on Emerging Trends in Electrical Engineering and Energy Management (ICETEEEM)*, 93–97. IEEE.

14 Gonen, T. (2015). *Electric Power Distribution Engineering*. CRC Press.

15 Winograd, S. (1978). On computing the discrete Fourier transform. *Mathematics of Computation* 32 (141): 175–199.

16 Harris, F.J. (1978). On the use of windows for harmonic analysis with the discrete Fourier transform. *Proceedings of the IEEE* 66 (1): 51–83. 138 Bibliography.

17 Girgis, A.A., Chang, W.B., and Makram, E.B. (1991). A digital recursive measurement scheme for online tracking of power system harmonics. *IEEE Transactions on Power Delivery* 6 (3): 1153–1160.

18 Chang, G., Chen, C., Liu, Y., and Wu, M. (2008). Measuring power system harmonics and interharmonics by an improved fast Fourier transform-based algorithm. *IET Generation, Transmission and Distribution* 2 (2): 192–201.

19 Rogoz, M. and Hanzelka, Z. et al. (2007). Power system harmonic estimation using neural networks. In: 2007 9th International Conference on Electrical Power Quality and Utilisation, EPQU 2007, pp. 1–8. IEEE.

20 Melo, I.D., Pereira, J.L., Variz, A.M., and Garcia, P.A. (2017). Harmonic state estimation for distribution networks using phasor measurement units. *Electric Power Systems Research* 147: 133–144.

21 Ujile, A. and Ding, Z. (2016). A dynamic approach to identification of multiple harmonic sources in power distribution systems. *International Journal of Electrical Power & Energy Systems* 81: 175–183.

22 Pires, I.A., Machado, A., Murta, M., and de Braze, Cardoso Filho, J. (2016). Experimental results of a thyristor switched series reactor for electric arc furnaces. In: *2016 IEEE Industry Applications Society Annual Meeting*, 1–5. IEEE.

23 Dekka, A.R., Beig, A.R., and Poshtan, M. (2011). Comparison of passive and active power filters in oil drilling rigs. In: *11th International Conference on Electrical Power Quality and Utilisation*, 1–6. IEEE.

24 Rohouma, W., Balog, R.S., Peerzada, A.A., and Begovic, M.M. (2020). D-STATCOM for harmonic mitigation in low voltage distribution network with high penetration of nonlinear loads. *Renewable Energy* 145: 1449–1464.

25 (2014). *Introductory Guide to ETAP for Power Students*. Irvine, CA: ETAP.

26 Khan, R.A.J., Junaid, M., and Asgher, M.M. (2009). Analyses and monitoring of 132 kV grid using ETAP software. In: *2009 International Conference on Electrical and Electronics Engineering, ELECO*. IEEE, Bursa, Turkey, pp. I-113, I-118 (5–8 November 2009).

27 Bevrani, H., Golpîra, H., Messina, A.R. et al. (2021). Power system frequency control: an updated review of current solutions and new challenges. *Electric Power Systems Research* 194: 2021. https://doi.org/10.1016/j.epsr.2021.107114.

28 Bevrani, H., Watanabe, M., and Mitani, Y. (2014). *Power System Monitoring and Control*. Hoboken, NJ, USA: IEEE Wiley.

29 Yoosefian, D. and Chabanloo, R.M. (2020). Protection of distribution network considering fault ride through requirements of wind parks. *Electric Power Systems Research* 178 (2020): 106019. https://doi.org/10.1016/j.epsr.2019.106019.

30 Bayrak, G., Ghaderi, D., and Sanjeevikumar, P. (2021). Chapter 4 - Fault ride-through (FRT) capability and current FRT methods in photovoltaic-based distributed generators. In: *Power Quality in Modern Power Systems* (ed. P. Sanjeevikumar, C. Sharmeela, J.B. Holm-Nielsen and P. Sivaraman), 133–149. ISBN: 9780128233467. https://doi.org/10.1016/B978-0-12-823346-7.00008-6. Academic Press.

31 Meri, A., Amara, Y., and Nichita, C. (2018). Impact of fault ride-through on wind turbines systems design. In: 2018 7th International Conference on Renewable Energy Research and Applications (ICRERA), IEEE, Paris, France, pp. 567–575. https://doi.org/10.1109/ICRERA.2018.8566910.

32 Li, R., Geng, H., and Yang, G. (2016). Fault ride-through of renewable energy conversion systems during voltage recovery. *Journal of Modern Power Systems and Clean Energy* 4: 28–39. https://doi.org/10.1007/s40565-015-0177-0.

33 El-Moursi, M. (2011). Fault ride through capability enhancement for self-excited induction generator-based wind parks by installing fault current limiters. *IET Renewable Power Generation* 5: 269–280.

34 Ngamroo, I. and Karaipoom, T. (2014). Cooperative control of SFCL and SMES for enhancing fault ride through capability and smoothing power fluctuation of DFIG wind farm. *IEEE Transactions on Applied Superconductivity* 24: 1–4.

35 Chen, L., Deng, C., Zheng, F. et al. (2015). Fault ride-through capability enhancement of DFIG-based wind turbine with a flux-coupling-type SFCL employed at different locations. *IEEE Transactions on Applied Superconductivity* 25: 1–5.

36 Hosseinzadeh, N., Aziz, A., Mahmud, A. et al. (2021). Voltage stability of power systems with renewable-energy inverter-based generators: a review. *Electronics* 10 (115): 2021. https://doi.org/10.3390/electronics10020115.

37 Kundur, P., Paserba, J., Ajjarapu, V. et al. (2004). Definition and classification of power system stability. *IEEE Transactions on Power Systems* 19 (2): 1387–1401.

38 Van Cutsem, T. and Vournas, C. (1998). *Voltage Stability of Electrical Power Systems*. New York: Springer Science.

39 Taylor, C.W. (1994). *Power System Voltage Stability*. New York: McGraw-Hill.

40 Zhu, T.X. and Tso, S.K. (2011). An investigation into the OLTC effects on voltage collapse. *IEEE Transactions on Power Systems* 15 (2): 515–521.

41 Vournas, C.D. and Karystianos, M. (2004). Load tap changers in emergency and preventive voltage stability control. *IEEE Transactions on Power Systems* 19 (1): 492–498.

42 Xu, F. and Wang, X. (2011). Determination of load shedding to provide voltage stability. *Electric Power System Research* 33 (3): 515–521.

43 Van Cutsem, T. and Vournas, C.D. (2007). Emergency voltage stability controls: an overview. In: *Proceedings of IEEE/PES General Meeting*, Tampa, FL, USA (24–28 June).

44 Kumudini Devi, R.P., Somsundaram, P., and Anbuselvi, S.V. (2014). *Power Evacuation Studies for Grid Integrated Wind Energy Conversion Systems*. Centre for Wind Energy Technology. http://www.cwet.res.in.

3

Multilevel Cascaded Boost Converter Fed Multilevel Inverter for Renewable Energy Applications

Marimuthu Marikannu[1], Vijayalakshmi Subramanian[1], Paranthagan Balasubramanian[1], Jayakumar Narayanasamy[2], Nisha C. Rani[2], and Devi Vigneshwari Balasubramanian[2]

[1]*Department of EEE, Saranathan College of Engineering, Trichy, India*
[2]*Department of EEE, The Oxford College of Engineering, Bommanahalli, Bangalore, India*

3.1 Introduction

Power electronics uses electronics for the control of high power using high power tubes like ignitron, mercury arc rectifiers, and thyratrons. The recent improvements of power semiconductor devices like SCR, IGBT, TRIAC, Power transistors, and Power MOSFET play an important role in handling larger power. Power electronics offers different converters namely rectifiers, inverters, chopper, and Cyclo converter. The main function of power electronic devices is to act like a switch [1–4].

DC chopper provides fixed DC voltage to varying DC voltage without using a transformer which is employed in many other applications like battery chargers, trolley cars, speed control of DC motor, electric traction, etc. This operation can be attained by controlling a switch that periodically turns ON and OFF. The arrival of fast self-quenching devices like power transistors, GTOs, and power MOSFETs enabled them to be normally used as a switch [5–7].

The non-linear choppers alter one level of DC potential to a different stage by varying the duty cycle ratio of the switch in the circuit. Hence choppers are compact in structure, highly effective, and provide an efficiency of 70–97%. Hence, they are commonly used in cellular phones, Laptop, computer peripherals, and medical equipment to generate various stages of DC voltages. The advantages of DC to DC converters are simplicity in design, put the least stress on the switch and necessitates reduced output filter for low output ripple [8–11].

Electricity has become the basic need in day-to-day life; hence to fulfill the requirement of electricity and meet the demands, an alternate energy resource called renewable sources are the most desirable ones. These renewable resources are classified into PV, Biomass, and Wind energy system. Choppers are broadly applied to interface the input sources (PV, fuel cell, and ultra-capacitor) to the grids or standalone loads through inverters is shown in Figure 3.1. Normally PV panel, fuel cells, and ultra-capacitors output voltage are at low level in the order of 20–60 V. Hence, high ratio gain converters are commonly used to interface such input sources with inverter. Higher level DC to DC converters are needed to enhance the low output level (20–60 V) from PV panels to a higher level of around 380 V to match the input of single phase 240 V bridge inverter in AC network connected in the electrical systems [9–13].

The DC choppers of high gain are widely categorized as isolated and non-isolated DC–DC converters. In the category of isolated converters, there is a necessity to use a transformer whereas in the non-isolated category inductors are used instead of transformers.

In order to acquire the rated high gain in isolated DC to DC converters, it requires high transformer ratio which in turn affects the total system performance. Switches in isolated converters are also suffered by large voltage

Artificial Intelligence-based Smart Power Systems, First Edition.
Edited by Sanjeevikumar Padmanaban, Sivaraman Palanisamy, Sharmeela Chenniappan, and Jens Bo Holm-Nielsen.
© 2023 The Institute of Electrical and Electronics Engineers, Inc. Published 2023 by John Wiley & Sons, Inc.

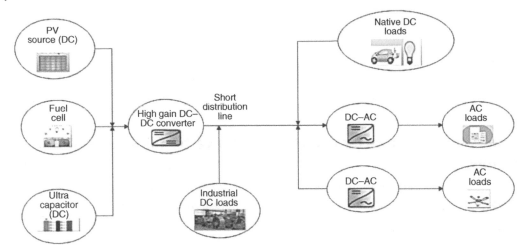

Figure 3.1 Requirement of high gain converters.

spikes, consequently raising the switching losses due to the transformer's leakage reactance. In a non-isolated topology, passive components are dominant to produce an enhanced gain in the circuit. Hence, non-isolated high boost converters are denser and further structured. The efficiency of the total grid-connected PV system is based on the number of converter stages and thus the DC to DC non-isolated converter are the most important portion for the applications of PV-connected system [14, 15].

To rise the voltage, gain in a normal boost converter, many derived topologies are available such as cascaded boost converters (CBCs), multilevel boost converters (MLBCs), boost converters with coupled inductors, center source MLBCs, interleaved boost, and hybrid converters. A traditional boost converter provides twice the input voltage when operated with 50% duty cycle. The current and potential stress across the diodes and MOSFETs switches and diodes are within the limits when the boost converter duty cycle ratio is 50%. However, for an increased duty cycle, the stress level of voltage and current increases which exceed the limits and deteriorates the voltage gain and performance.

CBC provides the essential enhanced gain but it agonizes from high voltage switching stress. Interleaving converters have minimum ripple input current, and current stress across the switch, resulting in improved power transmit efficacy but low voltage gain performance owing to the circulating current problem. With a coupled inductor correct turn ratio, and extreme duty ratio, it is possible to attain a larger voltage conversion ratio, but the leakage inductances effect reduces the efficiency slightly. Thus, the cascaded converters are not appropriate for non-conventional energy applications expecting superior voltage increase, and large power management system.

To magnify the voltage alteration ratio and reduce the voltage stresses in the switching device, concept of MLBC or multilevel cascaded boost converter (MCBC) is acquired from the boost converter [16]. The suggested topology yields the voltage output and that is N times the output voltage comprises of normal boost DC–DC converter, with $2N$ diodes and $2N-1$ capacitors. Multilevel boost and MCBCs are constructed in a modular way, so that any number of capacitors and diode can be clamped without affecting the basic circuit to attain the required voltage gain. MCBC is preferred to produce high voltage gain since its simple structure such that without disturbing the main circuit, any levels could be generated by inserting capacitors, and diodes clamped to the conventional DC–DC converters. Furthermore, the switching stress of multilevel converter can be reduced when compared with cascaded boost topology. The suggested topology holds a number of merits such as: (i) high conversion ratio of output voltage for reduced duty cycle, (ii) non-isolation converter without transformer, and (iii) higher efficiency with less voltage stress.

A multilevel inverter (MLI) is a circuit that produces a preferred multilevel AC voltage output as many times the rated input DC voltages. The high-power inverter should produce less total harmonic distortion (THD), high

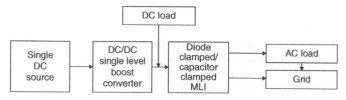

Figure 3.2 Multilevel inverter scheme with one DC source.

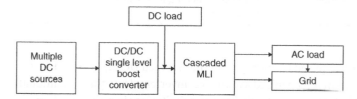

Figure 3.3 Multilevel inverter scheme with multiple DC sources.

quality sinusoidal AC voltage and power. To achieve the important terms in inverter, it should be operated with multilevel concept so that it can generate sinusoidal output with high efficiency. Almost all the non-conventional energy sources are DC, in order to attain pure sinusoidal AC, it is required to design filters which may raise the effective price of the system. Therefore, in such applications, the high-cost filters can be replaced with MLIs. There are single source MLIs, and multiple source MLIs are available, which are broadly classified into H-bridge multi-source, and single source inverters using huge condenser banks.

Conventional multilevel converter is exposed in Figure 3.2 consumes larger number of switches. For normal photovoltaic applications single DC source is stepped up using a boost DC–DC converter and the obtained output potential is fed to the diode clamped or capacitor clamped configuration for obtaining the multilevel AC output voltage.

MLIs block diagram constitutes more DC sources are shown in Figure 3.3. Every source is coupled to a step-up converter and the boost DC–DC converter outputs are coupled in cascaded fashion to meet the multilevel AC output potential. In this structure, there are too many numbers of boost converters used which shoot up the total price of the scheme.

The schematic diagram shown in Figure 3.4 represents the proposed MLI output in which a single source is converted to a source of multilevel output which is fed to a level circuit and H-bridge inverter type to achieve the necessary output.

In large power implementations, the output of the renewable input sources is needed to boost by step up converters. The suggested converter is used to build for low- and high-power applications. The preferred topology comprises of single source with multi-stage step-up output, whose voltage output is set to the H-bridge inverter source input through the multilevel circuit. Thus, the suggested MLI uses single voltage source, least count of switches, capacitors, and diodes for obtained for M-level stepped sinusoidal output voltage waveform.

Applications of the extended MLI circuit are: Solar powered Telecom equipment in remote areas, battery operated vehicles, home inverters supply, and renewable energy micro-grid.

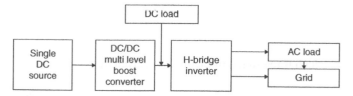

Figure 3.4 Multi boost converter fed MLI scheme with single DC source.

Table 3.1 Difference between conventional inverter and MLI.

S. No.	Conventional inverter	Multilevel inverter (MLI)
1.	High THD content	Reduced THD content
2.	Stress in switch is high	Reduced switching stress
3.	It is not suggested to high voltage systems	Suggested to high voltage systems
4.	Increased switching losses and very high switching frequency	Reduced switching losses and low switching frequency

An inverter normally produces an AC output with required frequency from the supplied DC input. The inverter which produces two different level of voltages or current is recognized as two level inverters. The square traditional inverter functions of high frequency switching suffer from losses of switching and limitations in rating for elevated applications of power and potential. High level of THD is additional problem. These difficulties collectively added make the normal inverters not suitable for interfacing renewable systems with grid. Hence, the researchers are very much interested in requirement for a diverse system of MLI.

A MLI is a circuit that produces a preferred voltage output as many times the rated input DC voltages. The high-power inverter should produce less THD, high quality sinusoidal AC voltage and power. To achieve the important terms in inverter, it should be operated with multilevel so that it can generate sinusoidal output.

The preferred topologies comprise of single source with multi-stage step-up topology whose voltage output is given to the conventional H-bridge inverter input through the level circuit. With the help of level circuit, M-level inverter performance is also achieved in this proposed circuit. Thus, the use of single voltage source, least count of switches, capacitors, and diodes can be obtained M-level stepped sinusoidal output voltage waveform.

MCBC fed MLI DC–DC converters are interfaced with level circuit and H-bridge inverter circuit to generate multilevel AC output voltage [17]. The difference between conventional inverter and MLI is given in Table 3.1.

3.2 Multilevel Cascaded Boost Converter

The converter topology is built with a single switching device with $2K$ diodes and $(2K-1)$ capacitors for K level of output voltage and generalized diagram is shown in Figures 3.5 and 3.6. The MCBC modules are designed with similar equations which are defined for CBC. The gain ratio (M) for the conventional CBCs output voltage is defined as [17].

$$M = \frac{1}{(1-d)^2} \tag{3.1}$$

K – number of potential voltage levels of converters,
M – Voltage gain ratio.

The formula for output voltage of CBC is

$$V_{\text{o}} = \frac{V_{\text{in}}}{(1-d)^2} \tag{3.2}$$

Considering K, the output voltage of five level CBC which is registered in Figure 3.6 is given by

$$V_{\text{o}} = K \frac{V_{\text{in}}}{(1-d)^2} \tag{3.3}$$

Figure 3.5 Generalized cascaded boost converters with *K*-multilevel output voltage.

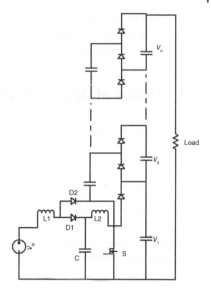

Figure 3.6 MCBC with five level structure.

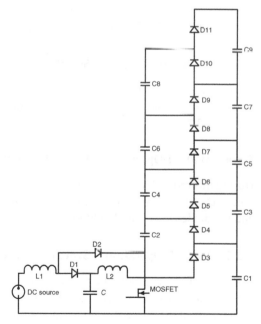

The inductances L_1, and L_2 are devised based on the subsequent equation, with suitable duty cycle ratio and power output

$$L_1 = \frac{d(1-d)^4 R}{2f} \tag{3.4}$$

$$L_2 = \frac{d(1-d)^2 R}{2f} \tag{3.5}$$

The value of the Capacitor can be obtained by,

$$\in = \frac{1-d}{2RfC} \tag{3.6}$$

Conventional multilevel converter consumes larger number of switches. For normal photovoltaic applications single DC source is stepped up using a boost converter and the obtained output voltage is fed to the diode clamped or capacitor clamped configuration for obtaining the multilevel AC output voltage.

3.3 Modes of Operation of MCBC

Operation of MCBC has 10 Modes. For the first five modes the switch S_A is closed. For the next five modes the switch S_A is opened.

3.3.1 Mode-1 Switch S_A Is ON

During Mode 1 operation switch S_A is made to conduct. Input voltage charges the inductor L_1 and capacitor C_A charges the inductor L_2. As an outcome, the current through the inductors starts increasing throughout this time of operation is revealed in Figure 3.7a.

3.3.2 Mode-2 Switch S_A Is ON

In mode-2 (Switch S_A is closed), if voltage across C_B is lesser than the potential across C_1, then the potential across C_B is clasped to a level equivalent to that on C_1 with the aid of D_4 diode and S_A switch. It is defined in Figure 3.7b.

3.3.3 Mode-3-Operation – Switch S_A Is ON

During mode-3-operation, if the sum of voltages $C_1 + C_2$ is greater than that across $C_B + C_C$, then the potential across C_B and C_C are fixed to that level of voltages over C_1 and C_2 through the switch S_A and diode D_6. It is offered in Figure 3.7c.

3.3.4 Mode-4-Operation – Switch S_A Is ON

During this mode, if the potential across $C_1 + C_2 + C_3$ are greater than that across $C_B + C_C + C_D$, then the potential on C_B, C_C, and C_D are clasped to that the potential across C_1, C_2, and C_3 correspondingly through the switch S_A and the diode D_8. It is highlighted in Figure 3.7d.

3.3.5 Mode-5-Operation – Switch S_A Is ON

During mode-5-operation, the potential on $C_1 + C_2 + C_3 + C_4$ are greater than that across $C_B + C_C + C_D + C_E$, then the potential across C_B, C_C, C_D, and C_E are clasped to that stage across C_1, C_2, C_3, and C_4 with the aid of the main switch S_A and the diode D_{10}. It is demonstrated in Figure 3.7e.

3.3.6 Mode-6-Operation – Switch S_A Is OFF

Mode 6 starts with switch S in open condition. Hence, the current flowing through the inductors L_1 and L_2 close the diode D_3. As an outcome, charge on the capacitor C_1 turn to discharge hence the potential on C_1 is clasped to a stage identical to that of the sum of the input supply inductor voltages $(V_{L1} + V_{L2})$. It is presented in Figure 3.7f.

3.3.7 Mode-7-Operation – Switch S_A Is OFF

In this mode D_5 closes, capacitor voltage C_B clamped to C_2 and input supply voltage plus the sum of inductor voltages $(V_{L1} + V_{L2})$ is clasped to the voltage across C_1 is explained in Figure 3.7g.

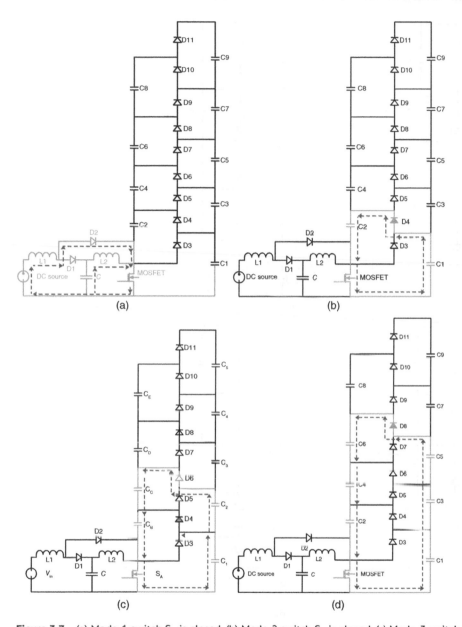

Figure 3.7 (a) Mode-1 switch S_A is closed, (b) Mode-2 switch S_A is closed, (c) Mode-3 switch S_A is closed, (d) Mode-4 switch S_A is closed, (e) Mode-5 switch S_A is closed, (f) Mode-6 switch S_A is opened, (g) Mode-7 switch S_A is opened, (h) Mode 8 switch S_A is opened, (i) Mode-9 switch S_A is opened, (j) Mode 10 Switch S_A is opened.

3.3.8 Mode-8-Operation – Switch S_A Is OFF

In this mode D_7 closes. Hence the capacitor voltage C_C is clamped to C_3. Capacitor voltage C_B is clamped to C_2. Input supply voltage plus sum of the inductor voltages $(V_{L1} + V_{L2})$ are clasped the voltage across C_3 and C_1, respectively, which is shown in Figure 3.7h.

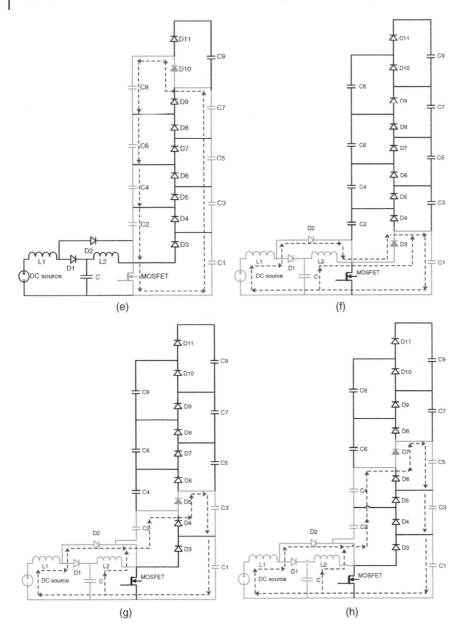

Figure 3.7 (*Continued*)

3.3.9 Mode-9-Operation – Switch S$_A$ Is OFF

In Mode-9, D$_9$ is closed. Hence, the capacitor voltage across C$_D$, C$_C$, C$_B$ are clasped to voltage across C$_4$, C$_3$, C$_2$, respectively. Input supply voltage plus sum of inductor voltages ($V_{L1} + V_{L2}$) are clasped the voltage across C$_1$ which is exposed in Figure 3.7i.

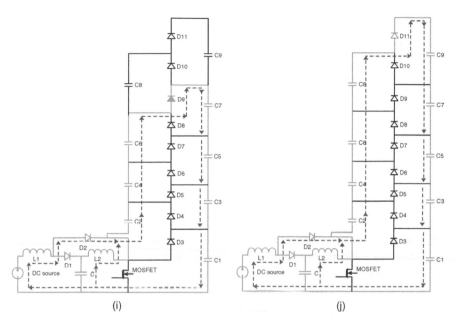

(i)   (j)

Figure 3.7 (Continued)

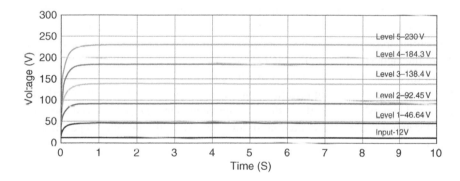

Lorem ipsum

Figure 3.8 Voltage obtained across R (load) for the suggested five level MCBC.

3.3.10 Mode 10-Operation – Switch S_A is OFF

In Mode-10, D_{11} closes. The capacitor voltages across C_E, C_D, C_C, C_B are clasped to C_5, C_4, C_3, C_2. The input supply voltage plus sum of inductor voltages $(V_{L1} + V_{L2})$ are clamped across C_1 respectively which is portrayed in Figure 3.7j.

3.4 Simulation and Hardware Results

The proposed non-isolated five stage MCBC is constructed with the components as given in the table and simulation output has been obtained with the aid of MATLAB/Simulink is portrayed in Figure 3.8.

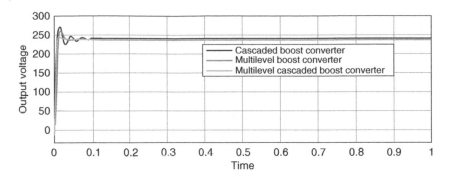

Figure 3.9 Output voltage obtained from CBC, MLBC, and MCBC.

In general, this converter is functioning as a traditional boost converter, and its acquired voltage output from an input is 12 V. By raising the number of diodes and capacitors in stages, the output voltages are further extended to 47, 93, 139, 185, and 230 V respectively. The components used in simulation and hardware are given in Table 3.2.

Further it is possible to achieve 230 V output voltage from 12 V input voltage using the following three converters namely CBC, MLBC, and suggested MCBC which is shown in Figure 3.9.

For achieving this output voltage in CBC, it should be operated with higher duty cycle. For obtaining 230 V from 12 V in MLBC, 10 levels have to be added in the modular structure. But the same output voltage is obtained in MCBC with five levels added to modular structure and also with minimum duty cycle when compared to CBC.

Figure 3.10 depicts potential across and current through an inductor L_1, L_2, capacitor C_A, and the switch S_A. From this waveform, 46.6 V is the potential stress in the switch S_A for five levels multilevel cascaded boost converter (MLCB) converter which is 19.33% of the highest voltage of the MCBC converter. Since the potential stress is less, lower rating MOSFET are enough for this greater output voltage converter circuits.

This converter produces a stiff output voltage with insignificant ripple across each leveland it is noteworthy to mention that there are no overshoots and undershoots and the system straighten outs more rapidly. It is important that, the level progression holds better and the capacitors potential differences are well controlled regardless of the differences in R (load) and the source.

Table 3.2 Components table for MCBC.

Parameter	Values
Source voltage (V_{in})	12 V
Frequency	10 kHz
Inductor L_1	0.6 mH
Inductor L_2	0.8 mH
Capacitor C_A	2 μF
Inductor ($L_1 = L_2$)	52 μH
Capacitors C_B, C_C, C_D, C_1, C_2, C_3, C_4, and C_5	300 μF
Duty cycle ratio (d)	50%
Switch-MOSFET	IRF540
Aurdino controller	1

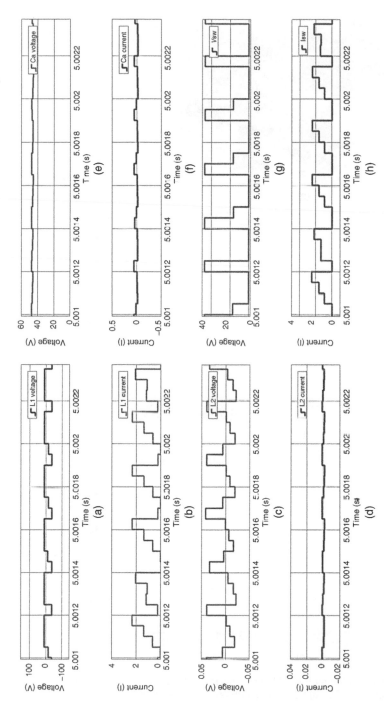

Figure 3.10 Current through and voltage across L_1, L_2, C_A, and switch of MCBC.

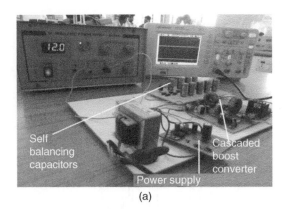

(a)

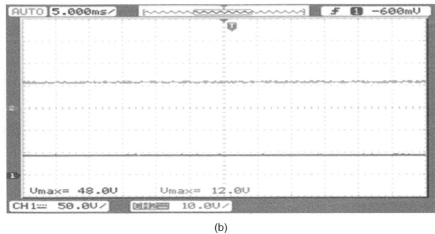

(b)

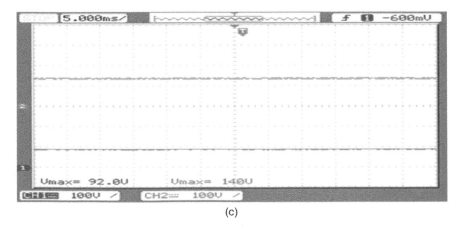

(c)

Figure 3.11 (a) Prototype of cascaded multilevel boost converter. (b) Input voltage and first level voltage of MCBC. (c) Second and third level voltage of MCBC. (d) Fourth and fifth level voltage of MCBC.

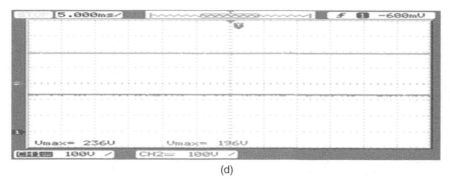

(d)

Figure 3.11 (*Continued*)

The hardware circuit is also constructed for the same five level MCBC with the input of 12 V, which is registered in Figure 3.11a. The obtained stage 1 output voltage is 48.1 V that is shown in Figure 3.11b, stage 2 and stage 3 output voltages are 93 and 141 V, respectively, which are shown in Figure 3.11c. The stage 4 and stage 5 output voltages are 197 and 237 V that are illustrated in Figure 3.11d.

3.5 Prominent Structures of Estimated DC–DC Converter with Prevailing Converter

Additionally, simulation circuit has been broadened to examine completely the enactment of the MCBC by comparing the output gain and it is presented in assessment with the other topologies in Table 3.3. In Table 3.3, recent converter topologies suggested in the references [3–8], respectively are associated against the offered non isolated MCBC.

3.5.1 Voltage Gain and Power Handling Capability

It is clear from the table that the recommended converter has been intended for 200 W yields a highest voltage gain of 19.33 which generates output voltage 236 V for an input voltage of 12 V with 50% duty cycle. Here, large boosting of voltage is attained with a single switch configuration itself. Hence, the recommended MCBC evidences its high-gain authority over the other conventional converters which are shown in Figures 3.12 and 3.13.

3.5.2 Voltage Stress

Usually voltage stress is stated that as a percentage of voltage across the output, while comparing from Table 3.3, it is proved that [3, 7] and the proposed converter are having the capacity to reduce voltage stress less than 20%. But references [3, 7] are subjected to a smaller amount voltage stress only for small potential gain ratios when related with the suggested converter. Henceforth, from Table 3.3 it is demonstrated that the recommended converter alone has low stress for high gain ratio which is shown in Figure 3.14. Therefore, it has noteworthy that the potential stress is more reduced when enhances the stage of converter.

3.5.3 Switch Count and Geometric Structure

From Table 3.3, it is clear that the number of switches count needed is more than one for all the converters excluding [7, 8] and the suggested converter is presented in Figure 3.15.

Table 3.3 Assessment of proposed MCBC with appropriate converter.

Review	[3]	[4]	[5]	[6]	[7]	[8]	Proposed
V_{in}	15	60	24	60	15	24	12
V_o	200	590	380	1100	200	220	236
Duty cycle	0.65	0.615	0.56	0.55	0.62	0.6	0.5
Gain	14	10	16	18	13	9	20
Power management capability (kW)	0.4	0.87	0.56	3	0.10	0.125	0.2
Gain formula	$\dfrac{4n}{1-D}$	$\dfrac{3n+1}{1-D}$	$\dfrac{4n+1}{1-D}$	$\dfrac{3(1+n)}{1-D}$	$\dfrac{3+D(n-1)}{1-D}$	$\dfrac{3+2D}{1-D}$	$N\dfrac{1}{(1-D)^2}$
Gain extension technique	Inter-leaved, CI and voltage quadrupler	Inter-leaved 3 winding CI and VMC	CI, diode, capacitor stages	Inter-leaved, lift capacitor CI and VMC	Super-lift	Switched-CI, diode capacitor multiplier	Cascaded, self-balancing capacitor with multilevel concept
Peak-voltage stress on switch	$\dfrac{V_o}{4n}$	$\dfrac{V_{cr}}{n}$	$\dfrac{V_o}{4n+1}$	$\dfrac{V_o}{1+nK}$	$\dfrac{V_o+V_{in}(n-1)}{n+2}$	$\dfrac{1}{1-D}V_{in}$	$\dfrac{V_{in}}{(1-D)^2}$
Voltage stress (% of V_o)	12.5	24.4	51.55	36.36	16.11	27.27	19.49
Switch count	6	2	2	3	1	1	1
Diodes	4	8	4	10	4	4	11
Capacitor	7	5	5	8	4	4	10
Total component	20	17	13	24	10	10	24

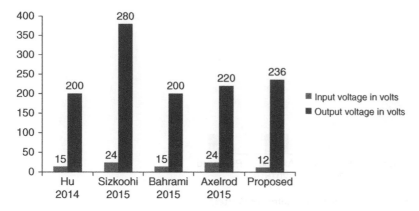

Figure 3.12 Output voltage assessment of suggested converter with latest converters.

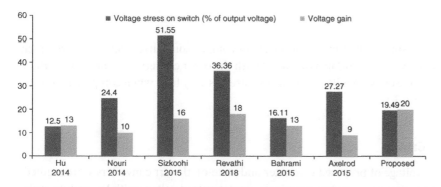

Figure 3.13 Comparison of MCBC converter with latest converters voltage gain.

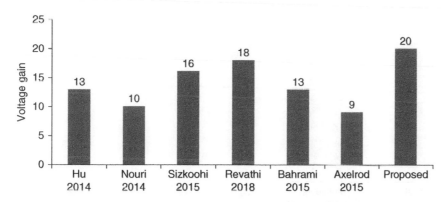

Figure 3.14 Voltage stress assessment of suggested converter with latest converters.

Despite the fact that [7, 8] only contains one switch, the voltage gain obtained when compared to the suggested converter is somewhat low. Switch count is higher in other converters, which directs switching losses to the source. The number of voltage levels can be raised without affecting the main circuit or the number of switches because the recommended converter has a modular design.

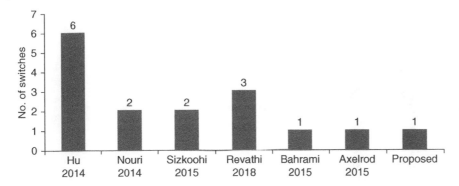

Figure 3.15 Switch count assessment of suggested converter with latest converters.

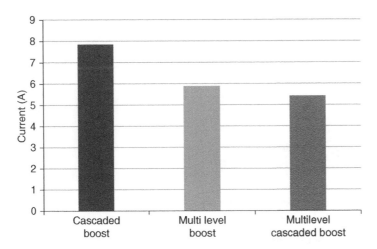

Figure 3.16 Assessment of current stress with various converter.

3.5.4 Current Stress

When investigating the current stress using simulation of converters in comparison, current stress is very much abridged by combining the merits of CBC and MLBC in assessment with the other conservative boost converters. When they are operated as the CBC alone or as a MLBC alone, whose stress is very high when compared with the planned MCBC. It is described in Figure 3.16.

3.5.5 Duty Cycle Versus Voltage Gain

The duty cycle is varied and output voltage of proposed converter and state-of–the-art converters were studied. From Figure 3.17, it is proved that recommended converter achieves greater output voltage with lesser duty cycle. When the duty ratio is reduced, the stress on the switch is also reduced, which improves the performance of MCBC.

3.5.6 Number of Levels in the Planned Converter

The suggested MCBC is fabricated in a modular way, hence any number of output levels can be added to the circuit. Figure 3.18 illustrates that the number of levels are raised, and an output potential gain of MCBC is also enlarged.

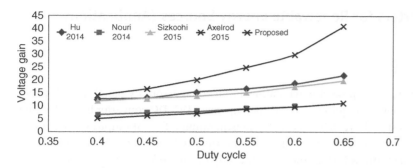

Figure 3.17 Duty cycle V_S voltage gain assessment of the recommended MCBC converter with recent converters.

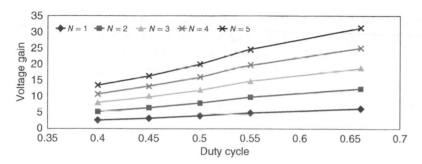

Figure 3.18 Output voltage V_s number of levels in proposed MCBC.

Table 3.4 Level circuit operation of MCBC.

S1	S2	S3	S4	S5	Output from level circuit
ON	–	–	–	–	$+kV_{dc}$
–	ON	–	–	–	$+2kV_{dc}$
–	–	ON	0	–	$+3kV_{dc}$
–	–	–	ON	–	$+4kV_{dc}$
–	–	–	–	ON	$+5kV_{dc}$

The level circuit in the prescribed topology is used to build the multilevel AC with the help of an H-bridge inverter. Totally five switches S1, S2, S3, S4, and S5 are employed to build up the level circuit which is tabulated in Table 3.4.

The switch S1 turned ON to carry the $+kV_{dc}$ of capacitor C_1 to the H-bridge inverter and it generates the positive half cycle by switching ON the S7 and S8 of the H-bridge inverter.

A similar process will be carried out with the next four levels thereby corresponding switches will be turned ON to achieve the particular multilevel AC positive half output voltage. For the negative half multilevel AC output, the switch S9, S10, and the corresponding level switches will be triggered.

3.6 Power Electronic Converters for Renewable Energy Sources (Applications of MLCB)

3.6.1 MCBC Connected with PV Panel

The multilevel DC–DC converters are very much powerful in handling input voltage from PV panel. The PV panel voltage is at low level and hence these levels are required to be stepped up using DC–DC converter before feeding to an inverter circuit. The multilevel boost and MCBCs are capable of generating different output voltage levels for the given PV panel input voltage. Since multilevel circuit produces multiple DC voltage levels, these multiple DC voltage levels are easily converted to multilevel AC output with the help of simple level changing circuit and H bridge circuit.

3.6.2 Output Response of PV Fed MCBC

Software models are performed using MATLAB. Figure 3.19 shows the interconnection of PV panel with MCBC. The PV input voltage of 12 V supplied to the MCBC is shown in Figure 3.20a and five output voltages are shown in Figure 3.20b. Level 1 voltage is 46 V, level 2 voltage is 91 V, level 3 voltage is 137 V, level 4 voltage is 182 V, and level 5 voltage is 227 V. These five-voltage level may be interfaced with inverter circuit to form a multilevel AC output voltage.

From the simulated waveforms, it is clear that MCBC is a good choice for PV-tied standalone system/grid-connected systems in the future.

3.6.3 H-Bridge Inverter

Figure 3.21 shows a traditional H-bridge inverter that is responsible for AC output at the load. Switches S11 and S41 are responsible for a positive cycle in AC output voltage, similarly switches S21 and S31 are responsible for the negative cycle. Table 3.5 and Figure 3.20 represent the selected level for inverter operation.

The output voltage level of the inverter is determined in relation to the input voltage level. The inverter's output voltage levels are denoted by M, while the MCBCs output voltage level is denoted by n.

$$\text{The switch count} = 5 + n \tag{3.7}$$

$$\text{Number of diodes} = \frac{3M - 7}{2} \tag{3.8}$$

$$\text{Number of capacitors} = M - 2 \tag{3.9}$$

$$\text{Inverter level } M = 2n + 1 \tag{3.10}$$

The potential difference across the output of the MCBC is

$$V_{0DC} = \sum_{i=1}^{\frac{M-1}{2}} V_i \tag{3.11}$$

where V_i is the individual capacitor voltage $= \dfrac{V_{in}}{1 - D} \times M \tag{3.12}$

D is the switching ratio of the boost converter

$$V_{0,\max} = \begin{cases} +\sum_{i=1}^{\frac{N-1}{2}} V_i, S_{11} = S_{41} = \text{ON} \\ -\sum_{i=1}^{\frac{N-1}{2}} V_i, S_{21} = S_{31} = \text{ON} \end{cases} \tag{3.13}$$

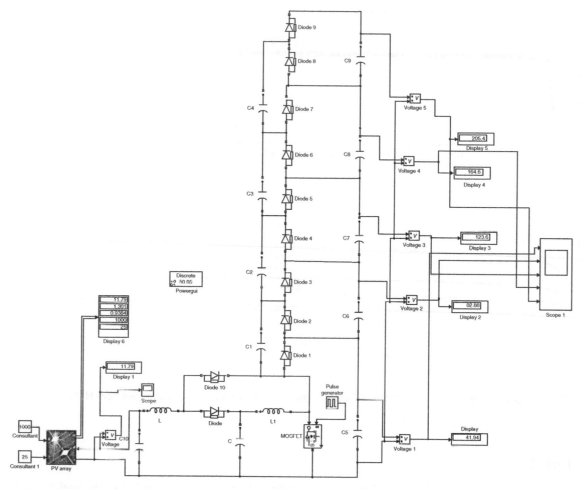

Figure 3.19 Simulation diagram of MCBC with PV panel.

3.7 Modes of Operation

MCBC-fed MLI modes of operations are represented in Figure 3.22a–j.

3.7.1 Mode 1

During the mode 1 operation, Switch S_2 is closed. The output voltage of the MCBC $+ kV_{dc}$ is supplied to the R through level switches S_2, S_7, and S_8 of the inverter which is shown in Figure 3.22a.

3.7.2 Mode 2

During the mode 2 operation, switch S_3 is closed. The output voltage $+ 2kV_{dc}$ is supplied to the load through switches S_7 and S_8 of the inverter circuit which is shown in Figure 3.22b.

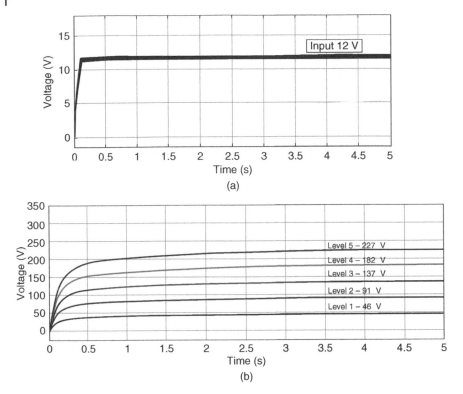

Figure 3.20 (a) Input supply voltage obtained from PV panel. (b) Five level output voltages obtained from MCBC.

3.7.3 Mode 3

During the mode 3 operation, switch S_4 is closed. The converter voltage $+3kV_{dc}$ is supplied to the load through switches S_7 and S_8 of H-bridge inverter which is given in Figure 3.22c.

3.7.4 Mode 4

During the mode 4 operation, switch S_5 is closed. The output voltage $+4kV_{dc}$ is applied to the load through switches S_7 and S_8 of H-bridge inverter which is shown in Figure 3.22d.

3.7.5 Mode 5

During the mode 5 operation, switch S_6 is closed. The converter voltage $+4kV_{dc}$ is applied to the resistive load through switches S_7 and S_8 of inverter circuit which is displayed in Figure 3.22e.

3.7.6 Mode 6

During the mode 6 operation, switch S_2 is closed. The output voltage $-kV_{dc}$ is applied to the load through switches S_9 and S_{10} of H-bridge inverter which is shown in Figure 3.22f.

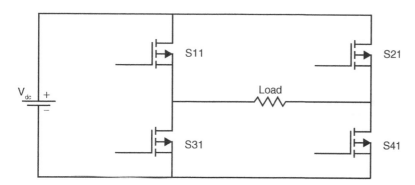

Figure 3.21 H-bridge inverter.

Table 3.5 Switching table for multilevel AC output voltage.

	Levels	Switching sequence					Voltage levels
		S2	S3	S4	S5	S6	
Positive half cycle S7, S8 ON	0	OFF	OFF	OFF	OFF	OFF	0
	1	ON	OFF	OFF	OFF	OFF	$+kV_{dc}$
	2	OFF	ON	OFF	OFF	OFF	$+2kV_{dc}$
	3	OFF	OFF	ON	OFF	OFF	$+3kV_{dc}$
	4	OFF	OFF	OFF	ON	OFF	$+4kV_{dc}$
	5	OFF	OFF	OFF	OFF	ON	$+5kV_{dc}$
	4	OFF	OFF	OFF	ON	OFF	$+4kV_{dc}$
	3	OFF	OFF	ON	OFF	OFF	$+4kV_{dc}$
	2	OFF	ON	OFF	OFF	OFF	$+2kV_{dc}$
	1	ON	OFF	OFF	OFF	OFF	$+kV_{dc}$
	0	OFF	OFF	OFF	OFF	OFF	0
Negative half cycle S9, S10 ON	1	ON	OFF	OFF	OFF	OFF	$-kV_{dc}$
	2	OFF	ON	OFF	OFF	OFF	$-2kV_{dc}$
	3	OFF	OFF	ON	OFF	OFF	$-3kV_{dc}$
	4	OFF	OFF	OFF	ON	OFF	$-4kV_{dc}$
	5	OFF	OFF	OFF	OFF	O	$-5kV_{dc}$
	4	OFF	OFF	OFF	ON	OFF	$-4kV_{dc}$
	3	OFF	OFF	ON	OFF	OFF	$-3kV_{dc}$
	2	OFF	ON	OFF	OFF	OFF	$-2kV_{dc}$
	1	ON	OFF	OFF	OFF	OFF	$-kV_{dc}$
	0	OFF	OFF	OFF	OFF	OFF	0

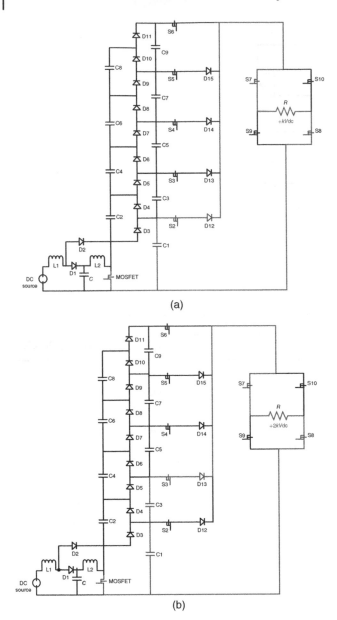

(a)

(b)

Figure 3.22 (a) Mode 1, (b) Mode 2, (c) Mode 3, (d) Mode 4, (e) Mode 5, (f) Mode 6, (g) Mode 7, (h) Mode 8, (i) Mode 9, (j) Mode 10.

3.7.7 Mode 7

During the mode 7 operation, switch S_3 is closed. The converter voltage $-2kV_{dc}$ is supplied to the resistor through switches S_9 and S_{10} of inverter circuit which is shown in Figure 3.22g.

3.7.8 Mode 8

During the mode 8 operation, switch S_4 is closed. The output voltage $-3kV_{dc}$ is applied to the load through switches S_9 and S_{10} of inverter circuit which is shown in Figure 3.22h.

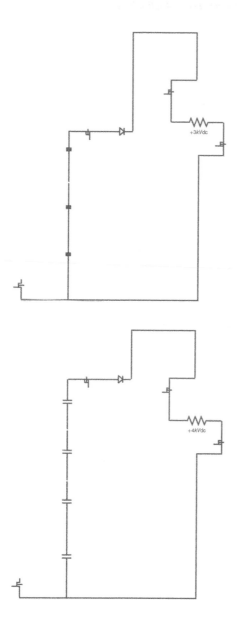

Figure 3.22 (*Continued*)

3.7.9 Mode 9

During the mode 9 operation, switch S_5 is closed. The converter voltage $-4kV_{dc}$ is applied to the load through switches S_9 and S_{10} of inverter circuit which is given in Figure 3.22i.

3.7.10 Mode 10

During the mode 10 operation, switch S_6 is closed. The converter voltage $-5kV_{dc}$ is applied to the load through switches S_9 and S_{10} of H-bridge inverter which is shown in Figure 3.22j.

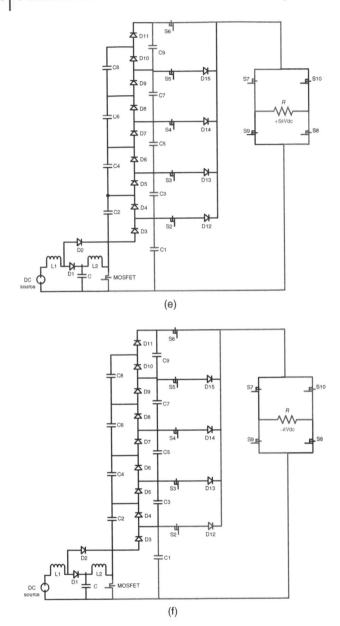

(e)

(f)

Figure 3.22 (*Continued*)

3.8 Simulation Results of MCBC Fed Inverter

Simulations only are performed in MCBC fed inverter which is shown in Figure 3.23. First stage in this inverter circuit is MCBC and it produces a five-level output voltage from given 24 V input the five levels are 65, 130, 195, 260, and 325. Using level circuit and H-bridge inverter combination these output voltages are converted into 11 level AC output across the load. For positive multilevel output voltage switches S_7 and S_8 are triggered and for negative multilevel output voltage switches S_9 and S_{10} of H-bridge inverter are in conduction. Thus a11-level multilevel output voltage is obtained using a single DC source.

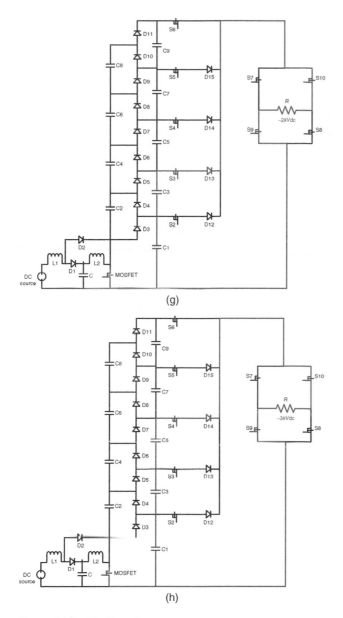

(g)

(h)

Figure 3.22 (*Continued*)

Figure 3.24 shows the THD spectrum for the proposed MCBC. THD value has been decreased in this topology when compared to previous structure because of increasing number of levels in output side and by increasing the output voltage amplitude without increasing the number of switches.

3.9 Power Electronic Converter for E-Vehicles

In an E-vehicle, the role of power electronic converter is the most important one. The role of power converter in E-vehicle has two modes: In the first mode, the power converter is responsible for energy management of the

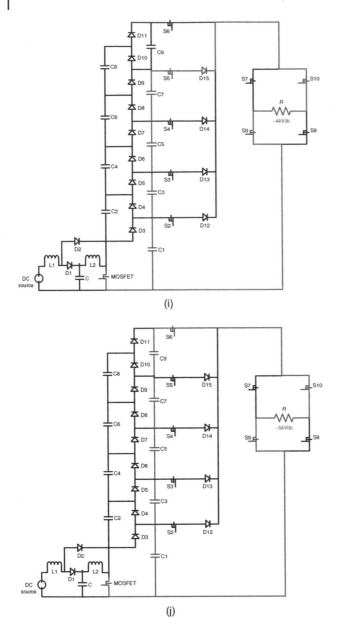

Figure 3.22 *(Continued)*

batteries in charging/discharging intervals. The second mode is an inverter. The main focus of this chapter is MLI that can be supported for second mode of E-vehicles. The MLI can be obtained from multilevel converter and H-bridge inverter is used in load side.

3.10 Power Electronic Converter for HVDC/Facts

In a high voltage DC (HVDC) transmission, both converter and inverter are used. In this chapter, both multilevel converter and inverter are explained. This kind of multilevel converter is used in transmission end and MLI is used

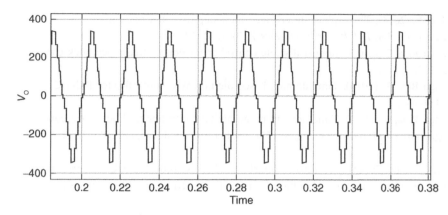

Figure 3.23 Simulated output of MCBC fed inverter.

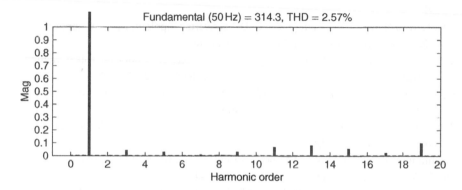

Figure 3.24 THD spectrum for MCBC fed inverter.

in receiving end of the HVDC transmission. FACTS devices are used in AC transmission. FACTS devices such as UPFC and GPFC power flow controllers are used. In a power flow controller, multilevel converter has a vital role in AC transmission system.

3.11 Conclusion

Thus, the MCBC is presented with modes of operation and simulated multilevel AC output voltage. From the waveform it is proved that with five level MCBC outputs we are able to achieve 11 levels in AC output voltage with a peak voltage of 320 V which is higher when compared to MLBC. To achieve more levels in multilevel AC output, by just increasing the number of levels in MCBC since it constructed in a modular way.

References

1 Zeng, J., Liu, N., Wu, J., and Goe, H. (2017). A quasi-resonant switched capacitor multi-level inverter with self-voltage balancing. *IEEE Transactions on Industrial Informatics* 13 (5): 2669–2679.

2 Xiang, X., Zhang, X., Luth, T. et al. (2018). A compact modular multilevel DC–DC converter for high step-ratio MV and HV use. *IEEE Transactions on Industrial Electronics* 65 (9): 7060–7071.

3 Hu, Y., Xiao, W., Li, W., and He, X. (2014). Three-phase interleaved high-step-up converter with coupled-inductor-based voltage quadrupler. *IET Power Electronics* 7 (7): 1841–1849.2.

4 Vijayalakshmi, S. and Sree Renga Raja, T. (2015). Robust discrete controller for double frequency buck converter. *Automatika* 56 (3): 303–317.

5 Vijayalakshmi, S. and Sree Renga Raja, T. (2014). Time domain based digital PWM controller for DC–DC converter. *Automatika* 55 (4): 434–445.

6 Vijayalakshmi, S. and Sree Renga Raja, T. (2014). Time-domain based digital controller for buck-boost converter. *Journal of Electrical Engineering and Technology* 9 (5): 1551–1561. http://dx.doi.org/10.5370/JEET.2014.9 .5.1551, ISSN: 2093-7423.

7 Nouri, T., Hosseini, S.H., Babaei, E., and Ebrahimi, J. (2014). Interleaved high step-up DC–DC converter based on three-winding high-frequency coupled inductor and voltage multiplier cell. *IET Power Electronics* 8 (8): 175–189.

8 Sizkoohi, H.M., Milimonfared, J., Taheri, M., and Salehi, S. (2015). High step-up soft-switched dual-boost coupled-inductor-based converter integrating multipurpose coupled inductors with capacitor-diode stages. *IET Power Electronics* 8 (9): 1786–1797.

9 Revathi, B.S. and Prabhakar, M. (2016). Non-isolated high gain DC–DC converter topologies for PV applications – a comprehensive review. *Renewable and Sustainable Energy Reviews* 66: 920–933.

10 Vijayalakshmi, S. and Sree Renga Raja, T. Design and implementation of a discrete controller for soft switching DC–DC converter. *Journal of Electrical Engineering* 12 (25): 23–28. Edition 3, ISSN: 1582-4594.

11 Marimuthu, M., Sait, H., Vijayalakshmi, S., and Paranthagan, B. (2018). Time domain based digital controller for boost converter. *Journal of Electrical Engineering* 18 (45): 35–41. Edition 4.

12 Shenbagalakshmi, R., Vijayalakshmi, S., and Geetha, K. (2020). Design and analysis PID controller for Luo converter. *International Journal of Power Electronics* 11 (3): 283–298.

13 Ponraj, R., Sigamani, T., and Subramanian, V. (2021). A developed H-bridge cascaded multilevel inverter with reduced switch count. *Journal of Electrical Engineering and Technology* 16: 1445–1455.

14 Bahrami, H., Iman-Eini, H., Kazemi, B., and Taheri, A. (2015). Modified step-up boost converter with coupled-inductor and super-lift techniques. *IET Power Electronics* 8 (6): 898–905.

15 Axelrod, B., Beck, Y., and Berkovich, Y. (2015). High step-up DC–DC converter based on the switched-coupled-inductor boost converter and diode-capacitor multiplier: steady-state and dynamics. *IET Power Electronics* 8 (8): 1420–1428.

16 Marimuthu, M. and Vijayalakshmi, S. (2020). Symmetric multilevel boost converter with single DC source using reduced number of switches for multilevel inverter. *Technickivjesnik (Technical Gazette)* 27 (5): 1585–1591.

17 Marimuthu, M. and Vijayalakshmi, S. (2020). A novel non-isolated single switch multilevel cascaded DC–DC boost converter for multilevel inverter application. *Journal of Electrical Engineering and Technology* 15 (5): 2157–2166.

4

Recent Advancements in Power Electronics for Modern Power Systems-Comprehensive Review on DC-Link Capacitors Concerning Power Density Maximization in Power Converters

Naveenkumar Marati[1], Shariq Ahammed[1], Kathirvel Karuppazaghi[1], Balraj Vaithilingam[1], Gyan R. Biswal[2], Phaneendra B. Bobba[3], Sanjeevikumar Padmanaban[4], and Sharmeela Chenniappan[5]

[1] PES, Valeo India Private Limited, Chennai, India
[2] Department of Electrical and Electronics Engineering, Veer Surendra Sai University of Technology (VSSUT), Burla, Odisha, India
[3] Department of Electrical and Electronics Engineering, Gokaraju Rangaraju Institute of Engineering and Technology (GRIET), Hyderabad, India
[4] Department of Electrical Engineering, Information Technology, and Cybernetics, University of South-Eastern Norway, Porsgrunn, Norway
[5] Department of Electrical and Electronics Engineering, Anna University, Chennai, India

4.1 Introduction

Power electronics is a revolutionary element in the present-day tech world that leads to compact and reliable power units that could deliver the desired power as an application requires. Foreseeing the future in power electronics at present power density will be a key factor associated with thermal management and high efficiency. High power density converters are an important trend in power electronics for many modern applications, including electric vehicles and renewable energy [1]. The development of recent technologies such as wide band gap (WBG) devices has led to a new era in power density improvement in power electronics. WBG power semiconductor device technology has become a key enabler for the miniaturization of power electronic systems [2]. All these add up to as described earlier power density being the major part contributor to future power electronics. DC-Link capacitors tend to be bulky and occupy large space as compared to other components as capacitor technology is not mature as the solid-state switches.

One can see the future with power electronics to tend with immense technological revolution reliability comes as in when the practical market aspects are considered for power electronic products or products with power conversion units. While capacitors contribute 30% to the total failure count of practical products in the market [3], they also stand to be the second-most failing unit in power converters while semiconductor switches being first [4]. Thus, practical reliability in power electronic products should also be a factor in regard to future technological pathways. In light of these aspects in power converters, DC-Link capacitors provide a significant threat to reliability. Reliability also comes to a brink line when thermal management comes up for electrolytic capacitors utilized in DC Links. Dissipation of the heat between nearby electrolytic capacitors can have an adverse effect on thermal management if proper mapping for the heat is not performed and which may cause a serious failure and reduction in the reliability of the product. Thus, a reduced lifespan is observed for electrolytic capacitors which are being utilized in DC Link.

Over the timeline, there are several reviews proposed now. Different power decoupling techniques for solar photovoltaic inverter systems with photovoltaic (PV) panel capacitance and leakage currents as a constraint is being discussed in ref. [5]. The review is very much keen on efficiency and cost for the solar photovoltaic inverter as an entity. Much less description and arrivals are made in regard to power density. Active power decoupling techniques in single-phase systems are being reviewed in ref. [6]. Even though the review proposes a structural and cost-effective side of each of the active techniques for decoupling paper discusses very less regard for power density and around 10% of the total topologies reviewed are being discussed for power density. Low-frequency decoupling techniques in single-phase systems are reviewed and skeletal analysis to the root is being carried out in ref. [7], yet the paper lags in power density comparison; however, it describes power density improvisation proposals that are being discussed in some kinds of literature that it reviews. A generic topology derivation with a review of existing topologies is being carried out in ref. [8]. However, demanding to have improved power density no particular comparison is provided to prove the same by the authors. Cost assessment and cost-effective proposal of topologies is being carried out in ref. [9] as described earlier to all reviews established, this one also lags with power density considerations and its comparisons. A much more in-depth analysis and root skeletal structural analysis of all topologies as per timeline is being carried out in ref. [10] with an extensive analysis of efficiency, control, cost, and the number of components further proposing a generalized topology is also being carried out.

In this review, limited research findings have been reported upon power density, while the major focus is given on generic topology derivation. As this analysis has described with viewpoints on future power electronics roadmap for converters, it is observed that power density does play a crucial role for the power conversion units. Thus, a review in regard to this aspect under all power conversion techniques where DC-Link capacitor has the least power density contribution, low thermal management capability, and lifetime limitation is of at most importance as an improvement over this can bring in better power density to a greater extent with enhanced thermal management with better lifespan. The review constitutes of six parts wherein the first describes the application of power converters, the second covers the general classification of DC-Link capacitors, the third over the briefing of various topologies for DC-Link, the fourth is a comparison between the different classifications and themselves within, and the fifth through the future and research gaps where this research can contribute in the future and then concluded over the final section.

4.2 Applications of Power Electronic Converters

With the increase in the technological innovations toward the sustainable innovations in the modern days, there exists a demand for the power electronic converters usage in the same proportion or high. For the next few decades, there is a huge demand for efficient power converters. One of the ways to improve the converter's efficiency is to select the appropriate configuration of the converter for the required application, and its DC-Link topology selection and knowing its limitation. The objective of this manuscript is to present in the detailed view of this different topologies.

4.2.1 Power Electronic Converters in Electric Vehicle Ecosystem

In the present decade most of the automotive players are moving toward the sustainable energy features in the power train architecture either modifying their power train architecture from conventional internal combustion engine to introducing electrification in the architecture through hybridization/hybrid electric vehicle (HEV), battery powered electric vehicles (BEVs), fuel cell electric vehicle (FCEV). Conversion of power train from conventional to EV/HEV/FCEV facilitates the introduction of the power electronic converters in to the architecture in different component levels visibly in DC–DC converters, motor controller units/inverters, on-boarding charging systems, etc. A typical block diagram of power train architecture of HEV/EV is shown in the Figures 4.1 and 4.2, respectively. Apart from the power train architecture there are other places in the EV ecosystem where one will also use the power converters in the applications of charging infrastructure requirements as well. Especially bi-direction converters are hugely used the applications of Vehicle-to-Grid (V2G) or Grid-to-Vehicle (G2V).

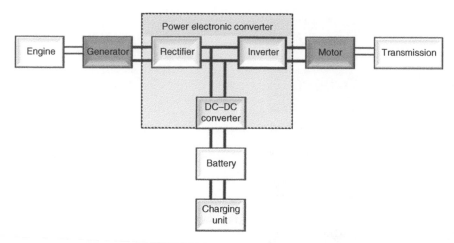

Figure 4.1 Typical power train architecture of hybrid electric vehicle with power electronic controllers.

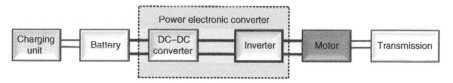

Figure 4.2 Typical power train architecture of electric vehicle with power electronic controllers.

4.2.2 Power Electronic Converters in Renewable Energy Resources

The other important aspect of usage of power electronic converters is in integration of the renewable energy resources like solar, wind to the grid and utility supplies. In this application of the renewable integration, the power electronic controller plays a key role in maintaining the power quality, grid stability, etc. One such typical example of usage of power electronic controllers in the applications of the solar PV integration with the grid integration is shown in the block diagram of Figure 4.3. The new topologies of the power electronic converters miniaturize the overall subsystems to bring down the cost and size of the systems. This introduction of renewable energy sources along with the power electronic controllers also makes easier to overcome the some of the challenges in the rural area power by bringing the concepts of micro-grid and smart-gird concepts. In today's modern day world, a very common feature in smart-grid or micro-grid environment is bringing up the artificial intelligence or digitalization of the power grid as though grid automation. So to do job easier in digitalization one should also involve the intelligent power converters in it. So to select these different topologies with more efficient converters

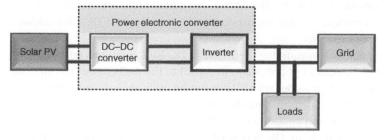

Figure 4.3 Block diagram for soar PV integration to the grid.

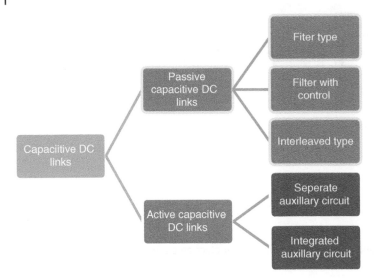

Figure 4.4 Classification of capacitive DC Links.

in this manuscript the authors have presented different converters with DC-Link topologies in improving its power density factors in subsequent Sections 4.3–4.6.

4.3 Classification of DC-Link Topologies

A generic classification in DC Link has been discussed in ref. [10], which is considered all present-day DC-Link systems available. Refreshing with the same classification layout concludes the DC Links to be either capacitive or inductive at the higher level. As this research discussion revolves more around capacitive, it closely describes the capacitive classification tree. Overall picture in capacitive DC-Link technologies is being depicted in Figure 4.4. Capacitive DC-Links can be further classified as passive and active where passive constitutes of filter type, filter with control, interleaved type. Active constitutes only two separate auxiliary and integrated auxiliary.

Passive filter type only constitutes a capacitive filter mechanism that filters out the AC component and provides a steady DC in the output in rectifiers as well as DC–DC converters. Filter type with control implies for the capacitive filter arrangement with an implemented control mechanism reduces the stress over the filter or DC-Link capacitor thus providing improved lifespan for the capacitors. Interleaved types of DC Link constitute interleaved structures that constitute capacitive filters. Active capacitive techniques employ active devices like MOSFETs or other solid state switches to divert or control the DC-Link currents and to improve the output characteristics. Separate auxiliary active circuit proposes active switching circuit utilized in addition to existing converter circuit which reduces DC-Link stresses and improves the lifespan for the device. Integrated auxiliary systems tend to propose a more complex as these utilize power switches within the converter circuit to suppress DC-Link stresses and improve output characteristics of the power converter.

Further classification proposed in ref. [10] includes considering the whole of DC-Link circuit as an entity and classifying the same on the basis of position of placement of the circuit in regard to the main power circuit. The proposal has been utilized to propose a generic topology as per ref. [10] but can be utilized in this review to categorize the whole DC-Link circuit as an entity and then determining the amount of power density compensation required or acquired by the addition of the compensation circuit to the converter as a whole. In accordance with placement the DC Links can be classified as four which are DC Series, DC Parallel, AC Series, and AC Parallel which are being depicted in Figure 4.5. Even though the review classifies this as an attribute of divergence in the single phase rectifier units, this classification can actually be attributed in DC–DC or AC–DC systems as the placement of the circuit over DC side or AC side will decide this classification. Here DC Series describes when the compensation of whole circuit is connected series to the DC side of the power converter. In DC Parallel the DC-Link circuit is connected in parallel to the DC side. In the AC–DC system the DC side stands strict for the respective side while in DC–DC converters this can be attributed as the output side where there is lower ripple is desired. Now, AC Series and parallel also contribute to a similar architecture implying if it is AC series then the DC-Link circuit will be series to the AC side while in parallel it will be parallel to the AC side.

4.4 Briefing on DC-Link Topologies

In this section, an initiation to provide a brief about each of the DC-Link topology that will be considered in this review. The section aims to give the researcher a better understanding on each of the topology while the comparison is carried out for different traits. As per the classification depicted in Figure 4.5, in capacitive DC Links the primary branch describes passive capacitive DC Links which are further classified into filter, filter with control, and interleaved types. Active being the secondary branch of classification consists of separate auxiliary and integrated auxiliary types of DC-Link topology. Each of these is discussed in detail below Sections 4.4.1 and 4.4.2.

4.4.1 Passive Capacitive DC Link

Passive component engaged balancing or filter arrangement using these passive components for achieving DC-Link property is generally referred to as a passive capacitive DC Link. The most primitive model of DC Link is a resistor balanced topology for two capacitors in series. The model is described as the most primitive and hence attributed to several backdrops. It consists of two resistors connected with each resistor in parallel to each capacitor in the DC-Link branch. Depicted in Figure 4.6 is a typical resistive balancing arrangement for

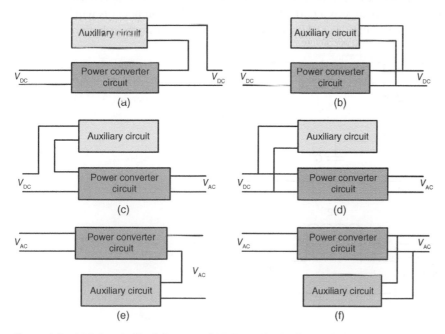

Figure 4.5 (a) DC series for DC output, (b) DC parallel for DC output, (c) DC series for AC output, (d) DC parallel for AC output, (e) AC series for AC output, (f) AC parallel for AC output.

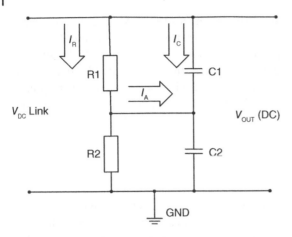

Figure 4.6 Resistive balancing for capacitive DC Link (passive method).

passive capacitive DC Link. Typical current observed and direction for the same is also marked. As it can be clearly acknowledged from the circuit that there exists a quiescent current through the resistor leg which tends to generate a statutory loss under all circumstances proving to be a major setback to this design which is a simple control topology.

4.4.1.1 Filter Type Passive Capacitive DC Links

In passive capacitive DC-Links filter type capacitive DC Links are the basic topology upon which all the other passive topologies were developed. Filter type capacitive DC Links are essentially filter capacitor filtering any ripple or high frequency undesired component from the required output. Several topologies have been proposed in time for this type of DC-Link topology in literature. The beginning in regard to this series can be attributed to the smoothing choke presented in Figure 4.7. A two-port strategy can be applied in this where the smoothing transformer and the capacitor for blocking forms the filter part in a DC circuit which forms the two-port network. Thus, a simple conclusion is the brief about this topology as a port where both AC and DC components will be present and the other port where the AC component will be canceled out as it gets transferred from the other port inverted

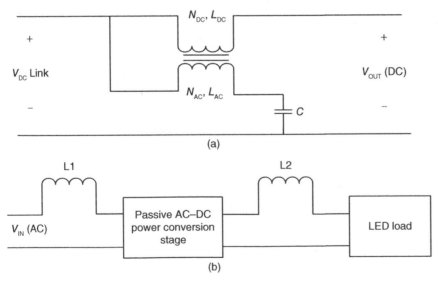

Figure 4.7 (a) Smoothing transformer filter capacitor choke proposed in [11], (b) LED load proposal for optimized DC Link [12].

over the smoothing transformer windings. While the advantages can be described as a mature technology and modularity with any converter, the disadvantages include reduced power density and electromagnetic interference (EMI)/electromagnetic compatibility (EMC) issues. A technique or concept applicable in LED loads is being discussed in ref. [12]. The idea proposed is very straight and clear, viz., in LED loads the ripple current attenuation from the supply side is not required to a perfect extent as in the case of other sensitive critical loads. This implies certain amount ripple current can be allowed in the supply to the load provided the load is LED. By utilizing this concept a reduced requirement for DC Link can be proposed and hence improved power density can be obtained. This disadvantage in major is that the method is not efficient as a DC Link nor is modular as this method only can be employed for LED loads. Two more literature can be classified under passive balancing in filter type [13] is more densely relying on hardware technique for suppression of DC-Link surges while ref. [14] relies on more of control strategy which can suppress requirement over DC Link there by optimizing it.

A combined rectifier inverter AC/DC/AC power module is being discussed in ref. [13], the key idea being increased power density by reduction of switches and optimized design of DC Link. In order to achieve a six-switch topology is being proposed where two switches are being shared to achieve both rectification and inverting action. Switching pattern is followed in such a manner so as to reduce the DC-Link capacitor surge minimization hence obtaining an optimized DC-Link capacitor thus improving the power density. The existing topology with the improved six-switch topology is being depicted in Figure 4.8.

Improved power density achieved in ref. [13] as desired by view can be attributed as a major advantage but as the proposal also searches for a modular technique the idea proposed is not modular to be applied in general to

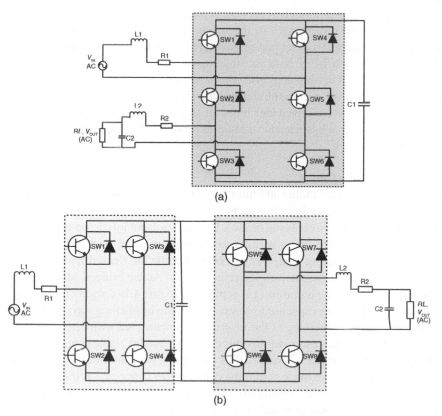

Figure 4.8 (a) Improved six switch high power density AC/DC/AC power module, (b) Conventional eight switch topology for AC/DC/AC power module [13].

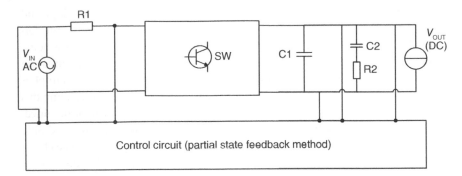

Figure 4.9 Single phase rectifier unit with control block scheme for partial state feedback control.

power conversion units. Control strategy focused topology for optimizing DC-Link capacitor and improved power density is being discussed in ref. [14]. Proposed technique employs a partial state feedback control which gives the component currents in rotating frame of reference. This primarily helps in contributing to reactive and active power control of the power module toward the grid side. Inner loop of control for current also actively monitors output current hence the capacitor current thereby contributing to control of DC-Link current, this control strategy monitoring for the DC-Link current helps to optimize the capacitor design and hence optimize the size leading to high power density. Design and evaluation of the control loop hence provides a better idea on DC-Link stresses thus contributes to early-stage design and optimization of DC Link which can be counter-verified with simulation. Single-phase rectifier with the control block proposed in ref. [14] is being depicted in Figure 4.9.

Shortcomings in regard to [14] topology can be directly seen in the control loop depicted in Figure 4.9. The sheer complexity with respect to the control loop can be described as the biggest demerit of this topology despite achieving an optimization in DC Link without any change in hardware topological aspects. However, a conceivable part in regard to this topology can be described as, if in a topology grid connected synchronization is not necessary especially for standalone applications in any rectification units the control loop proposed in ref. [14] can be adapted to the requirement with much less complexity. This conceived idea however requires a bit of work to be adapted into the existing topology and control loop which may again lead to a complex process but when processed at once as a generic one can be modular to many converters.

4.4.1.2 Filter Type Passive Capacitive DC Links with Control

Moving on to passive filter techniques with control technique and interleaved type, as discussed for passive filter capacitive technique there are several literatures available in this section also. Some of these literatures which are key contributors to the classification discussed here are being briefed. Both control and interleave types prove to be efficient for DC-Link optimization yet each contributes with its own limitation to the design. In general, it can be described a control technique scheme proposes complexity in control while interleaved type proposes complexity in change of topology to classical topologies such as H-bridge or Dual Active bridges. Moving into control technique for passive filters first proposal under literature [15] depicted in Figure 4.10 does not modify the topology of classical H-bridge but splits in the DC-Link capacitor to compensate there by opening ways for hybrid DC-Link model. The control part is contributed as each leg of the inverter is compensated with one capacitor. Net capacitance is derived as equivalent of capacitive power required for compensation for ripple components in the output. However, the control loop for each of them is individual, thus the topology is classified under control strategy passive capacitive filters. Significant note to be made is that the capacitances connected at the output is assumed to be equal, which is a critical bottleneck as the disparity in capacitance for these two capacitors can be lethal to the converter. Any circulating current if present undetected can lead to failure of the power module.

In ref. [16] proposed is a high-power density transformer-less inverter attributed with SiC MOSFETs the topology already projects in a strong vision of high-power density. Apart from this aspect change in position of placement

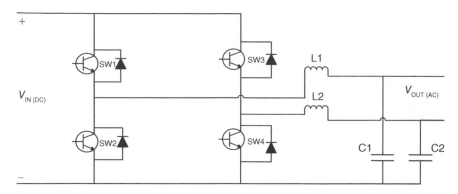

Figure 4.10 Power decoupling with H-bridge and split capacitor [15].

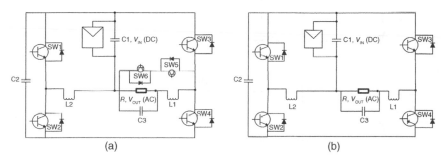

Figure 4.11 (a) High power density transformer-less PV inverter [16], (b) High power density active device removed transformer-less PV inverter [17].

in DC-Link capacitor is proposed. The topology claims to provide even better power density with this specific arrangement. An inherent three-level inverter is depicted in Figure 4.11a. This topology leads to a reduction in the ripple current in the inductor which further reduces DC-Link stresses. The power density improvement proposed associated with the DC-Link capacitor even though described as high doesn't seem to reflect proper results or comparison in the paper. Further, the SiC MOSFETs associated with significant switching losses are said to have switching loss control but this is evident only at the unity power factor.

A continued topology to that of ref. [16] is being proposed by the same authors in ref. [17] where the active switches in the three-level inherent leg part are removed to propose an improved efficiency at the cost of the DC-Link stresses which can lead to simple conclusions such as the power density efficiency trade-off is being done and established in a viable manner. Further, the removal of this current path can also cause instability to the topology by common-mode leakage currents, which are extremely harmful to the PV in non-isolated topologies of inverters. Figure 4.11a and b presents a clear-cut comparison between topologies [16, 17]. A similar modification to H-bridge topology is also being proposed in ref. [18]. In this literature, a decomposition and equivalent reduced DC-Link capacitor generation by reduced stress in DC Link through improved switching control are being discussed. Reduced switching losses and improved efficiency have been described in regard to the Zero Voltage Switching (ZVS) implemented in the topology which is accomplished by the same filter capacitor forming LC tank as depicted in Figure 4.12.

A major drawback in ref. [18] is the utilization of filter capacitors as both DC-Link and part of the LC-tank circuit. This can lead to critical design complications that need to be met through compromises made at the ripple voltage and capacitance value. Hence efficiency and power density come with a cost of trade-off to be made at the output end in terms of ripple voltage and capacitance. Multiple capacitors-based distributed energy compensation techniques are being discussed in ref. [19]. In conventional type four capacitors of H-bridge, where two capacitors

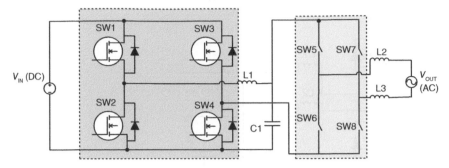

Figure 4.12 Dual DC–DC converter with reduced DC-Link capacitor [18].

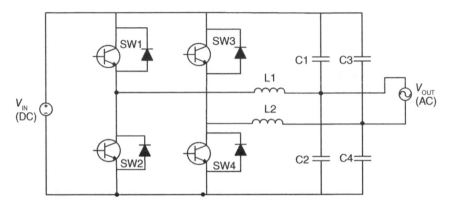

Figure 4.13 Four capacitor based transformer-less PV inverter [19].

are provided for the compensation purpose of each leg in the H-bridge. The logic is straightforward and simple but the imbalance in this capacitor network can be lethal to the topology circuit. Further, the thermal placements of these capacitors as it is four in place can be very challenging because capacitor deterioration is also contributed by the temperature of operation. The advantage is the same as [16–18] wherein H-bridge is utilized making the design more robust as there is much expertise in this classical topology. Topology in ref. [19] is depicted in Figure 4.13.

4.4.1.3 Interleaved Type Passive Capacitive DC Links

Interleaved topologies are the last ones in passive capacitive DC-Link techniques. Literature from [20–22] describes such differential arrangements where interleaved structures form a differential pair thus eliminating the unwanted ripple component. Fuel cell current is drawn and ripple elimination is being discussed in ref. [20]. The objective is obtained by putting in two DC–DC Converters in differential arrangement and canceling out the low-frequency ripples. Differential arrangement demands two capacitors in series which brings us back to the requirement that actually needed which is balancing of these series capacitors. Hence, the idea of these capacitors in parallel can lead to a worst-case when there are dissimilarities in these capacitors leading to reduced performance of the topology. The idea behind the topology is to eliminate the low-frequency components through optimal control of each output DC–DC Converter thereby generating appropriate counterparts to the ripple in any output of the DC–DC Converters. Topology and waveform control as described earlier are depicted in Figure 4.14.

Differential converter architecture is continued in ref. [21]. Conventional H-bridge is made into two DC–DC Converters with the DC-Link capacitor split into two capacitors each supporting one DC–DC Converter. Individual control on each DC–DC Converter allows the user to define whether to have a buck, boost, or marginal areas where buck-boost systems can be applied. The major drawback is similar to ref. [20] where the entire design and

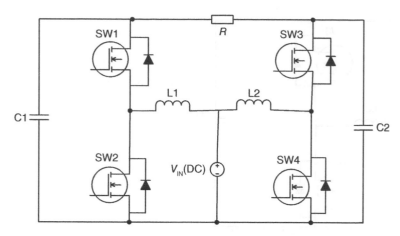

Figure 4.14 Differential converter arrangement for fuel cell drawn current stabilization [20].

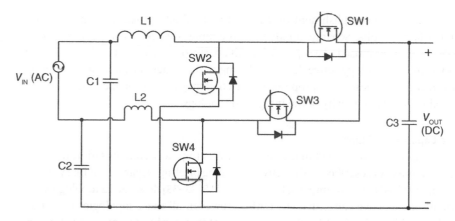

Figure 4.15 Differential converter for double line frequency elimination [21].

estimates depend on the capacitors placed in connection to each of the converters in the differential arrangement. Any variations in these capacitor values can be lethal as for capacitors lethality in ref. [20]. The topology for ref. [21] is shown in Figure 4.15.

More general characterization can be made at this point toward differential converters as the capacitors employed under the DC-Link part are always split into two capacitors but more importantly, tolerance in values of these capacitors can be deadly to the topology. The last in literature for interleaved topology is also a differential variant but more of a modified H-bridge where the source is kept at the center and the split capacitors are kept one parallel to every other leg as depicted in Figure 4.16. A similar profile of failure as in refs. [20, 21] with capacitors can be seen in this topology also where the failure of the topology can happen is tolerance capacitance value is present among the two capacitors.

4.4.2 Active Balancing in Capacitive DC Link

Active balancing of DC Link refers to employing active devices for balancing the DC Link from stresses and ripple. The active devices can also work in cooperation with other secondary energy storage devices such as inductors, capacitors, or DC Sources to contribute to power requirements as required. Classification of active balancing

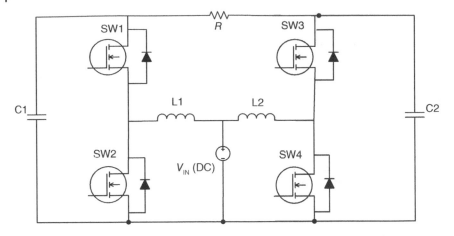

Figure 4.16 Differential converter on modified H-bridge [22].

branches to two specific topologies for the accomplishment of these active methods which are separate auxiliary circuits and the integrated auxiliary circuits. Separate auxiliary as the name suggests says the active device arrangement for balancing the DC Link is comprised of a separate circuit apart from the main topology while the integrated one tries to involve it within the topology active devices. The major drawback for integrated systems is that it tends to be more complex in converter design yet proposes better efficiency and in some cases, it can vary as the converter topology restricted is by the complexity of the integrated topology.

4.4.2.1 Separate Auxiliary Active Capacitive DC Links

Separate auxiliary capacitive DC Links can be described many close relatives to power compensation units in a power line or power factor correction units. As described in the active balancing briefing, separate auxiliary circuits apart from the converter topology are utilized for this compensation. In refs. [15, 23–33] describe 12 topologies with respect to separate auxiliary circuit-based active capacitive DC Links. However, all these topologies essentially described a similar topology in the auxiliary circuit utilized. The placement in regard to the converter topology may vary from topology to topology but as said the components or idea of compensation remains the same. All of this literature also has a prominent classification which can be made out as H-bridge based auxiliary support, Flyback/PFC-based auxiliary support, and Generic DC–DC Converter based auxiliary support. The classification tree depicted below with respective literature is shown in Figure 4.17.

In ref. [23] discussed is the primary aspect of active balancing with separate auxiliary circuits. Two switches and an energy transition element are accompanied by an energy storage element. Switches help in maintaining the compensation as required by desired switching pattern with appropriate duty. Whereas, an inductor which is an energy delivery element; it is usually utilized to couple the compensation circuit with the converter circuit. Energy storage elements which can be either a capacitor or a source will do management of energy. Garcia et al. [23] discusses a PFC-based circuit that is compensated using the above-described compensation circuit. The power delivery element which is an inductor will be the limiting point of this topology as the frequency will decide the power density for this element which can severely affect the prime target of power density maximization. Concepts in ref. [24] provide a temporary solution to this aspect by allowing the hardware designer to decide the placement of the auxiliary circuit thereby helping to decide the switching frequency which can, in turn, contribute to the improved power density. The scope of the idea is still limited as the main converter switching sides and their respective frequency will determine the power density for the circuit. Literature exactly follows the architecture and topology as followed in ref. [23] and basically supports the activity of a PFC Converter. A bidirectional converter with the same components but a different arrangement is followed for ref. [25]. Literature depicts a clear

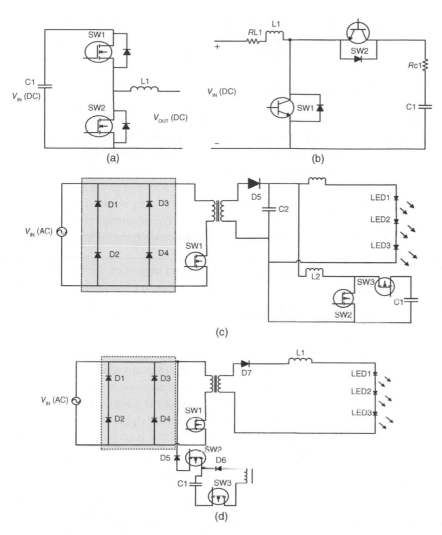

Figure 4.17 Flyback/PFC-based auxiliary support topologies. (a) Harmonic reducer converter [23], (b) Generalized technique of compensating low-frequency component of load current with a parallel bidirectional DC/DC converter [24], (c) A flicker-free electrolytic capacitor-less AC–DC LED driver [25], Generic DC–DC Converter based auxiliary support: (d) LED driver achieves electrolytic capacitor-less and flicker-free operation with an energy buffer unit [27].

improvement in capacitor lifespan however comprising efficiency and power density. Significant loss of efficiency is observed which should be taken care of as compared to the same PFC compensation as in refs. [23, 24] as the change in switches has contributed to a significant loss of efficiency which should be analyzed in this category of classification tree for auxiliary circuit based topologies.

Research in ref. [25] also tries another control design as in ref. [26] where this efficiency drop has been observed and tried to compensate yet to a limit as for expectation required. Feed-forward control implemented in the topology under ref. [26] limits the ripple current further which then eliminates the additional losses generated and thus is eliminated as compared to ref. [25]. Yet the losses pertain to a certain level that is a shortcoming for the topology. Fang et al. [27] is not entirely entitled under the separate auxiliary type of DC Link in the active category as the parallel circuit itself contributes to some modes in the operation of the basic converter. However, the majority of operational modes ranging up to 90% of the converter are separate in aspect, hence, it compensates for the PFC circuit that it can be considered as a PFC tree of classification for active separately connected DC Links. A coupled inductor is involved in this separate auxiliary circuit that not only helps in momentary power delivery but also contributes to the power storage in addition to the capacitor that acts as the storage element in this compensation circuit.

Moving on to the modified H-bridge-based auxiliary circuits (shown in Figure 4.18) for separate active balancing of DC-Link literature, [28–31] discuss this topology. The primary circuit of compensation remains intact from the literature [23] yet the combination of it with classical H-bridge topology is the streamline for this specific classification of separate auxiliary circuit-based active DC Link. The distinguishing feature of this particular topology is the utilization of a virtual capacitor. The control loop for this topology incorporates a virtual capacitor component that delivers the control logic with the effect of a capacitor added to the control. The consequence is good for higher frequencies but the undesired low-frequency components minimization is not visible as the actual capacitor is replaced by a virtual capacitor logic. Even though the control loop incorporates a capacitor to be present in the topology, the actual lack of this capacitor is significant in low-frequency areas. The advantage of the topology is improved power density as the electrolytic capacitor is eliminated and in alternate path reduced with the virtual capacitor in the control loop.

The high power density PWM rectifier described in ref. [29] seems to have a good impact on the research area of power density where the focus is. This positive impact is clearly depicted by the power density carried out by the authors. An adequate comparison between the two major bottlenecks of this research which are efficiency and power density has been very clearly compared with some of the topologies in this classification to prove the improvement brought in by their topology. The topology incorporates a basic compensation circuit for a separate auxiliary DC-Link system with the conventional H-bridge topology hence making it easy as compared to some of the other literature proposed under the same classification. The shortcoming in regard to any topology is the same as it is the deficiency in efficiency observed. The relative reduction in this amount of decrease of efficiency cannot be quantified to be high as compared to others but the overall control and design of the circuit can be described as simpler in part to other literature. Cao et al. [30] presents a multi-source-based active ripple compensator. This can be directly compared to harmonic compensators in power lines. The same objective is accomplished by using this active ripple compensator but with a change as said the compensation is for the ripple content or unwanted AC components observed in the output. The utilization of two energy storage elements makes it vulnerable to the fact that the amount of power density improvisation proposed is not effective.

The split capacitor technique discussed in the literature has been employed in ref. [31] but with a different control mechanism that can compensate for any variations or changes in the capacitor parameters thus possessing a great advantage of the elimination of circulation currents and failure in DC-Link capacitors. But the control complexity with the power delivery inductor for each of the capacitors makes the power density lag far behind compared to most of the literature on active capacitive DC Link already described. Current pulsation smoothing filter-based ripple reduction is observed in ref. [32] consisting of high frequency and low-frequency filter arraignment possessed by two capacitors and one power delivery inductor. This makes the system fall back in power density further the efficiency would take a massive hit if the input voltage goes beyond a minimum lower level thereby increasing the switching losses for active devices. Jang and Agelidis [33] discusses a generic compensation circuit of active DC Link as described in regard to the converter that it uses however an additional inductor which has an impact on power density has. Reddy and Narasimharaju [34] is a type of PFC compensation yet can be described as generic compensation a Flyback module and added advantage of support for discontinuous current operation mode.

4.4.2.2 Integrated Auxiliary Active Capacitive DC Links

Figure 4.19 discusses the integrated auxiliary circuit-based active DC-Link circuits and balancing techniques [35–43]. All of this literature is based on H-bridge compensation techniques as integration on this topology is comparatively easy in regard to other topologies where achieving an integrated operation would be very difficult.

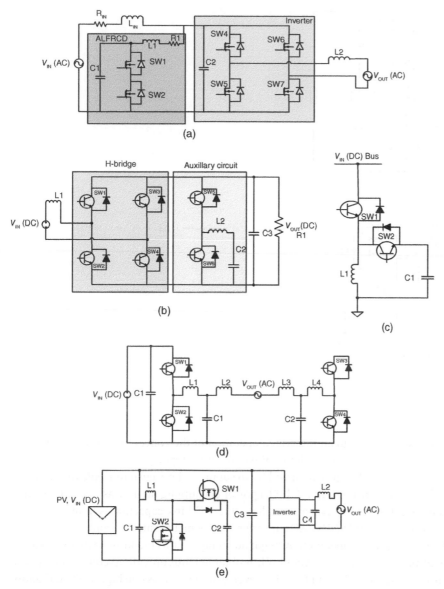

Figure 4.18 H-bridge based auxiliary support. (a) An active low-frequency ripple control method based on the virtual capacitor concept for BIPV systems [28], (b) A high power density single-phase PWM rectifier with active ripple energy storage [29], (c) Ripple eliminator to smooth DC-bus voltage and reduce the total capacitance required [30], (d) Improved power decoupling scheme for a single-phase grid-connected differential inverter with realistic mismatch in storage capacitances [31], (e) A novel parallel active filter for current pulsation smoothing on single stage grid-connected AC-PV modules [32], (f) A minimum power-processing-stage fuel-cell energy system based on a boost-inverter with a bidirectional backup battery storage [33], (g) Single-stage electrolytic capacitor less non-inverting buck-boost PFC based AC–DC ripple free LED driver [34], ALFRCD = dual active low-frequency ripple control.

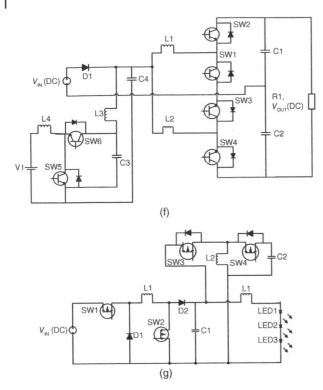

Figure 4.18 (*Continued*)

Some of these just control techniques contribute to an integrated operation of the converter where the suppression of the ripple current and stabilization of DC Link is done. While the other topologies try to add another leg in the H-bridge or alternate active switch-based loops which will try to suppress these current ripples at the same time while supporting the operation of the converter. In ref. [35] a power decoupling method as in any converter is discussed but the variation is imposed when this paper considered the DC-Link capacitor as another port in operation thus making the H-bridge a three-port DC/DC/AC converter arrangement. Thus the control scheme for power decoupling is designated for three ports of operation. A major advantage of this literature is that no change in efficiency is observed while an improved DC-Link current profile is obtained thus providing reduced stress over the DC Link. Reduced power density is the fallback point for this converter as it employs two high-value inductors for power delivery with the filter arrangement. Chen et al. [43] employs complex space vector pulse width modulation technique (SVPWM) and tries to eliminate electrolytic DC-Link capacitor by eliminating double frequency component. Even though the system accompanies a high power density, it employs a complex control mechanism that requires significant-good control hardware and efficient control technique. Dedicated control loop design with closed-loop results are displayed in literature [36] which in turn uses a generic topology of active compensation as discussed in separate auxiliary active DC Links. Even though the topology employs a generic compensation circuit for the ripple compensation, the topology combined the advantages of high energy-storage efficiency and the low requirement on control bandwidth [36]. No significant power density improvement rather than the proposed of those in separate auxiliary circuit section is observed hence cannot be considered to be any supreme. Rather than going for an active switch-based decoupling mechanism, ref. [37] utilizes active filter integration with a simple single-phase rectifier unit and multiple control loops with ripple voltage reference to control the compensation energy storage elements.

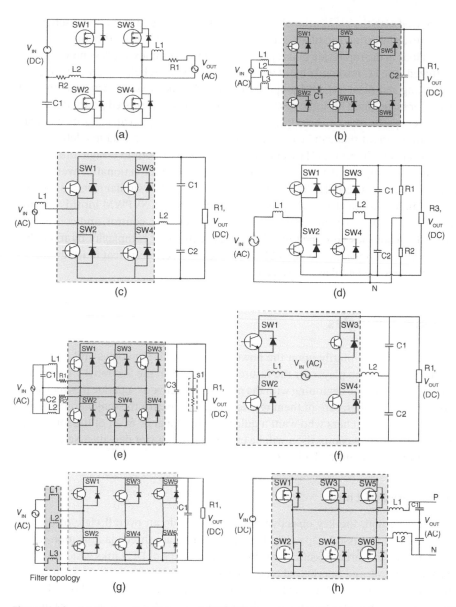

Figure 4.19 (a) Four-switch three-port DC/DC/AC converter [35], (b) Active power decoupling for high-power single-phase PWM rectifiers [36], (c) Active-filter-integrated AC/DC converter with a symmetrical half-bridge type of active filter [37], (d) Single-phase rectifier with a neutral leg [38], (e) Unity power factor PWM rectifier with DC ripple compensation [39], (f) Single-phase active power decoupling circuit for DC-Link capacitance reduction [40], (g) Three-leg AC–DC converters with generic filter topology [41], (h) Enhanced single-phase full bridge inverter circuit [42].

Ming and Zhong [38] discusses adding the support active leg for compensation in most of the literature but for a half-bridge. The control design and implementation are done for a half-bridge. Efficiency improvement is proposed in comparison to differential compensation techniques but none of the simple techniques as literature [23–43] are discussed or compared in the paper. Further power density comparison or any description is not provided thus making us assume the reduction of power density because of the power delivery inductor. Literature [39] is one of the oldest techniques in integrated auxiliary DC Link, where the generic compensation topology for the separate auxiliary technique is employed with the conventional H-bridge and an integrated operational scheme is formulated thus making the whole into an integrated auxiliary active DC-Link system. However, the compensation energy sources required are kept in a loop at the input side making the system more complex and less power-dense as power delivery inductors are utilized. The LC resonant circuit with the active switches helps in diminishing the current stresses over the DC Link. Tang and Blaabjerg [40] proposes a split DC-Link capacitor arraignment that requires no additional active devices or inductors. This indicates no additional cost and improved power density in regard to the previous larger DC-Link capacitor-based H-bridge. But as always for any split capacitor mechanism if there exists an undetected circulating current that can be outside of the control loop can lead to a lethal point of the circuit which should be taken care of. Literature [41] seems to be very complex both in design and control as two variants displayed in this literature represent three legs and four leg-based bidirectional converters for nano-grid applications with three-port architecture and SVPWM. The idea of three ports is similar to ref. [35] while three legs and four-leg basis form similar active device-based control for ripple current. SVPWM for this specific arrangement is very complex thus proving this literature to fall back on power density, control, and complexity in design. Literature [42] is one of the topologies described in ref. [41] but with lesser complexity by removing the SVPWM technique from the control technique. Zhu et al. [42] also possesses a serious threat of irregularities between the output side capacitors that may be lethal as described earlier.

4.5 Comparison on DC-Link Topologies

The comparison in this research literature is primarily upon the power density of all topologies and respective converters reviewed here. This comes in as a keen player for the future as described in the Section 4.1 with other respective properties. Comparative analysis is done beginning with the passive and active as separate in which each individual classification is compared with the literature in them respectively. Henceforth the major contribution of this part of the comparison is for those researchers who want a quick and brief glance at all of the topologies described earlier. Later Section 4.5.1–4.5.3 in this comparison will be exclusively for power density assessment as the longer roadmap for power electronics in the future depicts so.

4.5.1 Comparison of Passive Capacitive DC Links

As said in the earlier briefing on this section, the key objective will be to provide an overall view of the different topologies and techniques available in passive capacitive DC Links. There are three classification tables (Tables 4.1–4.3) in this section that discuss the three classifications in the passive type of DC Links which are passive filter type, passive filter type with control, and interleaved type. Each of these is being which are keen on DC Links in converters and their overall characteristics. Over the literature review that is conducted and based on all the analysis that has been observed is how these traits have been described and fixed as a point of comparative analysis. Table 4.1 is depicting the comparison of passive filter type DC Links.

Moving on to literature [15–19] discussed passive filter type with any advanced control technique employed for ripple suppression is compared next. Parameters of discussion and comparison remain the same to filter type passive capacitor DC Links as both share the same topological features concerning the DC-Link part. Similar key objective as that of the above comparison is shared with this category also henceforth not many descriptive decisions upon the comparison is being given but as said earlier Table 4.3 will give an overall view and layout of each literature is clearly depicted.

The final category of interleaved passive DC Links marks the end of the overview comparison for passive capacitive DC-Link topologies. As followed earlier for Passive capacitive DC-Link with control, same properties are followed by overview comparison with interleaved topologies as all the three are essentially passive capacitive DC Links. Literatures [20–22] discuss interleaved topologies. Even though they employ passive techniques they also

Table 4.1 Classification of separate auxiliary active DC Links with literature.

Flyback/PFC based auxiliary support	H-bridge based auxiliary support	Generic DC–DC converter based auxiliary support
Harmonic reducer converter [23]	An active low-frequency ripple control method based on the virtual capacitor concept for BIPV systems [29]	A novel parallel active filter for current pulsation smoothing on single stage grid-connected AC–PV modules [32]
Generalized technique of compensating low-frequency component of load current with a parallel bidirectional DC/DC converter [24]	A high power density single-phase PWM rectifier with active ripple energy storage [30]	A minimum power-processing-stage fuel-cell energy system based on a boost-inverter with a bidirectional backup battery storage [33]
A flicker-free electrolytic capacitor-less AC–DC LED driver [25]	Ripple eliminator to smooth DC-bus voltage and reduce the total capacitance required [31]	Single-stage electrolytic capacitor less non-inverting buck-boost PFC based AC–DC ripple free LED driver [34]
Feed-forward scheme for an electrolytic capacitor-less AC/DC LED driver to reduce output current ripple [26]	Improved power decoupling scheme for a single-phase grid-connected differential inverter with realistic mismatch in storage capacitances [32]	–
LED driver achieves electrolytic capacitor-less and flicker-free operation with an energy buffer unit [27]	–	–

incorporate interleaved operation with the converter to obtain an improved DC-Link ripple profile so as to improve DC-Link profile. Table 4.4 describes the comparison of the different topologies for interleaved capacitive DC Link.

4.5.2   Comparison of Active Capacitive DC Links

The second category for comparison is the generalized comparison for active capacitive DC Links. Similar to the passive capacitive DC-Link comparison discussed before the aim of this comparison also remains the same to give a whole idea of active capacitive DC Links. There will be two comparison sections that depict each of the classification groups in active capacitive DC Links. These are separate auxiliary circuit-based and integrated auxiliary circuit-based active capacitive DC Links. Apart from the six traits discussed in the passive capacitive DC-Link comparison an additional seventh trait which is active devices utilized in the circuit is also being discussed in this comparison. As the name indicates the requirement of this trait of comparison is required as active devices are employed in the topologies.

First section/paragraph of 4.5.2 active capacitive DC Links which is the separate auxiliary circuit group has three comparison groups which are as per the sub-classification developed in this review. These are H-bridge-based auxiliary support circuits, Flyback/PFC-based auxiliary support circuits, and Generic DC–DC Converter-based auxiliary support circuits. Corresponding literatures [15, 23–33] have been classified and compared in the three comparison tables (Tables 4.5–4.7) based on the traits described earlier.

The remaining category in active capacitive DC Link constitutes the integrated auxiliary circuit-based active capacitive DC Links which follows in literature [35–43]. All of these as described earlier refer to a compensation circuit or control technique that is embedded within the existing converter itself. Most of them rather look in for control techniques that can suppress the ripples in DC Link while others who propose integrated topology resort to H-bridge generic made with compensation to form modified H-bridge as integration to this architecture

Table 4.2 Comparison of filter type passive DC Links.

Literature	A zero ripple technique applicable to any DC converter [11]	A novel passive offline LED driver with long lifetime [12]	Six switches solution for single-phase AC/DC/AC converter with capability of second-order power mitigation in DC-Link capacitor [13]	Dynamic analysis and state feedback voltage control of single-phase active rectifiers with DC-Link resonant filters [14]
Electrical specifications	$V_{in} = 12\,V$ (DC), $V_{out} = 5\,V$ (DC), Prated = 20 W	$V_{in} = 230\,V$ (AC), $V_{out} = 158\,V$ (DC), Prated = 50 W	$V_{in} = 230\,V$ (AC), $V_{out} = 400\,V$ (DC), Prated = 80 kW	$V_{in} = 100\,V$ (AC), $V_{out} = 100\,V$ (DC), Prated = 1 kW
Ripple component and attenuation	2 V reduced to 0.02 V (reduction by $\cong$100 times)	No reduction in ripple, ripple is allowed as LED load	4 V ripple observed (1% of output voltage)	Ripple reduction or attenuation is less discussed while system stability during different conditions is discussed
DC Link capacitance specifications	4.4 μF no specified material described	Four 10 μF polypropylene capacitors each parallel pair forming 20 μF	470 μF no specified material described (simulation only carried out)	1000 μF (high as compared to the power level employed)
Control technique	Fourth order low pass mode operating for filter (coupled)	No specific control technique employed as ripple current allowed	PR (proportional and resonant) controller – inner inductor current loop and the PI (proportional and integral) controller – DC-Link voltage loop	Voltage controller – partial state feedback control (complex of the all four topologies discussed)
Modularity in regard to topology	Superior (as literature described integration with different DC–DC converters)	Inferior (as only applicable to LED loads)	Inferior (as only applicable AC/DC/DC specific topology)	Average (as can be integrated to most of the high power AC–DC units)
Efficiency comparison as per literature	No much consideration on efficiency impact as no additional active device used	High efficiency (~94%)	No much consideration on efficiency impact as only simulation carried out	No much consideration on efficiency impact as system studied for load condition response
Converter topology and configuration	Buck topology in DC–DC	Diode bridge with current source converter (AC–DC)	Specific AC–DC–DC topology with six switch	AC–DC generic high power LLC

Table 4.3 Comparison of filter type passive DC Links with control.

Literature	Power decoupling method for single-phase H-bridge inverters with no additional power electronics [15]	A high performance T-type single phase double grounded transformerless photovoltaic inverter with active power decoupling [16]	Adaptive DC-Link voltage control scheme for single phase inverters with dynamic power decoupling [17]	A low capacitance single-phase AC–DC converter with inherent power ripple decoupling [18]	Transformerless active power decoupling topologies for grid connected PV applications [19]
Electrical specifications	$V_{in} = 400\,V$ (DC), $V_{out} = 230\,V$ (AC), Prated = 1 kW	$V_{in} = 200\,V$ (DC), $V_{out} = 120\,V$ (AC), Prated = 1 kW	$V_{in} = 250\,V$ (DC), $V_{out} = 120\,V$ (AC), Prated = 500 W	$V_{in} = 500\,V$ (DC), $V_{out} = 230\,V$ (AC), Prated = 1 kW	$V_{in} = 600\,V$ (DC), $V_{out} = 325\,V$ (AC), Prated = 1 kW
Ripple component and attenuation	Ripple current of 1.25 A reduced to 0.17 A (7% reduction)	Output ripple minimized while input maintained low at 3% of input voltage (50% reduction)	Output ripple minimized and DC-Link voltage reduced by 24%	Reduction in ripple has not been quantified but a reduction is described through ZVS linked filter arrangement	Reduction even though not quantified the ripple has been limited to 1.67% of DC-Link voltage which is appreciable for this power level
DC-Link capacitance specifications	Two 60 µF film capacitor utilized for decoupling	30 µF film capacitor utilized for decoupling	45 µF film capacitor utilized for decoupling	38 µF capacitor, no specific material described as only simulation carried out	Four 60 µF film capacitor utilized for decoupling (very inferior property as number is four and value is around 60 µF)
Control technique	Cascaded voltage–current control	Voltage control with maximum power point tracking (MPPT) for input PV panel	Voltage control with MPPT and specific adaptive DC-Link voltage control	Feed forward control with ZVS and ZCS	Grid current control for output and voltage loop control for input side
Modularity in regard to topology	Superior (as employed with generic H-bridge)	Average (as can be utilized only in modified H-bridge with DC Link parallel to source)	Superior (easy integration as no modification in H-bridge)	Average (integration is only possible for dual active bridge or dual bridge topology)	Superior (easy integration with generic H-bridge)
Efficiency comparison as per literature	Overall efficiency decrease by 2%	Efficiency increase of 24% observed	Efficiency increase of ~2% observed	No efficiency as simulation only carried out	No efficiency as simulation only carried out
Converter topology and configuration	H-bridge generic (DC–AC)	H-bridge modified (DC–AC)	Generic H-bridge (DC–AC)	Dual bridge topology (DC–AC)	Generic H-bridge with split DC Link (DC–AC)

Table 4.4 Comparison of filter type passive DC Links with interleaved type.

Literature	Mitigation of low-frequency current ripple in fuel-cell inverter systems through waveform control [20]	Direct AC/DC rectifier with mitigated low-frequency ripple through inductor-current waveform control [21]	A dual mode operated boost inverter and its control strategy for ripple current reduction in single-phase uninterruptible power supplies [22]
Electrical specifications	V_{in} = 90 V (DC), V_{out} = 110 V (AC), Prated = 170 W	V_{in} = 110 V (AC), V_{out} = 200 V (DC), Prated = 50 W	V_{in} = 500 V (DC), V_{out} = 230 V (AC), Prated = 50 W
Ripple component and attenuation	45.1% of ripple reduction is stated with ripple less than 13% of the input	Ripple mitigation is not done but the same ripple at a reduced capacitance is achieved	Second order harmonics is eliminated almost completely attaining superior ripple mitigation
DC-Link capacitance specifications	Two 15 µF film capacitor of 800 V	Two 15 µF film capacitor of 600 V and one 30 µF film cap of 600 V	Two 60 µF film capacitor of 800 V
Control technique	Differential converter control with inner current loop control and waveform control	Waveform control based differential converter control	Differential mode control for inverter with common mode control for ripple elimination
Modularity in regard to topology	Superior (as buck, boost and buck-boost can be utilized for differential arrangement)	Inferior (as literature only substantiate the data only for modified boost converter in differential arrangement)	Average (differential arrangement is only applicable to modified H-bridge for inverter architecture with split capacitor at output)
Efficiency comparison as per literature	Superior efficiency has been depicted as per test results for the specific topology with waveform control	Lower efficiency at lower power and higher observed to compare topologies at higher than rated power	No efficiency comparison is made as the research is toward the control technique employed
Converter topology and configuration	Differential DC–DC converter to obtain inverter operation (DC–AC)	Differential arrangement for rectification (AC–DC)	Differential modified H-bridge to obtain inverter operation (DC–AC)

is much less complex. Below the nine topologies described in the literature have been compared (Table 4.8) to give an overall view of these with the selected traits for active capacitive DC Links as did with the earlier comparisons.

4.5.3 Comparison of DC Link Based on Power Density, Efficiency, and Ripple Attenuation

The major focus of this review is on the power density maximization for different topologies of DC-Link compensation with their respective converters included as shown in Figure 4.20. This section tries to depict different comparative analyzes in regard to the power density of each of these topologies with key aspects such as Efficiency and Ripple Attenuation which are the traits in question to be proposed as an improvement as these are DC Links employed. As per the conducted review, observation of these three traits, i.e. Power Density, Efficiency, and Ripple Attenuation varies one-on-one and there needs to be a suitable trade-off made which can

Table 4.5 Comparison of separate auxiliary circuit active DC Links (H-bridge type).

Literature	An active low-frequency ripple control method based on the virtual capacitor concept for BIPV systems [29]	A high power density single-phase PWM rectifier with active ripple energy storage [30]	Ripple eliminator to smooth DC-bus voltage and reduce the total capacitance required [31]	Improved power decoupling scheme for a single-phase grid-connected differential inverter with realistic mismatch in storage capacitances [32]
Electrical specifications	$V_{in} = 400\,V$ (DC), $V_{out} = 220\,V$ (AC), Prated $= 2\,kW$	$V_{in} = 540\,V$ (DC), $V_{out} = 220\,V$ (AC), Prated $= 15\,kW$	$V_{in} = 230\,V$ (AC), $V_{out} = 400\,V$ (DC), Prated $= 1\,kW$	$V_{in} = 500\,V$ (DC), $V_{out} = 230\,V$ (AC), Prated $= 800\,W$
Ripple component and attenuation	Ripple attenuation from compared model is only of a 0.2 A observed in the results (superior for 2 kW power level)	Ripple attenuation from compared model is stated as 2% limited ripple to the DC bus voltage requirement	Ripple attenuation from compared model is stated as 20–25% limited ripple to the DC bus voltage requirement	Ripple at second order level of the fundamental 50 Hz is said to be eliminated while few fourth order remains
DC-Link capacitance specifications	DC-Link capacitor of 3 µF and energy storage capacitor of 100 µF	DC-Link capacitor of 140 µF and energy storage capacitor 200 µF	DC-Link capacitor of 110 µF and energy storage capacitor 165 µF	Two DC-Link capacitor of 60 µF film type
Control technique	Basic current control is employed for DC-Link current control but included is a virtual DC-Link capacitor which improves control profile	Discontinuous current control mode of operation is followed for auxiliary support circuit	PWM rectification control is provided for the H-bridge while discontinuous current control mode of operation is followed for auxiliary support circuit	Differential control is applied for the differential arrangement of the converter while common mode current control for ripple elimination applied for the auxiliary circuit
Modularity in regard to topology	Superior modularity is observed as auxiliary circuit can be easily integrated to any inverter topology for single phase applications	Superior modularity is observed as auxiliary circuit can be easily integrated to H-bridge configuration in generic mode	Superior modularity is observed as auxiliary circuit can be easily integrated to H-bridge configuration in generic mode	Average modularity is observed as a modified H-bridge is utilized for the topology implementation with split DC-Link capacitor
Efficiency comparison as per literature	Due to the addition of virtual capacitor equivalent series resistor (ESR) for actual capacitor is eliminated which improves efficiency to some extent	Efficiency decrease of 1.8% observed due to the addition of auxiliary circuit	Literature states there is an efficiency drop in the system but to what extent is not quantized	Lossless ideal conditions are described for the development of the topology and no relevant comparison is given stating support for efficiency
Active devices utilized in auxiliary circuit	Two extra MOSFETs are employed	Two extra MOSFETs are employed	Two extra MOSFETs are employed	No extra active devices utilized
Converter topology and configuration	Generic H-bridge with generic compensation based on two active devices (DC–AC)	Generic H-bridge with generic compensation based on two active devices (DC–AC)	Generic H-bridge with generic compensation based on two active devices (AC–DC)	Modified H-bridge with split DC capacitor at DC-Link side (DC–AC)

Table 4.6 Comparison of separate auxiliary circuit active DC Links (generic type).

Literature	A novel parallel active filter for current pulsation smoothing on single stage grid-connected AC–PV modules [32]	A minimum power-processing-stage fuel-cell energy system based on a boost-inverter with a bidirectional backup battery storage [33]	Single-stage electrolytic capacitor less non-inverting buck-boost PFC based AC–DC ripple free LED driver [34]
Electrical specifications	$V_{in} = 34$ V (DC), $V_{out} = 230$ V (AC), Prated $= 110$ W	$V_{in} = 26$ V (DC), $V_{out} = 100$ V (AC), Prated $= 1.2$ kW	$V_{in} = 30$–90 V (AC), $V_{out} = 40$ V (DC), Prated $= 7$ W
Ripple component and attenuation	Full ripple elimination is described to be achieved eliminating any rippled DC components in the expected DC output	Ripple attenuation can be said as ripple elimination in this topology as the attenuation levels are nearly 99%	Ripple attenuation of significant level is observed even though not quantified in literature as comparison
DC-Link capacitance specifications	Reduction in capacitance is not much observed in literature, energy storage capacitor of 500 µF is used and input DC Link of 2.78 mF	Output filter capacitance of 30 µF and DC Link and storage capacitors of 30 µF both categories are with metal film	Output filter capacitance of 6.6 µF and storage capacitors of 13.2 µF both categories are with metal film
Control technique	MPPT for the PV panel is followed in front end with current pulsation smoothing with the parallel active filter control	PI Control for both compensation circuit and converter with anti-windup technique on both control loop	Higher voltage is maintained for compensation circuit output which simplifies control complexity as in feed forward control [26]
Modularity in regard to topology	Reduced modularity for the generic compensation circuit is observed as the specific application is limited to PV converters	Primitive level of modularity is observed as specific compensation is only applicable to specified topology	Restricted modularity is observed as compensation circuit is only applicable in LED drivers
Efficiency comparison as per literature	No change in efficiency as compared to previous topology in literature is observed as the entire PV power is not handled by the compensation circuit	Higher efficiency to compare literature is observed and stated in the literature [32]	Reduced efficiency is observed for higher output voltages; average decrement being ~2%
Active devices utilized in auxiliary circuit	Two extra MOSFETs are employed	Two extra MOSFETs are employed	Two extra MOSFETs are employed
Converter topology and configuration	Generic compensation circuit is provided in front end which also contributes MPPT function with inverter in the output	Generic compensation circuit is provided in front end of four switch based boost inverter (DC–AC)	Modified compensation circuit topology employed with PFC based front end rectifier (AC–DC)

propose an optimum solution in each case. Each case is referred to here as the topology may be an inverter, rectifier, or DC–DC converter, or even an AC–AC conversion unit. As the comparison considered is a trade-off as described earlier this research utilizes radar charts to compare these traits thus providing a better comparison for the topologies. In the chart 0 indicates low or incapable while 2 represents high or excellent in trait and 1 indicates intermediate.

Table 4.7 Comparison of separate auxiliary circuit active DC Links (Flyback type).

Literature	Harmonic reducer converter [23]	Generalized technique of compensating low-frequency component of load current with a parallel bidirectional DC/DC converter [24]	A flicker-free electrolytic capacitor-less AC–DC LED driver [25]	Feed-forward scheme for an electrolytic capacitor-less AC/DC LED driver to reduce output current ripple [26]	LED driver achieves electrolytic capacitor-less and flicker-free operation with an energy buffer unit [27]
Electrical specifications	V_{in} = 230 V (AC), V_{out} = 430 V (DC), Prated = 200 W	V_{in} = 125 V (AC), V_{out} = 210 V (DC), Prated = 360 W	V_{in} = 230 V (AC), V_{out} = 48 V (DC), Prated = 34 W	V_{in} = 90–264 V (AC), V_{out} = 48 V (DC), Prated = 34 W	V_{in} = 89–132 V (AC), V_{out} = ~60 V (DC), Prated = 15 W
Ripple component and attenuation	Ripple attenuation of 75–80% observed for different harmonic levels duly compared with EN61000-3-2	Ripple attenuation can be said as ripple elimination in this topology as the attenuation levels are nearly 99%	Larger ripples are allowed for LED load because of which reduced ripple attenuation is observed	Ripple attenuation of superior quality is observed at a rate of ~22% from the previous topology [25]	Larger ripples are allowed for LED load because of which reduced ripple attenuation is observed
DC-Link capacitance specifications	No specification in regard to DC-Link capacitance given. Additional capacitor utilized for storage element in auxiliary circuit	Specification is reduced from 3.8 to 1.8 mF and can be brought down even to ~600 µF with a 2% ripple	DC-Link capacitor of 0.47 µF (metalized polyester film) and energy storage capacitor of 20 µF (metalized polyester film)	DC-Link capacitor of 0.47 µF (metalized polyester film) and energy storage capacitor of 4.7 µF (metalized polyester film)	DC-Link capacitor of 10 µF (multi-layer ceramic capacitor [MLCC]) and two energy storage capacitor of 3.3 µF (metalized polyester film)
Control technique	PFC control is achieved through current control loop while auxiliary circuit control through voltage control loop	PFC control is linked for front end converter control while current control loop is put in for ripple energy compensation	PFC control is clubbed in with bidirectional controller for control of ripple energy to a limit as LED load	PFC control is embedded with bidirectional converter control which is feed-forward mechanism to improve ripple attenuation	PFC control and auxiliary circuit control through current control for ripple power compensation compared with LED load power
Modularity in regard to topology	Superior modularity is observed as it can be easily integrated with any topology converter and control can be employed	Two configurations for this compensation circuit is given which makes it have superior modularity	Lower modularity observed as specific topology is only applicable as PFC use in LED load drivers	Lower modularity observed as specific topology is only applicable as PFC use in LED load drivers	Lower modularity observed as specific topology is only applicable as PFC use in LED load drivers

(Continued)

Table 4.7 (Continued)

Efficiency comparison as per literature	Reduction of efficiency is stated at lower powers contributed by the active devices in the auxiliary circuit	No efficiency statements are observed in the literature while additional active devices are there in auxiliary circuit which may cause reduced efficiency	~3% decreased efficiency is observed in regard to the additional two active devices utilized	~5% decreased efficiency is observed in regard to the additional two active devices utilized with the feed-forward control	~2% decreased efficiency is observed in regard to the additional two active devices utilized
Active devices utilized in auxiliary circuit	Two extra MOSFETs are employed	Two extra MOSFETs are employed	Two extra MOSFETs are employed	Two extra MOSFETs are employed	Two extra MOSFETs and two diodes are employed
Converter topology and configuration	Generic PFC employed with generic compensation circuit with two active devices (AC–DC)	Compensation circuit is put in upfront as a compensator with added PFC control (AC–DC)	Buck-boost compensation circuit with Flyback converter for LED load is observed (AC–DC)	Buck–Boost with feed forward compensation circuit with Flyback converter for LED load is observed (AC–DC)	Flyback converter for LED Driving with coupled inductor-based compensation to the Flyback transformer (AC–DC)

Table 4.8 Comparison of integrated circuit active DC Link.

Literature	A power decoupling method based on four-switch three-port DC/DC/AC converter in DC microgrid [35]	DC capacitor-less inverter for single-phase power conversion with minimum voltage and current stress [43]	Active power decoupling for high-power single-phase PWM rectifiers [36]	Integration of an active filter and a single-phase AC/DC converter with reduced capacitance requirement and component count [37]	A single-phase rectifier having two independent voltage outputs with reduced fundamental frequency voltage ripples [38]	A unity power factor PWM rectifier with DC ripple compensation [39]	A component-minimized single-phase active power decoupling circuit with reduced current stress to semiconductor switches [40]	Control and modulation of bidirectional single-phase AC–DC three-phase-leg sine pulse width modulation (SPWM) converters with active power decoupling and minimal storage capacitance [41]	Enhanced single-phase full-bridge inverter with minimal low-frequency current ripple [42]
Electrical specifications	$V_{in} = 200\,V$ (DC), $V_{out} = 110\,V$ (AC), Prated = 500 W	$V_{in} = 185\,V$ (DC), $V_{out} = 120\,V$ (AC), Prated = 1.5 kW	$V_{in} = 230\,V$ (AC), $V_{out} = 450\,V$ (DC), Prated = 4 kW	$V_{in} = 110\,V$ (AC), $V_{out} = 260\,V$ (DC), Prated = 210 W	$V_{in} = 110\,V$ RMS (AC), $V_{out} = 300/200$ (DC–500 V DC Bus), Prated = 1 kW	$V_{fr.} = 50\,V$ PEAK (AC), $V_{out} = 140\,V$ (DC), Prated = 100 W	$V_{in} = 120\,V$ (AC), $V_{out} = 350\,V$ (DC), Prated = 1 kW	$V_{in} = 110\,V$ (AC), $V_{out} = 220\,V$ (DC), Prated = 550 W	$V_{in} = 200\,V$ (DC), $V_{out} = 110\,V$ (AC), Prated = 50 W
Ripple component and attenuation	Double line frequency components are observed to be reduced by 2% thus improving ripple attenuation	Double line frequency components are observed to be reduced and limited to 2.5% of output voltage indicating improved ripple attenuation	Superior ripple component attenuation is observed reaching up to 70% in reduction to previous considered topology without active devices	Ripple in the output DC component is reduced to 1.1% of output with input current distortion at a controlled level	Ripple attenuation of considerable amount is stated and quantified with individual voltage outputs reduced from 20 to 3 V and 16 to 6 V	Ripple attenuation is improved by 10 times as compared to the previous uncompensated system as ripple current reduced by 1/10 to the previous one	Not much of ripple attenuation capability is demanded by the topology, literature more focuses on active device less decoupling	Ripple attenuation is quantified by stating output voltage ripple magnitude of 2.5 V which is 1.1% of the rated output	Superior improvement in ripple attenuation is quantified and justified with results with reduction from 87% to 14% in ripple current

(Continued)

Table 4.8 (Continued)

DC-Link capacitance specifications	Reduction in capacitance is not much observed in literature, 100 µF electrolytic capacitor is used	Huge reduction is capacitors are stated with DC-Link capacitance being reduced to 170 µF from 4.6 mF and output AC capacitor of 100 µF	DC-Link capacitance of 450 µF is being utilized which are film type replacing the electrolytic types	Two DC-Link capacitance of 50 µF is being utilized which are Film type replacing the 900 µF electrolytic types	Split capacitor is said to be utilized in output of values 1120 and 560 µF with no type specified however values indicates electrolytic type	DC-Link capacitance of 1200 µF is utilized which is not that significant reduction in addition to which two 165 µF in AC input	DC-Link capacitance of 90 µF of film type is utilized	DC-Link capacitance of 200 µF and AC capacitance of 147 µF which is film type	AC capacitance of 60 µF each is used in the output each of film type
Control technique	Outer voltage loop and inner current loop based control is done for the converter	Complex harmonic injection control using SVPWM is followed similar to STATCOM	Capacitor voltage control system consists of inner loop pole-placement control and outer loop proportional resonant control	PFC control with feed forward voltage control with two closed-loop voltage-ripple-based reference generation method is utilized	Rectification control and output voltage control is done through PI control while capacitor current control through repetitive current control	Current control with PFC control is employed for the topology rather than PFC control in uncompensated system	Inner current control with double loop voltage control is provided for control of the two identical capacitors	SPWM three leg control is provided with vector based estimation of grid synchronization	Dedicated control technique is developed stated as ripple confinement control where output voltage and capacitance voltage is measured
Modularity in regard to topology	Average modularity is observed as compensation can be integrated with any H-bridge topology alone	Reduced modularity is observed as applicable only to H-bridge and complex control via SVPWM is followed	Modularity is being restricted to H-bridge and PWM rectifiers	Average modularity can be demanded as the design only implies to H-bridge rectification topology	Restricted modularity can be demanded as the design only implies to modified H-bridge rectification topology	Average modularity can be demanded as the design only implies to H-bridge rectification topology	Restricted modularity can be demanded as the design only implies to modified H-bridge rectification topology	Average modularity can be described as the system integrates additional leg to three leg based inverter topology	Average modularity can be demanded as the design only implies to integrate one leg with H-bridge

Efficiency comparison as per literature	No change in efficiency is observed as no additional active devices are used	No stated record on efficiency in the literature hence makes it difficult to decide on efficiency	Topology is stated have improved efficiency than that of inductor based compensation circuits	Conventional H-bridge is observed to have higher efficiency at low load, while at higher loads proposed topology has higher efficiency	Superior efficiency at both low and higher loads are observed and quantified in the literature for specified topology	No stated record on efficiency in the literature hence makes it difficult to decide on efficiency	Drop in efficiency is observed and quantified in the paper. At higher powers this decrement in efficiency is less	Reduced efficiency of about 3% decrement to standard topology is observed and quantified in the literature with three-leg in comparison	No stated record on efficiency in the literature hence makes it difficult to decide on efficiency, additional switches may decrease efficiency
Active devices utilized in auxiliary circuit	No additional active devices	No additional active devices	Two extra MOSFETs are employed	Two extra MOSFETs are employed	No additional active devices	Two extra IGBTs are employed	No additional active devices	Two extra IGBTs are employed	Two extra MOSFETs are employed
Converter topology and configuration	Generic H-bridge converter with no auxiliary support (DC–AC)	Generic H-bridge converter with no auxiliary support (DC–AC)	Generic H-bridge converter with auxiliary support of single leg (half bridge) (AC–DC)	Generic H-bridge converter with additional two MOSFETs as auxiliary circuit (AC–DC)	Modified H-bridge topology (AC–DC)	Generic H-bridge converter with additional two IGBTs as auxiliary circuit (AC–DC)	Modified H-bridge topology (AC–DC)	Generic H-bridge converter with additional two IGBTs as auxiliary circuit (AC–DC)	Generic H-bridge converter with additional two MOSFETs as auxiliary circuit (DC–AC)

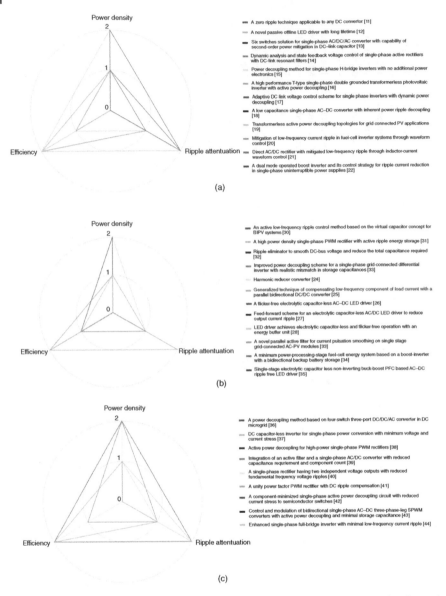

Figure 4.20 Radar chart for trade-off comparison between efficiency, power density and ripple attenuation for (a) passive balancing topologies in DC Link, (b) Active balancing topologies in DC Link except integrated topologies, (c) Active balancing topologies with integrated compensation in DC Link.

4.6 Future and Research Gaps in DC-Link Topologies with Balancing Techniques

The review performed on topologies in DC Link and their balancing techniques exposes a few areas where a requirement of immense research and improvisation exists. Further to the survey conducted literature [4, 9, 29, 44–60] also discusses the future of power electronics with power density, reliability, and failures in capacitors. Even though stated as few of these are the areas of the future power electronics as described in the Section 4.1. The three identified areas of improvement are (i) Power Density, (ii) Thermal Management, and (iii) Optimal trade-off numerical technique for prediction of DC Link. The first area power density as described in is one of the emerging trends in power electronics.

The rise of wide band gap devices using SiC and GaN requires the molding of DC Links also to suit the design and energy level optimization in these high power density converters. DC Link thermal management is the second area of future scope as the dissipation of heat in high-end power-rated converters can be challenging when optimization is done in regard to power density maximization. Layout constraints in high power density converters will have a significant impact on large DC-Link optimization techniques.

The third and final category is the optimal trade-off numerical techniques for the prediction of DC-Link design. There are several numerical techniques including machine learning and deep learning techniques which are being presently utilized for different estimation techniques. Such an estimation or optimization technique can be developed for the design of DC Link where the constraints can be efficiency, ripple attenuation, and power density. Generic topology derivation technique is being discussed in ref. [8] in continuation to this an optimization for the generic derived topology can be carried out which thus can project an impact layout on the trade-off made and suitable generic topology then after optimization is what can be the point of research in action. On a whole, these three traits can be concluded as the key points to be considered in the future rise of research in DC-Link topologies and balancing techniques.

4.7 Conclusion

A comprehensive review on different DC-Link topologies and their balancing techniques has been carried out in this research. All the topologies up till now are included in the literature, which has been reviewed in this research. Some of the literature includes the same topologies as the ones discussed with a different application. In this review, initial Section 4.4 are described in a generalized brief on the topologies for discussion and their comparison while at the end a trade-off comparison is given in regard to the power density of each category of topologies whether it be passive or active. This review tries to fill in the gap on the overall review for passive and active topologies for DC Links keeping a keen outlook on power density maximization also. Further research and scope areas required in this field are being identified and discussed to throw light and help build a track for researchers in this area which can contribute to a rich and sustainable future in power electronics.

References

1 Kim, K.A., Liu, Y.C., Chen, M.C., and Chiu, H.J. (2017). Opening the box: survey of high power density inverter techniques from the little box challenge. *CPSS Transactions on Power Electronics and Applications* 2 (2): 131–139.

2 Mantooth, H.A., Evans, T., Farnell, C. et al. (2017). Emerging trends in silicon carbide power electronics design. *CPSS Transactions on Power Electronics and Applications* 2 (3): 161–169.

3 Wang, H., Liserre, M., and Blaabjerg, F. (2013). Toward reliable power electronics: challenges design tools and opportunities. *IEEE Industrial Electronics Magazine* 7 (2): 17–26.

4 Yang, S., Bryant, A., Mawby, P. et al. (2011). An industry-based survey of reliability in power electronic converters. *IEEE Transactions on Industry Applications* 47 (3): 1441–1451.

5 Hu, H., Harb, S., Kutkut, N. et al. (2012). A review of power decoupling techniques for microinverters with three different decoupling capacitor locations in PV systems. *IEEE Transactions on Power Electronics* 28 (6): 2711–2726.

6 Sun, Y., Liu, Y., Su, M. et al. (2015). Review of active power decoupling topologies in single-phase systems. *IEEE Transactions on Power Electronics* 31 (7): 4778–4794.

7 Vitorino, M.A., Alves, L.F.S., Wang, R., and de Rossiter Corrêa, M.B. (2016). Low-frequency power decoupling in single-phase applications: a comprehensive overview. *IEEE Transactions on Power Electronics* 32 (4): 2892–2912.

8 Wang, H., Wang, H., and Zhu, G. et al. (2016). A generic topology derivation method for single-phase converters with active capacitive DC-Links. In: *2016 IEEE Energy Conversion Congress and Exposition (ECCE),* IEEE, Milwaukee, WI, USA, https://doi.org/10.1109/ECCE.2016.7854689 (September 2016), pp. 1–8.

9 Wang, H., Wang, H., and Zhu, G. et al. (2016). Cost assessment of three power decoupling methods in a single-phase power converter with a reliability-oriented design procedure. In: *2016 IEEE 8th International Power Electronics and Motion Control Conference (IPEMC-ECCE Asia)* (May 2016), IEEE, Hefei, China, https://doi.org/10.1109/IPEMC.2016.7512905, pp. 3818–3825.

10 Wang, H., Wang, H., Zhu, G.R., and Blaabjerg, F. (2020). An overview of capacitive DC links-topology derivation and scalability analysis. *IEEE Transactions on Power Electronics* 35 (2): 1805–1829.

11 Hamill, D.C. and Krein, P.T. (1999) A zero ripple technique applicable to any DC converter. In: *30th Annual IEEE Power Electronics Specialists Conference. Record (Cat. No. 99CH36321),* Charleston, SC, USA, IEEE, vol. 2, pp. 1165–1171, https://doi.org/10.1109/PESC.1999.785659.

12 Hui, S.Y., Li, S.N., Tao, X.H. et al. (2010). A novel passive offline LED driver with long lifetime. *IEEE Transactions on Power Electronics* 25 (10): 2665–2672.

13 Liu, X., Wang, P., and Loh, P.C. et al. (2011). Six switches solution for single-phase AC/DC/AC converter with capability of second-order power mitigation in DC-Link capacitor. In: *IEEE Energy Conversion Congress and Exposition* (September 2011), IEEE, Phoenix, AZ, USA, https://doi.org/10.1109/ECCE.2011.6063938, pp. 1368–1375.

14 Vasiladiotis, M. and Rufer, A. (2013). Dynamic analysis and state feedback voltage control of single-phase active rectifiers with DC-Link resonant filters. *IEEE Transactions on Power Electronics* 29 (10): 5620–5633.

15 Serban, I. (2015). Power decoupling method for single-phase H-bridge inverters with no additional power electronics. *IEEE Transactions on Industrial Electronics* 62 (8): 4805–4813.

16 Xia, Y., Roy, J., and Ayyanar, R. (2016). A high performance T-type single phase double grounded transformerless photovoltaic inverter with active power decoupling. In: *IEEE Energy Conversion Congress and Exposition (ECCE)* (February 2017), IEEE, Milwaukee, WI, USA, https://doi.org/10.1109/ECCE.2016.7854693, pp. 1–7.

17 Xia, Y. and Ayyanar, R. (2016). Adaptive DC Link voltage control scheme for single phase inverters with dynamic power decoupling. In: *IEEE Energy Conversion Congress and Exposition (ECCE)* (February 2017), IEEE, Milwaukee, WI, USA, https://doi.org/10.1109/ECCE.2016.7854824, pp. 1–7.

18 Gottardo, D., De Lillo, L., and Empringham, L. et al. (2016) A low capacitance single-phase AC–DC converter with inherent power ripple decoupling. In: *IECON 2016-42nd Annual Conference of the IEEE Industrial Electronics Society* (October 2016), IEEE, Florence, Italy, https://doi.org/10.1109/IECON.2016.7793634, pp. 3129–3134.

19 Kumar, V.P. and Fernandes, B.G. (2016). Transformerless active power decoupling topologies for grid connected PV applications. In: *IECON 2016-42nd Annual Conference of the IEEE Industrial Electronics Society*, IEEE, Florence, Italy, https://doi.org/10.1109/IECON.2016.7793349, pp. 2410–2419.

20 Zhu, G.R., Tan, S.C., Chen, Y., and Chi, K.T. (2012). Mitigation of low-frequency current ripple in fuel-cell inverter systems through waveform control. *IEEE Transactions on Power Electronics* 28 (2): 779–792.

21 Li, S., Zhu, G.R., Tan, S.C., and Hui, S.R. (2014). Direct AC/DC rectifier with mitigated low-frequency ripple through inductor-current waveform control. *IEEE Transactions on Power Electronics* 30 (8): 4336–4348.

22 Tang, Y., Yao, W., and Blaabjerg, F. (2019). A dual mode operated boost inverter and its control strategy for ripple current reduction in single-phase uninterruptible power supplies. In: *9th International Conference on Power Electronics and ECCE Asia (ICPE-ECCE Asia),* IEEE, Seoul, Korea (South), https://doi.org/10.1109/ICPE.2015.7168086, pp. 2227–2234.

23 Garcia, O., Martínez-Avial, M.D., Cobos, J.A. et al. (2003). Harmonic reducer converter. *IEEE Transactions on Industrial Electronics* 50 (2): 322–327.

24 Dusmez, S. and Khaligh, A. (2014). Generalized technique of compensating low-frequency component of load current with a parallel bidirectional DC/DC converter. *IEEE Transactions on Power Electronics* 29 (11): 5892–5904.

25 Wang, S., Ruan, X., Yao, K. et al. (2011). A flicker-free electrolytic capacitor-less AC–DC LED driver. *IEEE Transactions on Power Electronics* 27 (11): 4540–4548.

26 Yang, Y., Ruan, X., Zhang, L. et al. (2013). Feed-forward scheme for an electrolytic capacitor-less AC/DC LED driver to reduce output current ripple. *IEEE Transactions on Power Electronics* 29 (10): 5508–5517.

27 Fang, P., Sheng, B., Webb, S. et al. (2018). LED driver achieves electrolytic capacitor-less and flicker-free operation with an energy buffer unit. *IEEE Transactions on Power Electronics* 34 (7): 6777–6793.

28 Cai, W., Liu, B., Duan, S., and Jiang, L. (2013). An active low-frequency ripple control method based on the virtual capacitor concept for BIPV systems. *IEEE Transactions on Power Electronics* 29 (4): 1733–1745.

29 Wang, R., Wang, F., Boroyevich, D. et al. (2011). A high power density single-phase PWM rectifier with active ripple energy storage. *IEEE Transactions on Power Electronics* 26 (5): 1430–1443.

30 Cao, X., Zhong, Q.C., and Ming, W.L. Ripple eliminator to smooth DC-bus voltage and reduce the total capacitance required. *IEEE Transactions on Industrial Electronics* 62 (4): 2224–2235.

31 Yao, W., Wang, X., Loh, P.C. et al. (2016). Improved power decoupling scheme for a single-phase grid-connected differential inverter with realistic mismatch in storage capacitances. *IEEE Transactions on Power Electronics* 32 (1): 186–199.

32 Kyritsis, A.C., Papanikolaou, N.P., and Tatakis, E.C. (2007). A novel parallel active filter for current pulsation smoothing on single stage grid-connected AC-PV modules. In: *European Conference on Power Electronics and Applications* (September 2017), IEEE, Aalborg, Denmark, https://doi.org/10.1109/EPE.2007.4417545, pp. 1–10.

33 Jang, M. and Agelidis, V.G. A minimum power-processing-stage fuel-cell energy system based on a boost-inverter with a bidirectional backup battery storage. *IEEE Transactions on Power Electronics* 26 (5): 1568–1577.

34 Reddy, U.R. and Narasimharaju, B.L. Single-stage electrolytic capacitor less non-inverting buck-boost PFC based AC–DC ripple free LED driver. *IET Power Electronics* 10 (1): 38–46.

35 Cai, W., Jiang, L., Liu, B. et al. A power decoupling method based on four-switch three-port DC/DC/AC converter in DC microgrid. *IEEE Transactions on Industry Applications* 51 (1): 336–343.

36 Li, H., Zhang, K., Zhao, H. et al. Active power decoupling for high-power single-phase PWM rectifiers. *IEEE Transactions on Power Electronics* 28 (3): 1308–1319.

37 Li, S., Qi, W., Tan, S.C., and Hui, S.R. Integration of an active filter and a single-phase AC/DC converter with reduced capacitance requirement and component count. *IEEE Transactions on Power Electronics* 31 (6): 4121–4137.

38 Ming, W.L. and Zhong, Q.C. A single-phase rectifier having two independent voltage outputs with reduced fundamental frequency voltage ripples. *IEEE Transactions on Power Electronics* 30 (7): 3662–3673.

39 Shimizu, T., Fujita, T., Kimura, G., and Hirose, J. A unity power factor PWM rectifier with DC ripple compensation. *IEEE Transactions on Industrial Electronics* 44 (4): 447–455.

40 Tang, Y. and Blaabjerg, F. A component-minimized single-phase active power decoupling circuit with reduced current stress to semiconductor switches. *IEEE Transactions on Power Electronics* 30 (6): 2905–2910.

41 Wu, H., Wong, S.C., Chi, K.T., and Chen, Q. Control and modulation of bidirectional single-phase AC–DC three-phase-leg SPWM converters with active power decoupling and minimal storage capacitance. *IEEE Transactions on Power Electronics* 31 (6): 4226–4240.

42 Zhu, G.R., Wang, H., Liang, B. et al. Enhanced single-phase full-bridge inverter with minimal low-frequency current ripple. *IEEE Transactions on Industrial Electronics* 63 (2): 937–943.

43 Chen, R., Liu, Y., and Peng, F.Z. DC capacitor-less inverter for single-phase power conversion with minimum voltage and current stress. *IEEE Transactions on Power Electronics* 30 (10): 5499–5507.

44 Wang, H., Ma, K., and Blaabjerg, F. (2012). Design for reliability of power electronic systems. In: *Proceedings of IEEE Industrial Electronics Society Annual Conference (IECON)* (September 2012), IEEE, Montreal, QC, Canada, https://doi.org/10.1109/IECON.2012.6388833, pp. 33–44.

45 Koutroulis, E. and Blaabjerg, F. (2013). Design optimization of transformer-less grid-connected PV inverters including reliability. *IEEE Transactions on Power Electronics* 28 (1): 325–335.

46 Song, Y. and Wang, B. (2013). Survey on reliability of power electronic systems. *IEEE Transactions on Power Electronics* 28 (1): 591–604.

47 Biela, J., Waffler, S., and Kolar, J.W. (2009). Mission profile optimized modularization of hybrid vehicle DC/DC converter systems. In: *Proceedings of IEEE International Power Electronics and Motion Control Conference*, IEEE, Wuhan, China, https://doi.org/10.1109/IPEMC.2009.5157601, pp. 1390–1396.

48 Petrone, G., Spagnuolo, G., Teodorescu, R. et al. (2008). Reliability issues in photovoltaic power processing systems. *IEEE Transaction in Electronics* 55 (7): 2569–2580.

49 Ma, K., Blaabjerg, F., and Liserre, M. (2013). Thermal analysis of multilevel grid side converters for 10 MW wind turbines under low voltage ride through. *IEEE Transactions on Industry Applications* 49 (2): 909–921.

50 Imam, A.M., Habetler, T.G., and Harley, R.G. (2005). Failure prediction of electrolytic capacitors using DSP methods. In: *Proceedings of IEEE Application Power Electronics Conference*, IEEE, Austin, TX, USA, https://doi.org/10.1109/APEC.2005.1453106, pp. 965–970.

51 Maccomber, L.L. (2011). Aluminum electrolytic capacitors in power electronics. In: *Proceedings of IEEE Application Power Electronics Conference, Exposit*. Cornell Dubilier Special Session, IEEE, Cornell dubilier exposit, pp. 1–14.

52 Alwitt, R.S. and Hills, R.G. (1965). The chemistry of failure of aluminum electrolytic capacitors. *IEEE Transactions on Parts, Materials Packaging* PMP-I (2): 28–34.

53 Liu, D. and Sampson, M.J. (2012). Some aspects of the failure mechanisms in $BaTiO_3$ – based multilayer ceramic capacitors. In: *Proceedings of CARTS International*, Capacitors and Resistors Technology Symposium (CARTS) International, Las Vegas, NV; (https://ntrs.nasa.gov/citations/20120009286), pp. 59–71.

54 Minford, W.J. (1982). Accelerated life testing and reliability of high K multilayer ceramic capacitors. *IEEE Transactions on Components, Hybrids, and Manufacturing Technology* CHMT-5 (3): 297–300.

55 Kubodera, N., Oguni, T., Matsuda, M. et al. (2012). Study of the long term reliability for MLCCs. In: *Proceedings of CARTS International*, Capacitors and Resistors Technology Symposium (CARTS) International, Las Vegas, NV, pp. 1–9.

56 Pelletier, P., Guichon, J.M., Schanen, J.L., and Frey, D. (2009). Optimization of a DC capacitor tank. *IEEE Transactions Industrial Application* 45 (2): 880–886.

57 Gasperi, M.L. and Gollhardt, N. (1998). Heat transfer model for capacitor banks. In: *Conference Record IEEE IAS Annual Meeting*, IEEE, St. Louis, MO, USA, https://doi.org/10.1109/IAS.1998.730299, pp. 1199–1204.

58 Harada, K., Katsuki, A., and Fujiwara, M. (1993). Use of ESR for deterioration diagnosis of electrolytic capacitor. *IEEE Transactions on Power Electronics* 8 (4): 355–361.

59 Vogelsberger, M.A., Wiesinger, T., and Ertl, H. (2011). Life-cycle monitoring and voltage-managing unit for DC-link electrolytic capacitors in PWM converters. *IEEE Transactions on Power Electronics* 26 (2): 493–503.

60 Wang, H., Wang, H., and Zhu, G. et al. (2016). A generic topology derivation method for single-phase converters with active capacitive DC-Links. In: *Proceedings of IEEE ECCE' 2016* (September 2016), IEEE, Milwaukee, WI, USA, https://doi.org/10.1109/ECCE.2016.7854689, pp. 1–8.

5

Energy Storage Systems for Smart Power Systems

Sivaraman Palanisamy[1], Logeshkumar Shanmugasundaram[2], and Sharmeela Chenniappan[3]

[1] *World Resources Institute (WRI) India, Bengaluru, India*
[2] *Department of Electronics and Communication Engineering, Christ the King Engineering College, Coimbatore, India*
[3] *Department of Electrical and Electronics Engineering, Anna University, Chennai, India*

5.1 Introduction

Concerns about the environment, such as global warming, have become global issues. As a result, the use of renewable energy sources like solar and wind power, as well as Smart Grids that effectively use all sorts of power sources, are seen as a very promising technology [1, 2]. Electricity grids achieve reliable power supply by balancing supply and demand to the best of their abilities. However, when the use of solar power and other renewable energy sources, which have variable production, grows, the grid's power supply may become unstable [3]. This poses several difficulties. To overcome such difficulties, an energy storage system has been required in the power system [4]. In an emergency, energy storage devices can also be used as a backup power supply [5]. Energy storage systems are extremely adaptable, and they may be used to satisfy the needs of a wide range of customers and in a variety of industries [6, 7]. These include renewable energy power producers, grid equipment such as transmission and distribution equipment, as well as commercial buildings, factories, and residences [8–10].

The grid-connected energy storage systems are used for both energy management applications as well as power quality applications such as load balancing or leveling, reducing the intermittency and smoothing the renewable energy integration, peak demand shifting or shaving, providing the uninterrupted power supply (UPS) to end-user loads, voltage and frequency regulation, voltage sag mitigation, reactive power control, and management, black start and islanded mode of operation, Volt/var control, etc. in the power systems [11]. The amount of renewable energy penetration keeps increasing every year, which demands the requirements of grid-connected energy storage systems. The energy storage system (ESS) is used to smooth the power output from renewable energy sources such as solar photovoltaic (PV) and wind energy conversion systems and level the load pattern [12].

The ESS can be classified in many ways, but one of the most useful is one based on the duration and frequency of power delivery. They are (i) short-term (seconds to minutes), (ii) medium-term (day storage), and (iii) long-term ESS (weekly to monthly) [1, 13].

The short-term ESS (less than 25 minutes) [13] shall be used for spinning reserve, peak shaving, UPS, primary and secondary frequency control, electric vehicles, etc.

The medium-term ESS (1–10 hours) [13] shall be used for load leveling, tertiary frequency control, UPS, tertiary frequency control, etc.

Finally, the long-term ESS (from 50 hours to less than three weeks) [13] shall be used for long-term services during periods whenever power output from renewable power plants (solar and wind) is limited (known as "dark-calm periods").

Artificial Intelligence-based Smart Power Systems, First Edition.
Edited by Sanjeevikumar Padmanaban, Sivaraman Palanisamy, Sharmeela Chenniappan, and Jens Bo Holm-Nielsen.
© 2023 The Institute of Electrical and Electronics Engineers, Inc. Published 2023 by John Wiley & Sons, Inc.

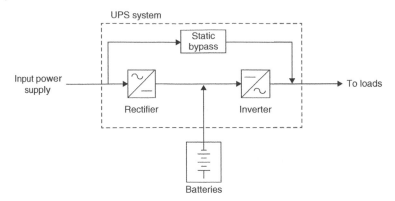

Figure 5.1 SLD of UPS system.

The short-term services possibly will be provided via flywheels, superconductive magnetic coils, and super-capacitors. The medium-term services possibly will be provided by pumped hydropower, compressed air ESS, thermoelectric storage, and electrochemical ESS, such as lithium-ion, lead acid, high temperature, and flow batteries [14–16]. Hydrogen or natural gas storage systems can provide long-term services. This chapter discusses the need for ESS in a different segment of power systems.

5.2   Energy Storage System for Low Voltage Distribution System

The ESS in low voltage distribution systems is employed in the UPS system for a couple of decades. This will provide the UPS to business-critical loads/emergency loads/group of loads that require a high input power supply reliability. It consists of the inbuilt rectifier and inverter and external battery systems. The typical single-line diagram (SLD) of the UPS system is shown in Figure 5.1.

During the normal operating time, loads are powered and parallelly batteries are also being charged. In case of emergency, energy stored in the batteries is used to power the loads without any break. The nature of the application of UPS systems is particularly used for business-critical loads/emergency loads/small groups of loads and not generally used for other non-critical loads. Here, the purpose of energy storage is to provide an UPS to the loads during the emergency time only.

In recent years, due to the factors like power supply reliability, utilizing the surplus power available from solar PV systems during the nonpeak hours, peak shaving, load shifting, etc., the ESS are being considered [17, 18]. The typical SLD of ESS used in a manufacturing plant is shown in Figure 5.2. The installed capacity of battery ESS is 500 kVA/500 kWh and is capable to supply plant loads for a minimum of 60 minutes duration during the grid power supply failure event.

Generally, in the past decades, diesel generator sets are considered for providing the backup power supply (standby power supply) to any small-scale commercial or industrial plants. These diesel generators act as a standby power source in addition to the grid power source. Whenever the grid power supply is not available, these diesel generators are turned ON and provide the power supply to those commercial or industrial plants. Automatic transfer switch (ATS) is used to transfer the operating source from grid power to diesel generator power and vice versa.

Nowadays, the rapid depletion of fossil fuels and drastic increase in diesel cost demand the users look for an alternative to a diesel generator. The ESS is one of the alternative options to these diesel generators. As shown in Figure 5.2, a 500 kVA/500 kWH battery energy storage system is used as an alternative option to a 650 kVA diesel generator in the manufacturing industry. The battery energy storage system consists of a bi-directional converter and battery banks. A bi-directional converter acts as a rectifier-cum-charger during the charging time (converting

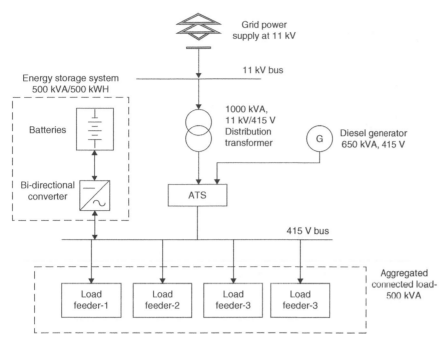

Figure 5.2 SLD of ESS used in a manufacturing plant.

the AC supply into DC supply and recharging the battery banks). It will act as an inverter during the discharging time (converting the DC supply into AC supply and feeding back to the grid).

During the normal operating condition, the grid power source is used to power the plant loads as well as charge the battery energy storage system as shown in Figure 5.3.

In case of failure of grid power supply, a battery energy storage system is used to power the connected loads till the grid power supply is resumed back as shown in Figure 5.4. The grid power supply interruption due to temporary faults is managed by using the battery energy storage system.

In case energy in the battery storage is drained and the grid power supply is not resumed, a diesel generator is used to power the plant loads as shown in Figure 5.5.

5.3 Energy Storage System Connected to Medium and High Voltage

The medium-scale and large-scale renewable power plants are connected to medium voltage and high voltage systems. Since the output from renewable energy sources is intermittent, the grid's stability is being affected. To smooth the power output, one of the possible solutions is adding the ESS into the grid. Solar PV and wind power generations, among the other options, now account for a major portion of the global energy requirements. However, in the situation of a higher amount of renewable energy integration into power systems, ESS is critical in combining a sustainable energy source with a reliable load dispatch and mitigating the effects of intermittency. As a result, ESS installation has expanded in recent years all over the world. Figure 5.6, the typical SLD of the grid network, shows the locations of ESS that can be employed.

The typical structure of battery energy storage system (BESS) connected to the medium voltage grid is shown in Figure 5.7. The battery packs, DC/DC conversion, and DC/AC conversion are taking place in this system. Generally, BESS connected to medium voltage is of two types, namely with transformer and transformerless. In the first

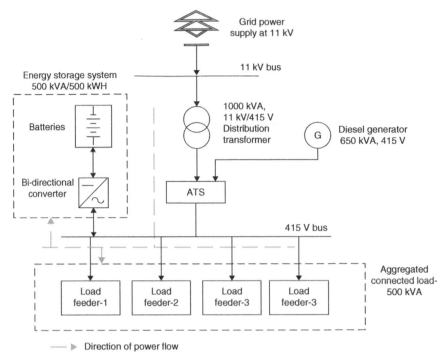

Figure 5.3 Direction of power flow during normal operating conditions.

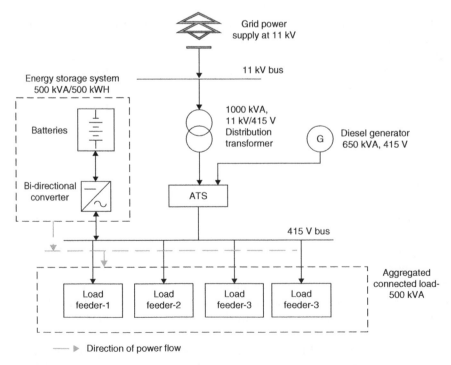

Figure 5.4 Direction of power flow during emergency conditions.

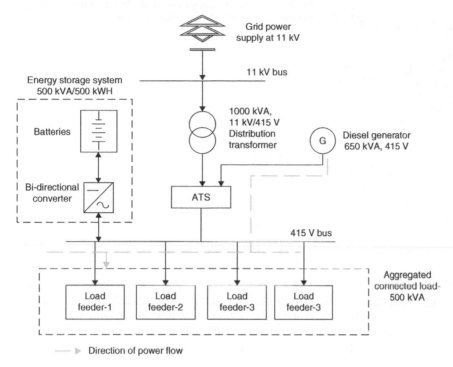

Figure 5.5 Direction of power flow during emergency conditions.

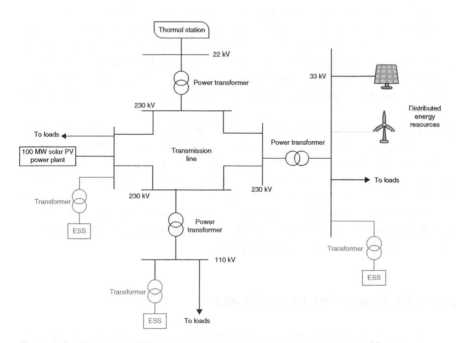

Figure 5.6 The typical SLD of the grid network shows the locations of ESS.

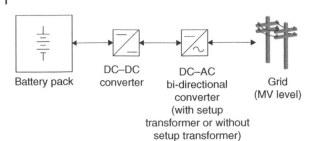

Figure 5.7 The typical structure of a battery energy storage system connected to the medium voltage.

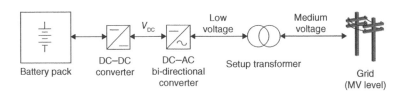

Figure 5.8 The typical step-up transformer-based BESS system connected to medium voltage.

type, a step-up transformer is used to connect the BESS to the medium voltage. The step-up transformer is applicable, where the BESS power conditioning system is in low voltage. On the other hand, transformerless AC/DC converter technologies have been used in BESS application and it is suitable for medium voltage interconnection. It has two levels with serial switches and modular multilevel converters (MMCs). A detailed analysis or study is required to examine the efficiency, complexity, and techno-economical benefits of with step-up transformer versus transformerless option.

The typical step-up transformer-based BESS system connected to medium voltage is shown in Figure 5.8. It consists of a step-up transformer, passive filter circuits, converter (power conditioning unit), and battery packs. The step-up transformer is used to step up the low voltage from the power conditioning unit to the grid's medium voltage (MV) and high voltage (HV). A low pass filter is generally used between the converter and transformer to reduce the harmonic current injection into the system. Mostly, inductor-capacitor (LC) or inductor-capacitor-inductor (LCL) passive filter combinations are widely used for this purpose [19]. The power conditioning system plays an important role in BESS application [20, 21]. Generally, this power conditioning system is a classic two-level converter either voltage source converter (VSC), Z-source converter (ZSI), and quasi-Z-source converter (qZSI) [22–27].

In case there is a fault in a BESS system of a single higher power rating, then the entire BESS system will be disconnected/isolated and it will be connected to the grid after the rectification of the fault. To increase reliability, a parallel combination of BESS systems has been used in place of a single BESS system with a higher rating. The typical parallel combination of the BESS system is shown in Figure 5.9. Another advantage of using multiple parallel combinations of the BESS system can be interconnected at a single location or multiple locations in the grid (transmission or distribution) based on the system requirement and provide necessary functions/support distributed across the system. This parallel combination of multiple BESS systems (power block) concepts nowadays is generally used in the system due to distributed energy resources integration.

5.4 Energy Storage System for Renewable Power Plants

Renewable energy sources like solar PV and wind energy conversion systems in a power system keep increasing over the last decades. The major issue associated with these renewable power plants is intermittency in the output power, i.e. power output from these renewable energy sources is extremely irregular (non-uniform) and varies

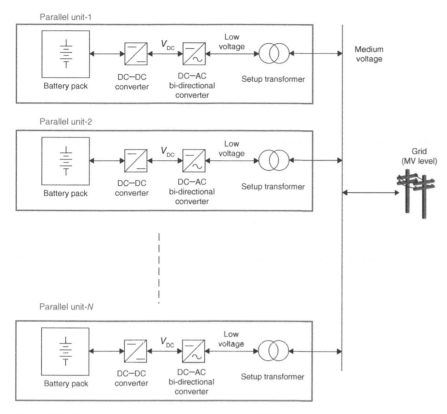

Figure 5.9 The parallel combination of multiple BESS systems.

based on climatic and environmental constraints/factors. This leads to problems like frequency variation, power balance, and stability issues in power systems.

Example 5.1   The SLD of the 15 MW solar PV power plant connected to a 110 kV transmission system is shown in Figure 5.10. The output power of the 15 MW solar PV power plant is shown in Figures 5.11 and 5.12. Figure 5.11 shows the output power for one typical day (day 1) and Figure 5.12 shows the output power for another typical day (day 2) of the monitoring.

Figures 5.11 and 5.12 clearly show the output power from the 15 MW solar PV power plant is highly fluctuating based on several environmental and climatic conditions. In day 1, the output power fluctuates highly between 01:00 p.m. and 02:30 p.m., and in day 2, across the day, the output power is highly fluctuating.

Example 5.2   The SLD of the 10 MW solar PV power plant connected to a 66 kV transmission system is shown in Figure 5.13. The output power of the 10 MW solar PV power plant is shown in Figures 5.14–5.16. Figure 5.14 shows the output power for one typical day (day 1), Figure 5.15 shows the output power for another typical day (day 2), and Figure 5.16 shows the output power for another typical day (day 3) of the monitoring.

Figures 5.14–5.16 shows the output power from the 10 MW solar PV power plant highly fluctuates across the day based on several environmental and climatic conditions.

From Examples 5.1 and 5.2, output power from the solar PV power plants (15 and 10 MW) is not uniform and varies across the day. This negatively affects the power system's performance, such as power imbalance (power generation ≠ power demand), frequency variation, stability problems, etc. Hence, the application of an energy

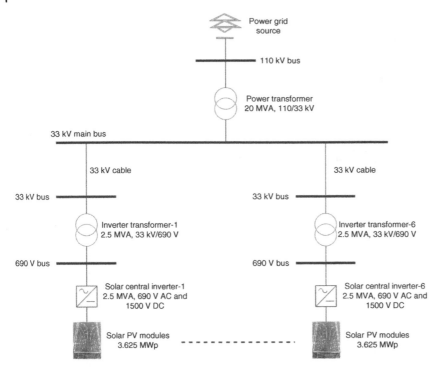

Figure 5.10 SLD of 15 MW solar PV power plant.

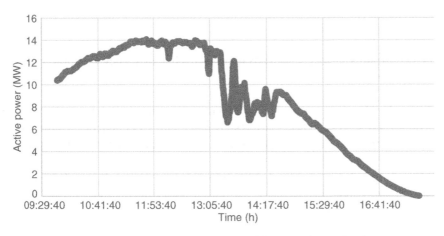

Figure 5.11 Output power from the 15 MW solar PV power plant – day 1.

storage system is essential to overcome the above-mentioned problems and smoothening the output power from renewable power plants, curtailment of renewable power, and providing ancillary services like peak shaving, load shifting, power quality enhancement, etc.

5.4.1 Renewable Power Evacuation Curtailment

The large-scale renewable power plants are connected to high voltage and extra-high voltage levels and communicated with the load dispatch center through fiber-optic communications. The function of the load dispatch center

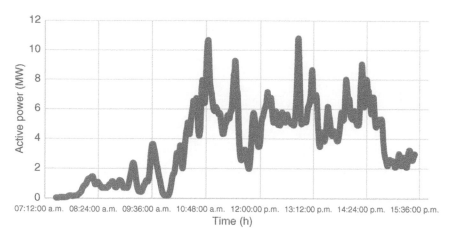

Figure 5.12 Output power from the 15 MW solar PV power plant – day 2.

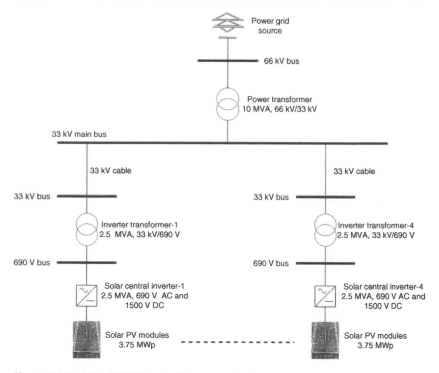

Figure 5.13 SLD of 10 MW solar PV power plant.

is to maintain the power balance (power demand is equal to power generation) and frequency. The grid operator in the load dispatch center will take care of this function. To maintain the power balance (power demand is equal to power generation), the grid operator will do necessary action to increase or decrease the power generation as well as increase or decrease the power demand. The large-scale renewable power plants also receive instructions from the load dispatch center for a decrease or increase in power generation, i.e. a decrease in power generation from the renewable power plant is reducing the output power delivery to the grid, and an increase in power generation from the renewable power plant is increasing the output power delivery to the grid. It has an impact on the energy output from the plant.

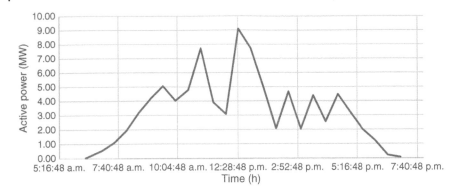

Figure 5.14 Output power from the 10 MW solar PV power plant – day 1.

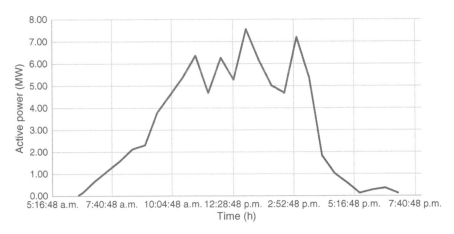

Figure 5.15 Output power from the 10 MW solar PV power plant – day 2.

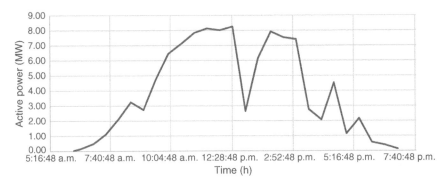

Figure 5.16 Output power from the 10 MW solar PV power plant – day 3.

Example 5.3 The 100 MW solar PV power plant is connected to the power grid through a 220 kV transmission system. Assume a situation, the plant is presently delivering the rated power (100 MW) to the grid. The typical power generation of the day is shown in Figure 5.17.

In case, if the grid is operating with excessive power generation than the load demand, the grid operator instructs the solar power plant to decrease the power generation from the rated capacity for some period of time (Figure 5.18).

Whenever the plant is operating at reduced capacity (80 MW in this case), energy output from the plant is also reduced, resulting in financial reduction.

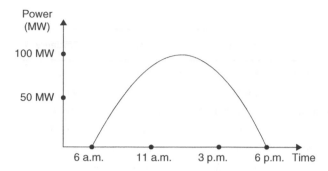

Figure 5.17 Typical power generation of the day.

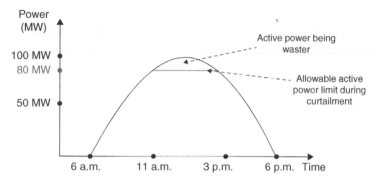

Figure 5.18 Power generation at reduced capacity due to curtailment.

The energy storage system is the alternative option to utilize the energy being wasted during the reduced operation (curtailment period) as demanded by the grid operator. This energy storage system will utilize the energy during this period by storing it and stored energy can be utilized later as required.

5.5 Types of Energy Storage Systems

The energy storage systems are classified into five categories based on their technology. They are:

☐ The mechanical energy storage system
☐ Hydrogen energy storage system
☐ The thermal energy storage system
☐ Batteries energy storage system
☐ The pumped hydro energy storage system

The power and energy densities of various energy storage technology are listed in Table 5.1 [2].

5.5.1 Battery Energy Storage System

The battery energy storage system is one of the energy storage systems with higher growth potential in the last decades. Some of the advantages of a battery energy storage system are listed below:

☐ Suitable for both high power and high energy applications.
☐ It is possible to install at any place as well as existing power plants.
☐ Overall charging and discharging process is instant and easily achievable.

Table 5.1 Comparison of various energy storage technologies.

S. no	Types of storage	Power	Energy density	Response time	Efficiency (%)
1	Pumped hydro	100 MW–2 GW	400 MWh–20 GWh	12 min	70–80
2	Compressed air energy storage	110–290 MW	1.16–3 GWh	12 min	90
3	BESS	100 W–100 MW	1 kWh–200 MWh	In seconds	60–80
4	Flywheels	5 kW–90 MW	5–200 kWh	12 min	80–95
5	Superconducting magnetic energy storage	170 kW–100 MW	110 Wh–27 kWh	In milliseconds	95
6	Supercapacitors	1 MW	1 Wh–1 kWh	In milliseconds	95

It consists of a bi-directional AC–DC converter, DC–DC converter, and battery packs, with a setup transformer or without a setup transformer based on power rating and interconnection voltage level [28–30]. The following types of battery chemistries are used for energy storage applications:

☐ Lead-acid
☐ Lithium-Ion [31]
☐ Sodium-sulfur
☐ Nickel-cadmium
☐ Zinc-bromine
☐ Iron-chromium
☐ Vanadium redox

5.5.2 Thermal Energy Storage System

The thermal energy storage system is one of the storage systems and is mostly used in solar thermal power plants. This system works on capturing and releasing heat or thermal and changes the state of storage from one form to another form, i.e. gas to a liquid, solid to liquid, and vice versa. Energy storage with molten salt and liquid air, as well as cryogenic storage, are examples of technologies.

5.5.3 Mechanical Energy Storage System

Mechanical energy storage methods are used to store the energy on rotational or gravitational kinetic forces and release it whenever required. Flywheels and compressed air systems are the most popular mechanical energy storage systems. In today's modern grid applications, feasibility demands the use of cutting-edge technologies, and this technology is not widely used.

5.5.4 Pumped Hydro

Pumped hydroelectric power plants are traditionally used as energy storage systems in the power systems. It pumps the water from the lower basin to the upper basin during the nonpeak hours at a reduced cost of electricity and discharges it during the peak hours at a higher cost and maintains the grid stability.

5.5.5 Hydrogen Storage

Hydrogen energy storage, which is still in its infancy, would include converting power to hydrogen by electrolysis and storing it in tanks. It can then be re-electrified or distributed to new applications such as transportation, industry, and residential as a supplement or alternative to gas. Fuel-cell vehicles require enough hydrogen to offer a driving range of more than 300 mi and the ability to refill as fast as possible.

5.6 Energy Storage Systems for Other Applications

The energy storage systems are used for other applications in the power system as well as it provides ancillary services. Some of the ancillary services provided by energy storage systems are explained in this section.

5.6.1 Shift in Energy Time

This is a pretty common use, and it is one of the main tasks that ordinary pumped-storage hydropower plants already handle. It requires "buying" energy at a lesser cost during the off-peak hours (by consuming energy from the grid, such as recharging batteries or pumping water to the upper basin or reservoir in the case of hydroelectric pumping) and "selling" it at a higher cost during the peak demand hours. The benefits of this application go beyond the financial gains from selling energy at a higher price. When energy demand is low, this "energy movement" helps to increase it, and when demand is high, it helps to decrease it.

Peak shaving lowers the influence of peaks in both the generation and load curves shown in Figure 5.19, resulting in a "smooth" curve form. This makes things easier to anticipate and handle.

5.6.2 Voltage Support

The reactive power regulation on each generator is commonly used to achieve voltage control in an electrical energy system depicted in Figure 5.20. An energy storage system could provide this service. The voltage regulation done by the energy storage system can also be classified as "power quality" because it is extremely valuable in improving the quality of service supplied by the power supply company.

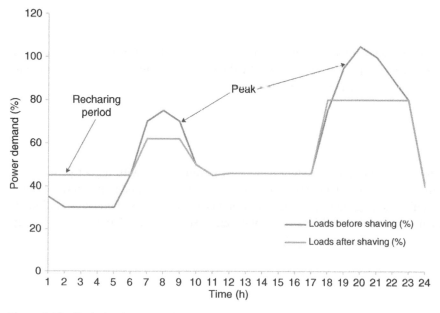

Figure 5.19 Peak shaving example.

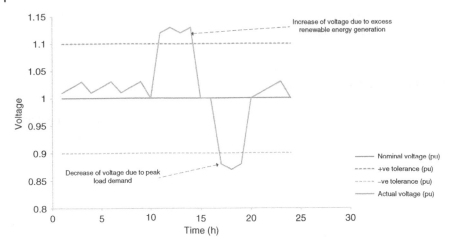

Figure 5.20 Voltage variation out of standard range.

5.6.3 Frequency Regulation (Primary, Secondary, and Tertiary)

Whenever the power generation of the system is not matched to the power demand, the frequency of the system gets affected. The ESS is used to balance the power generation equal to power demand by absorbing the power from the grid (charging the batteries) and supplying the power back to the grid (discharging the batteries). Hence, the ESS is improving the frequency regulation.

5.6.4 Congestion Management

When network components are nearing their maximum power limit, the energy storage system can be used to "cushion" the power transfer without shutting down generators or requiring further electrical network investment.

5.6.5 Black Start

The inconveniences caused by a potential blackout on portions of a network can be reduced by deploying an energy storage device that can deliver enough power to the users affected by the outage. The ESS might also be used to restart the entire electrical system in the event of a general blackout.

5.7 Conclusion

This chapter discusses the need for an energy storage system in a smart power system with higher penetration of renewable energy integration into the power system. Also, it discusses the ESS connected to the low, medium, and high voltage levels, as well as energy storage in renewable power plants and examples of power curtailment. Several examples are presented to understand the need for an energy storage system. Also, different types of energy storage systems are explained.

References

1 Baker, J.N. and Collinson, A. (1999). Electrical energy storage at the turn of the millennium. *Power Engineering Journal* 13 (3): 107–112.

2 Sivaraman, P., Sharmeela, C., Mahendran, R., and Thaiyal Nayagi, A. (2020). *Basic Electrical and Instrumentation Engineering*. Wiley.

3 Sivaraman, P., Sharmeela, C., and Kothari, D.P. (2017). Enhancing the voltage profile in distribution system with 40 GW of solar PV rooftop in Indian grid by 2022: a review. In: *First International Conference on Large Scale Grid Integration of Renewable Energy in India* (September 2017), New Delhi, pp. 1–5.

4 Lawder, M.T., Suthar, B., Northrop, P.W.C. et al. (2014). Battery energy storage system (BESS) and battery management system (BMS) for grid-scale applications. *Proceedings of the IEEE* 102 (6): 1014–1030.

5 Chen, H., Cong, T.N., Yang, W. et al. (2009). Progress in electrical energy storage system: a critical review. *Progress in Natural Science* 19 (3): 291–312.

6 Feehally, T., A. Forsyth, D. Strickland, et al. (2016). Battery energy storage systems for the electricity grid: UK research facilities. In: *8th IET International Conference on Power Electronics, Machines and Drives (PEMD 2016)*, pp. 1–6. April 2016. https://doi.org/10.1049/cp.2016.0257.

7 Gallardo-Lozano, J., Milanés-Montero, M.I., Guerrero-Martínez, M.A., and Romero-Cadaval, E. (2012). Electric vehicle battery charger for smart grids. *Electric Power Systems Research* 90: 18–29.

8 Xu, X., Bishop, M., Oikarinen, D.G., and Hao, C. (2016). Application and modeling of battery energy storage in power systems. *Journal of Power Energy Systems* 2 (3): 82–90.

9 Sivaraman, P. and Sharmeela, C. (2020). Introduction to electric distribution system. In: *Handbook of Research on New Solutions and Technologies in Electrical Distribution Networks* (ed. B. Kahn, H.H. Alhelou and G. Hayek), 1–31. Hershey, PA: IGI Global.

10 Sivaraman, P. and Sharmeela, C. (2020). Existing issues associated with electric distribution system. In: *Handbook of Research on New Solutions and Technologies in Electrical Distribution Networks* (ed. B. Khan, H.H. Alhelou and G. Hayek), 1–31. Hershey, PA: IGI Global.

11 Sivaraman, P. and Sharmeela, C. (2017). Battery energy storage system addressing the power quality issue in grid connected wind energy conversion system. In: *First International Conference on Large-Scale Grid Integration Renewable Energy in India* (September 2017), New Delhi, India, pp. 1–3.

12 Sanjeevikumar, P., Sharmeela, C., Holm-Nielsen, J.B., and Sivaraman, P. (2021). *Power Quality in Modern Power Systems*. Academic Press.

13 Ghiani, E. and Pisano, G. (2018). Impact of renewable energy sources and energy storage technologies on the operation and planning of smart distribution networks. In: *Operation of Distributed Energy Resources in Smart Distribution Networks* (ed. K. Zare and S. Nojavan), 25–48. Academic Press, https://doi.org/10.1016/B978-0-12-814891-4.00002-3.

14 Wang, G., Konstantinou, G., Twonsend, C.D. et al. (2016). A review of power electronics for grid connection of utility scale battery energy storage systems. *IEEE Transactions on Sustainable Energy* 7 (4): 1778–1790.

15 Vazquez, S., Lukic, S.M., Galvan, E. et al. (2010). Energy storage systems for transport and grid applications. *IEEE Transactions on Industrial Electronics* 57 (12): 3881–3895.

16 Logeshkumar, S. and Manoharan, R. (2014). Influence of some nanostructured materials additives on the performance of lead acid battery negative electrodes. *Electrochimica Acta* 144: 147–153.

17 D. Cicio, G. Product, M. Energy, and S. Solutions, "*EssPro™ – Battery Energy Storage the Power to Control Energy Challenges of the Future Power Grid Long-Term Drivers for Energy Storage,*" 2017, https://new.abb.com/docs/librariesprovider78/eventos/jjtts-2017/presentaciones-peru/(dario-cicio)-bess%2D%2D-battery-energy-storage-system.pdf?sfvrsn=2 (accessed 1 October 2022).

18 Abdi, H., Mohammadi-ivatloo, B., Javadi, S. et al. (2017). *Energy Storage Systems*. Elsevier BV.

19 Ota, J.I.Y., Sato, T., and Akagi, H. (2016). Enhancement of performance, availability, and flexibility of a battery energy storage system based on a modular multilevel cascaded converter (MMCC-SSBC). *IEEE Transactions on Power Electronics* 31 (4): 2791–2799.

20 Maharjan, L., Inoue, S., and Akagi, H. (2008). A transformerless energy storage system based on a cascade multilevel PWM converter with star configuration. *IEEE Transactions on Industry Applications* 44 (5): 1621–1630.

21 Krishnamoorthy, H.S., Rana, D., Garg, P. et al. (2014). Wind turbine generator–battery energy storage utility interface converter topology with medium-frequency transformer link. *IEEE Transactions on Power Electronics* 29 (8): 4146–4155.

22 Peng, F.Z. (2003). Z-source inverter. *IEEE Transactions on Industry Applications* 39 (2): 504–510.

23 Anderson, J. and Peng, F.Z. (2008). Four quasi-Z-source inverters. In: *IEEE Power Electronics Specialists Conference*, Rhodes, Greece, June 2008, pp. 2743–2749. https://doi.org/10.1109/PESC.2008.4592360.

24 Cintron-Rivera, J.G., Li, Y., Jiang, S. et al. (2011). Quasi-Z-source inverter with energy storage for photovoltaic power generation systems. In: *IEEE Applied Power Electronics Conference and Exposition*, Fort Worth, TX, USA, March 2011, pp. 401–406. https://doi.org/10.1109/APEC.2011.5744628.

25 Logeshkumar, S. and Vanathi, P.T. (2016). Modelling and simulation of low power polymer micro-heater for MEMS gas sensor applications. *Asian Journal of Research in Social Sciences and Humanities* 6 (7): 144–151.

26 Logeshkumar, S., Vanathi, P.T., and Menon, R. (2014). Modelling and simulation of MEMS based sensors for Microbotics. In: *2014 IEEE International Conference on Computational Intelligence and Computing Research*. IEEE.

27 Liu, Y., Ge, B., Abu-Rub, H., and Peng, F.Z. (2013). Control system design of battery-assisted quasi-z-source inverter for grid-tie photovoltaic power generation. *IEEE Transactions on Sustainable Energy* 4 (4): 994–1001.

28 Qian, H., Zhang, J., Lai, J.S., and Yu, W. (2011). A high-efficiency grid-tie battery energy storage system. *IEEE Transactions on Power Electronics* 26 (3): 886–896.

29 Zhou, H., Bhattacharya, T., Tran, D. et al. (2011). Composite energy storage system involving battery and ultracapacitor with dynamic energy management in microgrid applications. *IEEE Transactions on Power Electronics* 26 (3): 923–930.

30 Chatzinikolaou, E. and Rogers, D.J. (2017). A comparison of grid – connected battery energy storage system designs. *IEEE Transactions on Power Electronics* 32 (9): 6913–6923.

31 Horiba, T. (2014). Lithium-ion battery systems. *Proceedings of the IEEE* 102 (6): 939–950.

6

Real-Time Implementation and Performance Analysis of Supercapacitor for Energy Storage

Thamatapu Eswararao¹, Sundaram Elango¹, Umashankar Subramanian², Krishnamohan Tatikonda³, Garika Gantaiahswamy³, and Sharmeela Chenniappan⁴

¹Department of Electrical and Electronics Engineering, Coimbatore Institute of Technology, Coimbatore, India
²Renewable Energy Laboratory, Department of Communications and Networks, Prince Sultan University, College of Engineering, Riyadh, Saudi Arabia
³Department of Electrical and Electronics Engineering, JNTU Kakinada, Andhra Loyola Institute of Engineering and Technology, Vijayawada, Andhra Pradesh, India
⁴Department of Electrical and Electronics Engineering, Anna University, Chennai, India

6.1 Introduction

General Electric (GE) engineers began working with components for fuel cells and rechargeable batteries utilizing porous carbon electrodes in the 1950s. Becker invented the "low-voltage electrolytic capacitor with porous carbon electrodes" in 1957. Because it stores a significant quantity of energy, this capacitor is known as a supercapacitor (SC). Esha Khare, an Indian–American girl, designed a novel electrode for supercapacitors in 2013 [1, 2].

Supercapacitors are revolutionary technology with the potential to enable considerable advancements in energy storage. Supercapacitors use high-surface-area electrodes and thinner dielectrics to obtain higher capacitance [3, 4]. They are regulated by the same fundamental equations as those for normal capacitors. This enables higher energy densities than those possible using batteries. As a result, supercapacitors may become a more appealing power source for a growing range of applications [5, 6]. Numerous types of supercapacitors, related quantitative modeling domains, and the future of supercapacitor research and development are covered in this concise introduction [7]. Table 6.1 that compares the supercapacitor performance to that of the accumulator [8] quickly reveals the rationale for the interest in supercapacitors [9]. Another significant benefit of supercapacitors is their long life and minimal temperature sensitivity. Because a battery relies on a chemical reaction between the electrolyte and electrodes, each charge–discharge cycle deteriorates both the active materials and the electrolyte, resulting in a usable lifetime measured in hundreds or thousands of cycles when the battery is fully charged [10]. There are three different types of supercapacitors: double layer, pseudo, and hybrid capacitors. Figure 6.1 shows the different types of supercapacitors and their classification mentioned.

Supercapacitors are used in several applications. Some of these applications are given below.

1. Used in the start-up of diesel engines in submarines and tanks.
2. Used in hybrid electrical vehicle (HEV) and new trains to recover braking energy and deliver it during acceleration phases.
3. Global positioning system (GPS)-guided missiles can use it as a backup energy source.

Artificial Intelligence-based Smart Power Systems, First Edition.
Edited by Sanjeevikumar Padmanaban, Sivaraman Palanisamy, Sharmeela Chenniappan, and Jens Bo Holm-Nielsen.
© 2023 The Institute of Electrical and Electronics Engineers, Inc. Published 2023 by John Wiley & Sons, Inc.

Table 6.1 Comparison of storage devices under different parameters.

Condition	Supercapacitor	Capacitor (conventional)	Battery (lead acid)
Time during charging condition	0.3–30 s	10^{-3}–10^{-6} s	1–5 h
Time during discharging condition	0.3–30 s	10^{-3}–10^{-6} s	0.3–3 h
Energy (Wh/kg)	1–10	<0.1	10–100
Life cycle	>500 000	>500 000	1000
Power (W/kg)	<10 000	<100 000	<1000
Efficiency	0.85–0.98	>0.95	0.7–0.85
Temperature	−40–65 °C	−20–65 °C	−20–100 °C

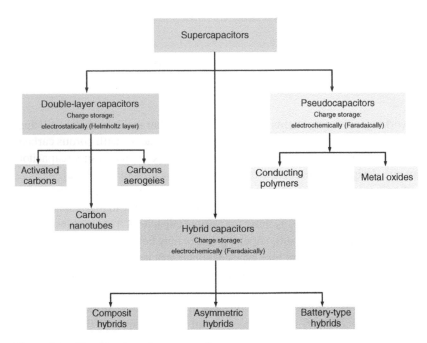

Figure 6.1 Classification of supercapacitors.

4. China is testing a new type of electric bus that operates without the use of power lines and is powered by enormous onboard ultracapacitors. In early 2005, a few prototypes were tested in Shanghai. Two commercial bus routes began using supercapacitor vehicles in 2006.
5. The supercapacitor's strongest suit is delivering or accepting power during short-duration events [11, 12].

Table 6.1 compares the various parameters of the supercapacitor, conventional capacitor, and battery. This chapter compared the charging and discharging of a supercapacitor using an field-programmable gate array (FPGA)-based pulse width modulation (PWM) technique, which is discussed in Section 6.4. The electrical structure and modeling of proposed supercapacitor is discussed in Section 6.2. The proposed bidirectional buck–boost converter and FPGA controller are discussed in Section 6.3.

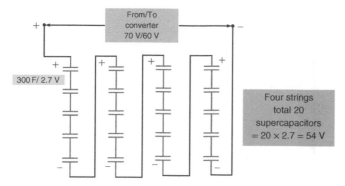

Figure 6.2   Proposed structure of supercapacitor.

6.2   Structure of Supercapacitor

The basic construction of a supercapacitor consists of aluminum current collectors and electrodes made of activated carbon soaked in an organic or aqueous electrolyte. To isolate the two electrodes, a separator was placed between them. The unit was constructed using classic capacitors [13]. Figure 6.2 depicts the proposed supercapacitor structure. A supercapacitor's operation is based on the storage of energy by the dispersion of ions from the electrolyte along the surface of the two electrodes. Accordingly, when a terminal voltage is applied to supercapacitors, a zone of space charge is created between the electrode and the electrolyte. As there is no electrochemical reaction, energy storage is electrostatic rather than faradic, as in the case of batteries [14, 15]. Table 6.2 lists the supercapacitor parameters.

6.2.1   Mathematical Modeling of Supercapacitor

In this section, a supercapacitor model is presented. This model was used in the simulation test, but it was not necessary for DC control law design. Two main types of models have been proposed in the literature. First, the model is based on a constant distribution of charge, as in the case of a transmission line [16]. The parameter calculation of this model is similar to that of a transmission line that consists of solving two partial differential equations that describe the variation in voltage and current along the line. Second, two resistance, capacitance (RC) branches with a nonlinear capacitance varying with the voltage on their terminals describe the electric behavior of the supercapacitor in power electronic applications [17]. The parameters of the model were determined experimentally. This is divided into three parts. The first part is composed of branch R1C1, known as the principal branch, and the

Table 6.2   Parameters of supercapacitor.

Parameter variable	Value
Total voltage of capacitor set	54 V
Capacitor voltage	2.7 V
Capacitor rating	300 F
Total supercapacitors	20
Total strings	4
Capacitors per string	5

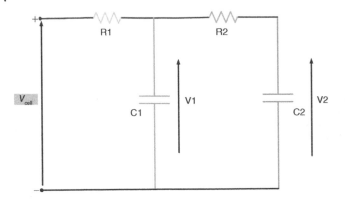

Figure 6.3 Equivalent circuit of supercapacitor.

capacitance is modeled as a voltage that depends on the differential capacitor C1 [18, 19]. The third (not depicted) is composed of *Rp* and the equivalent parallel resistance. The electrical equivalent circuit of the supercapacitor is illustrated in Figure 6.3. The total voltage across supercapacitors set is shown in Eq. (6.1) and the voltage across capacitor C2 of single supercapacitor cell is shown in Eqs. (6.2) and (6.3) [20].

Supercapacitors are a promising method for storing brake energy in rail vehicles, according to a market survey conducted by Montana [2, 21]. Diesel engines, hybrid electric cars, industrial uses, elevators, and pallet trucks are all examples of predicted technological advancements outside the railway sector. The development time frame was projected to be 5–10 years [22]. The major objectives for the development are as follows:

1. Long life time.
2. Increase in the rated voltage

$$V_{\text{Total}} = N_{\text{sc-s}} \cdot V_{\text{sc}} = N_{\text{sc-s}} \cdot \left[V_{\text{sc}} + R_1 \frac{I_{\text{sc}}}{N_{\text{sc-P}}} \right] \tag{6.1}$$

$$V_2 = \frac{1}{C_2} \int i_2 \cdot \mathrm{d}t \tag{6.2}$$

$$V_2 = \frac{1}{C_2} \int \frac{1}{R_2} (V_1 - V_2) \mathrm{d}t \tag{6.3}$$

V_{Total} = Total voltage across supercapacitors set
$N_{\text{sc-s}}$ = Number of supercapacitors in series
$N_{\text{sc-P}}$ = Number of supercapacitors in parallel
V_{sc} = voltage across the single supercapacitor

6.3 Bidirectional Buck–Boost Converter

The converter was placed between a supercapacitor and DC supply fixed at a constant voltage. A supercapacitor is regarded as power system with a high current and low voltage source [23]. The DC–DC converter is a bidirectional buck–boost converter, which acts as a boost converter when the supercapacitor provides the power requirement to load or DC link, and acts as a buck converter when the supercapacitor is charged from a DC source [24].

Figure 6.4 shows the circuit diagram for buck–boost converter. In the proposed system, the DC voltage is 70 V initially applied to the bidirectional converter's input terminals, and as the applied voltage decreases to a low value,

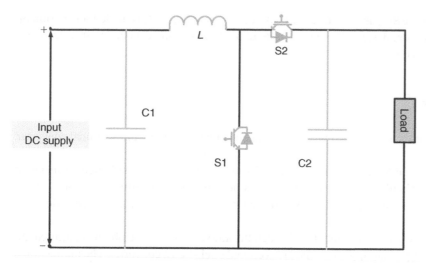

Figure 6.4 Bidirectional buck–boost converter.

the required value of super capacitor charging voltage is typically 2.7 V into play. At this point, switch S2 operates and switch S1 is in the off state; the inductor is in the charging condition, and the output buck voltage, which is 2.7 V, is applied to the supercapacitor [25].

In the proposed system, after full charging of the supercapacitor, the supercapacitor delivers power to the load through a bidirectional converter in the boost converter mode [26]. In this mode, switch S1 operates and switches to the S2 off state, and the inductor discharges through the diode of S2 to the load until the supercapacitor is fully discharged [27].

6.3.1 FPGA Controller

A FPGA controller is used in a proposed system: to sense the voltage and current of the actual values from the sensors and compare it with the reference values. The necessary action will be taken in the form of a gate pulse applied to the insulated gate bi-polar transistor (IGBT)/metal oxide semiconductor field effect transistor (MOS-FET) of the boost converter [28]. An FPGA SPARTAN6 controller (SP6 LX9 Rev-1.0) was used to implement the gate pulses applied to the IGBT/MOSFET of the bidirectional converter, as shown in Figure 6.5. The Xilinx

Figure 6.5 SPARTAN6 FPGA controller.

Spartan-6 FPGA [29] is included on the SP6, an easy-to-use FPGA development board. It is made for experimenting with FPGAs and researching system design. The Xilinx XC6SLX9 FPGA on this development board has a maximum of 100 user input/outputs (IOs).

6.4 Experimental Results

The line diagram of the supercapacitors is associated with the load and supply through a bidirectional converter, as shown in Figure 6.6.

The PWM pulses for switches S1 and S2 in a bidirectional buck–boost converter are implemented using an FPGA controller in the hardware. The hardware model consists of (i) supercapacitors, (ii) bidirectional buck–boost converter, (iii) rectifier, (iv) single-phase AC supply of 230 V provided by the auto transformer. Specifications are shown in Table 6.3.

Initially, a variable AC supply was applied to the rectifier through the auto transformer to obtain a required DC supply of 70 V. The input 70 V DC supply was applied to the bidirectional buck–boost converter. The bidirectional converter acted as a buck converter. Because the supercapacitor will charge up to a maximum voltage of 2.7 V, there are a total of 20 supercapacitors and the total charging voltage is 54 ($20 \times 2.7 = 54$) V, so in order to provide voltage from 70 to 54 V, the bidirectional converter acts as a buck converter.

Figure 6.7a shows the supercapacitor charging a voltage of 12 V and current of 0 A from a constant DC supply voltage of 70 V at a 10% duty cycle. Figure 6.7b shows the supercapacitor charging a voltage of 15 V, and current of

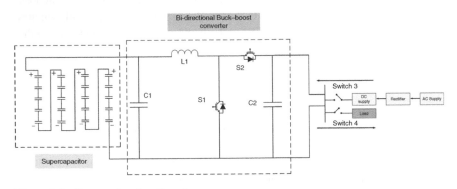

Figure 6.6 Proposed system line diagram.

Table 6.3 Parameters of buck–boost converter.

Parameter variable	Value
Output power	1000 W
DC input voltage (min)	300 V
DC output voltage (max)	100 V
Frequency at switching conditions	100 KHz
Inductor (L2)	5 mH
Capacitors (C3 and C4)	1500 µF
IGBT	1200 V/100 A

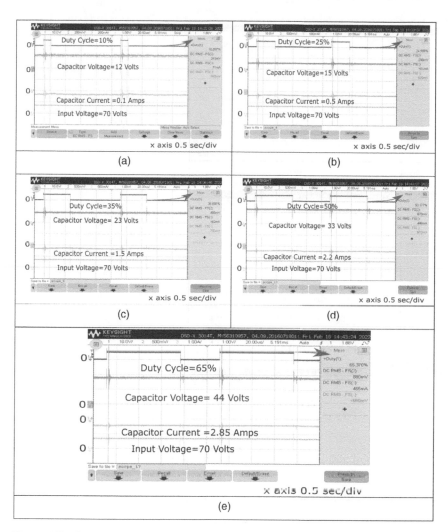

Figure 6.7 Supercapacitor charging under different duty cycles at constant input voltage of 70 V (a) 10% duty cycle, (b) 25% duty cycle, (c) 35% duty cycle, (d) 50% duty cycle, (e) 65% duty cycle.

0.5 A from a constant DC supply voltage of 70 V at a 25% duty cycle. Figure 6.7c shows the supercapacitor charging a voltage of 23 V, and current of 1.5 A from a constant DC supply voltage of 70 V at a 35% duty cycle. Figure 6.7d shows the supercapacitor charging a voltage of 33 V, and current of 2.2 A from a constant DC supply voltage of 70 V at a 50% duty cycle. Figure 6.7e shows the supercapacitor charging a voltage of 44 V, and current of 2.85 A from a constant DC supply voltage of 70 V at a 65% duty cycle.

After full charging of supercapacitors, that is, 51 V, in order to observe the discharging of the supercapacitor at different duty cycles, we need to connect a constant load of 250 W. In this condition, the input supply to the bidirectional buck–boost converter is provided by supercapacitors of 51 V. Therefore, to reach the required load demand, the converter acts as a boost converter.

Figure 6.8a shows the super capacitor delivers energy in order to meet the load demand of 250 W at 10% duty cycle and discharging voltage of 40 V and discharging current of 0.2 A. Figure 6.8b shows the super capacitor delivers energy in order to meet the load demand of 250 W at 20% duty cycle and discharging voltage of 37 V

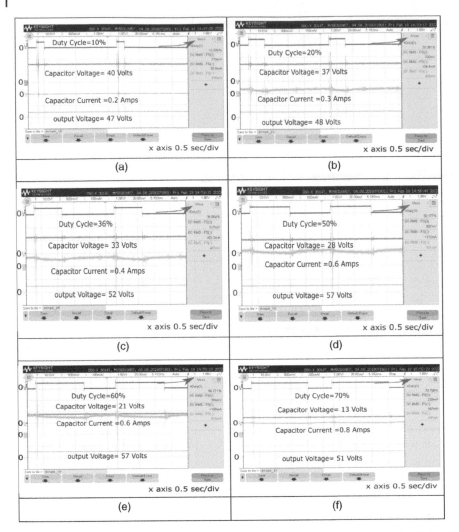

Figure 6.8 Supercapacitor discharging under different duty cycles at constant load of 250 W (a) 10% duty cycle, (b) 20% duty cycle, (c) 35% duty cycle, (d) 50% duty cycle, (e) 60% duty cycle, (f) 70% duty cycle.

and discharging current of 0.3 A. Figure 6.8c shows the super capacitor delivers energy in order to meet the load demand of 250 W at 35% duty cycle and discharging voltage of 33 V and discharging current of 0.4 A. Figure 6.8d shows the super capacitor delivers energy in order to meet the load demand of 250 W at 50% duty cycle and discharging voltage of 28 V and discharging current of 0.6 A. Figure 6.8e shows the super capacitor delivers energy in order to meet the load demand of 250 W at 60% duty cycle and discharging voltage of 21 V and discharging current of 0.6 A. Figure 6.8f shows the super capacitor delivers energy in order to meet the load demand of 250 W at 70% duty cycle and discharging voltage of 13 V and discharging current of 0.8 A. The supercapacitor is charged, which means it stores energy when the available renewable energy sources are supplied in excess of the load demand. For example, when the load demand is 250 W higher than this load demand while still being supplied by renewable energy sources, the supercapacitor will charge up to its maximum voltage of 54 V. Similarly, the super capacitor will discharge when the amount of renewable energy supplied falls short of the load demand; in this case, a load

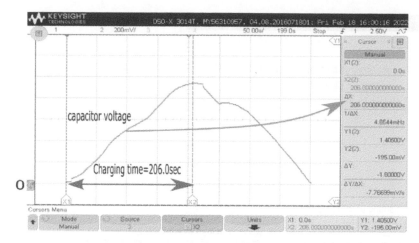

Figure 6.9 Supercapacitor total charging time at constant input voltage of 70 V and 75% duty cycle.

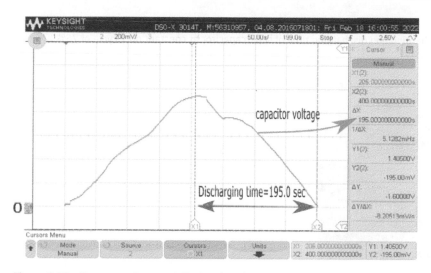

Figure 6.10 Supercapacitor total discharging time at constant load of 250 W and 75% duty cycle.

demand of 250 W to satisfy the load demand. This means that in the proposed system, the stored energy is delivered to the load up to its maximum voltage of 54 V.

Figure 6.9 shows the total charging time of the supercapacitor at a constant duty cycle of 75% at a constant input DC supply voltage. Figure 6.10 shows the total discharging time of the supercapacitor at a constant duty cycle of 75% at a constant load of 250 W. Figure 6.11 shows the total charging and discharging times of the supercapacitor at a constant duty cycle of 75%. The experimental setup for the proposed system is shown in Figure 6.12. The performance analysis of the supercapacitor during charging and discharging are listed in Tables 6.4 and 6.5.

6.5 Conclusion

In this chapter, an interleaved, multiphase, bidirectional, soft-switching multidevice, and FPGA-driven-based buck–boost converter is proposed with real-time FPGA-based modulation capability. The use of an FPGA in a

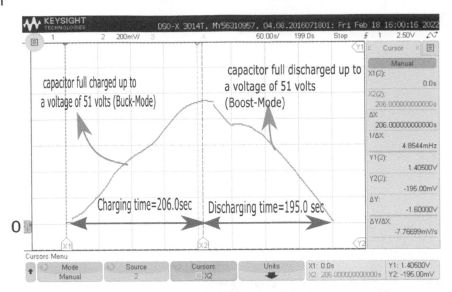

Figure 6.11 Representation of total charging and discharging time of supercapacitor at constant duty cycle of 75%.

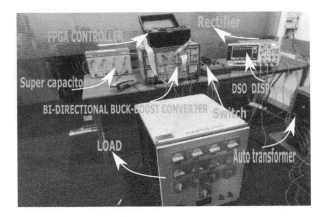

Figure 6.12 Experimental setup of the proposed system, DSO = digital storage oscilloscope.

multidevice technique provides for separate switching and inductor frequency, allowing for greater flexibility in optimizing various power conversion parameters such as inductor and power device losses, inductor ripple, and frequency at the switching node for electro magnetic interference (EMI) filters. A direct digital synthesizer (DDS) is relevant for non-parallelizable power devices since it is not parallelizable. It is well suited for IGBTs and is a technique for boosting power capabilities. A versatile technique for multidevice structures is provided using a digital FPGA control loop and a customizable PWM generator. A comparative analysis of the charging and discharging of a supercapacitor under different duty cycle conditions was performed using an FPGA-based PWM technique in a bidirectional converter hardware setup and the proposed hardware environment, which offers a good understanding of the supercapacitors.

Table 6.4 Performance of supercapacitor during charging.

Duty cycle (%)	Capacitor charging voltage (V)	Capacitor charging current (A)	Input voltage of converter (V)
10	12	0.1	70
25	15	0.5	70
30	18	1.3	70
35	23	1.5	70
45	30	2.0	70
50	33	2.2	70
60	39	2.7	70
65	44	2.85	70

Table 6.5 Performance of supercapacitor during discharging.

Duty cycle (%)	Capacitor discharging voltage (V)	Capacitor discharging current (A)	Output voltage of converter (V)	Output current of converter (A)
10	44	0.2	47	0.2
15	39	0.2	46	0.2
20	37	0.3	47	0.3
25	36	0.3	49	0.2
30	35	0.3	51	0.4
35	33	0.4	52	0.3
40	31	0.4	53	0.3
45	29	0.5	55	0.5
50	28	0.6	57	0.5
55	25	0.6	60	0.4
60	21	0.6	57	0.5
65	18	0.7	57	0.4
70	13	0.8	51	0.3

References

1 Andreev, M.K. (2021). An overview of super capacitors as new power sources in hybrid energy storage systems for electric vehicles. In: *National Conference with International Participation (ELECTRONICA)*, 1–4. IEEE.

2 Vrindavanam, J., Das, S., Vinay, S. et al. (2021). Design and implementation of super capacitor energy storage in satellite applications. In: *IEEE International Conference for Innovation in Technology (INOCON)*, 1–5. IEEE.

3 Ibrahim, T., Stroe, D., Kerekes, T. et al. (2021). An overview of super capacitors for integrated PV – energy storage panels. In: *19th International Power Electronics and Motion Control Conference (PEMC)*, 828–835. IEEE.

4 Jayasawal, K., Karna, A.K., Thapa, K.B. et al. (2021). Topologies for interfacing super capacitor and battery in hybrid electric vehicle applications: an overview. In: *International Conference on Sustainable Energy and Future Electric Transportation (SEFET)*, 1–6. IEEE.

5 Rahman, O., Robinson, D.A., Elphick, S. et al. (2021). Power sharing and energy management between super capacitor and battery in a hybrid energy system for EVs. In: *31st Australasian Universities Power Engineering Conference (AUPEC)*, 1–6. IEEE.

6 Prashar, A. and Kumar, N. (2021). Compensation of DVR with super capacitor based energy storage. *International Journal of Engineering Research & Technology, IJERT* 10: 1076–1080.

7 Patnaik, C., Lokhande, M.M., Pawar, S.B. et al. (2020). Hybrid energy storage system using super capacitor for electric vehicles. In: *Innovations in Power and Advanced Computing Technologies (i-PACT)*, 1–5. IEEE.

8 Zulfiqar, M.H., Riazand, K., Tauqeer, T. et al. (2020). Foldable, eco-friendly and easy go designed paper based super capacitor: energy storage device. In: *17th International Bhurban Conference on Applied Sciences and Technology (IBCAST)*, 40–43. IEEE.

9 Berrueta, A., Ursúa, A., Martín, I.S. et al. (2019). Super capacitors: electrical characteristics, modeling, applications, and future trends. *IEEE Access* 7: 50869–50896.

10 Guruvareddiyar, G. and Ramaraj, R. (2020). Super capacitor based energy recovery system from regenerative braking used for electric vehicle application. In: *International Conference on Clean Energy and Energy Efficient Electronics Circuit for Sustainable Development (INCCES)*, 1–3. IEEE.

11 Arunkumar, C.R. and Manthati, U.B. (2019). Design and small signal modeling of battery-super capacitor HESS for DC micro grid. In: *Region 10 Conference (TENCON)*, 2216–2221. IEEE.

12 Lan, T., Cao, S., Cheng, Z. et al. (2019). The wireless electric vehicle system based on super-capacitor power supply. In: *International Conference on Mechatronics and Automation (ICMA)*, 1818–1822. IEEE.

13 Selim, A.M., Wasfey, M.A., Abduallah, H.H. et al. (2019). Fabrication of super capacitor based on reduced graphene oxide for energy storage applications. In: *6th International Conference on Advanced Control Circuits and Systems (ACCS)*, 94–98. IEEE.

14 Cabrane, Z. and Lee, S.H. (2022). Electrical and mathematical modeling of supercapacitors: comparison. *Energies* 15: 01–12, MDPI.

15 Jha, D., Karkaria, V.N., Karandikar, P.B. et al. (2022). Statistical modeling of hybrid supercapacitor. *Journal of Energy Storage* 46: 103–109, Elsevier.

16 Hinov, N., Dimitrov, V., Vacheva, G. et al. (2020). Mathematical modelling and control of hybrid sources for application in electric vehicles. In: *24th International Conference Electronics*, 1–5. IEEE.

17 Naseri, F., Karimi, S., Farjah, E. et al. (2022). Supercapacitor management system: a comprehensive review of modeling, estimation, balancing, and protection techniques. *Renewable and Sustainable Energy Reviews* 155: 111–118, Elsevier.

18 Reigstad, M., Storebo, F., Steinsland, V. et al. (2021). Comparison of supercapacitor and battery transient response for DC-bus. In: *International Conference on Electrical, Computer, Communications and Mechatronics Engineering (ICECCME)*, 1–5. IEEE.

19 Ariyarathna, T., Kularatna, N., Gunawardane, K. et al. (2021). Development of supercapacitor technology and its potential impact on new power converter techniques for renewable energy. *IEEE Journal of Emerging and Selected Topics in Industrial Electronics* 2: 267–276.

20 Wang, Z., Lin, H., Guo, X. et al. (2021). Super capacitor energy storage system's charging design based on composite control mode. In: *6th International Conference on Power and Renewable Energy (ICPRE)*, 950–954. IEEE.

21 Zulfiqar, M.H., Riaz, K., Tauqeer, T. et al. (2020). Foldable, eco-friendly and easy go designed paper based supercapacitor: energy storage device. In: *7th International Bhurban Conference on Applied Sciences and Technology (IBCAST)*, 40–43. IEEE.

22 Malev, E. (2021). Study of bidirectional DC–DC converter for regenerating braking for ultralight electric vehicle. In: *National Conference with International Participation (ELECTRONICA)*, 1–7. IEEE.

23 Chang, X., Li, R., Gao, L. et al. (2022). Bidirectional buck/boost converter based on SiC MOSFET. In: *5th Conference on Energy Internet and Energy System Integration (EI2)*, 3704–3708. IEEE.

24 Chen, J.-Y., Chiou, G.J., Jiang, B.S. et al. (2021). Implementation of digital bidirectional buck–boost converter with changeable output voltage for electric vehicles. In: *International Future Energy Electronics Conference (IFEEC)*, 1–6. IEEE.

25 Rakesh, O. and Anuradha, K. (2021). Analysis of bidirectional DC–DC converter with wide voltage gain for charging of electric vehicle. In: *7th International Conference on Electrical Energy Systems (ICEES)*, 135–140. IEEE.

26 Lee, H.-S. and Yun, J.-J. (2018). High-efficiency bidirectional buck–boost converter for photovoltaic and energy storage systems in a smart grid. *IEEE Transactions on Power Electronics* 34: 4316–4328.

27 Makhdoomi, A., Ahamed, M.B., Deshmukh, K. et al. (2019). A review on recent advances in hybrid supercapacitors: design, fabrication and applications. *Renewable and Sustainable Energy Reviews*, Elsevier 101: 123–145.

28 Senthilvel, A., Vijeyakumar, K.N., Vinothkumar, B. et al. (2020). FPGA based implementation of MPPT algorithms for photovoltaic system under partialshading conditions. *Microprocessors and Microsystems*, Elsevier 77: 103–111.

29 Ilyas, A., Khan, M.R., Ayyub, M. et al. (2020). FPGA based real-time implementation of fuzzy logic controller for maximum power point tracking of solar photovoltaic system. *Optik*, Elsevier 213: 164–172.

7

Adaptive Fuzzy Logic Controller for MPPT Control in PMSG Wind Turbine Generator

Rania Moutchou[1], Ahmed Abbou[1], Bouazza Jabri[2], Salah E. Rhaili[1], and Khalid Chigane[1]

[1]*Department of Electrical Engineering, Mohammed V University in Rabat, Mohammadia School of Engineers, Rabat, Morocco*
[2]*Department of Physical, LCS Laboratory, Faculty of Sciences, Mohammed V University in Rabat, Rabat, Morocco*

7.1 Introduction

Given the global warming the world is facing in recent years due to the use of renewable energies to reduce the serious problems produced by other energies (environmental problems, depletion of fossil resources, …). Wind power is one of the clean energy sources that are growing significantly and considered to be one of the main sources of sustainable, dependable, and environmentally friendly energy. In general, only the wind turbine can provide access to electricity by changing the kinetic energy of the wind into electrical energy. The use of endless attraction coetaneous creator (permanent magnet synchronous generator [PMSG]) for a variable speed wind turbine has entered a lot of attention due to trust ability, simplicity, better perfection, lower conservation, and increased effectiveness leading to a factor of advanced power and high effectiveness [1].

Nowadays, experts are considered advantages to PMSG of working without mechanical gear diverse, which ensures the direct coupling of mechanical system and rotor, slow rotation speed, no rotor current and can be applied without gearbox. In addition, the synchronous generator gives less weight compared to the asynchronous generator, the high efficiency and low conservation will alleviate the operating cost which is the most inconvenient to spend on it. In addition, the variable speed of the generator must also be raised by the inverter [2].

In relation to the nonstop evolution of wind power technology, the effectiveness of the inverter instrument, fronting some delicate problems, completes an important part in perfecting the performance of the wind energy generation system. MPPT command charity can be attuned in order to release maximum wind power. A variety of maximum power point tracking (MPPT) ways have been used for the wind energy conversion system (WECS) [3].

The chain of conversion consists of a PMSG which rotates at variable speed, a thyristor rectifier, and an inverter. The output of the PMSG is alternating current which is converted to direct current using a diode controlled rectifier to remove the ripple present in the AC component and a smoothing capacitor is placed on the rectifier to minimize l ripple due to non-linearity [4]. As wind power varies, the fuzzy logic controller tracks the output voltage and current to generate an efficient duty cycle for converter operation. Thus, the maximum power is produced according to the available wind speed [5].

Due to the immediate variable speed of the wind turbine, it is necessary to decide the optimum speed of the creator which guarantees the maximum energy supplied. Generally, the MPPT system uses detectors to track the MPPT by controlling rotor speed and necklace [6]. The PMSG is a non-linear system, which operates at lagging variable operating points at time-varying wind speed. However, conventional MPPT control is not suitable to give optimum performance for different operating points. The feedback linearization control is designed for the PMSG [7].

The fuzzy logic controller strategy has been extensively applied in power electronics [8], endless attraction coetaneous motor [9], and low voltage lift-through of the PMSG- grounded WECS [10]. The fuzzy logic controller anticipates nonlinear systems with greater dynamic performances than the regulators described grounded on an approached direct mode and direct fashion.

This study analyzes the pitch angle regulator approach to describe the goods and effectiveness of PMSG for independent WECS. In addition, the MPPT regulator is also used to ameliorate the control strategy, connected to the network using non-linear adaptive control grounded on the observation of disturbances to increase the energy conversion effectiveness in the time-varying wind. The high gain bystander is incorporated into the original fuzzy logic controller to design the MPPT control scheme [11].

The performance of the fuzzy logic controller under variable wind speed is estimated. The proposed control strategy has the bettered capability to capture maximum wind power. Thus, the suggested control system is an affair bolster regulator, which is fluently enforced for a practical scheme. In this chapter, the effectiveness of the proposed regulator is vindicated and the performance of the system using the fashion bettered with regard to delicacy, quality of fitted power, set point shadowing, response time, static crimes, and minimization of total harmonious deformation (THD) and robustness [11, 12].

For large wind turbines in offshore sites, if the sector is a full expansion, the idea of using the fuzzy MPPT allows on the one hand the pursuit of the maximum point according to the variation of the wind in order to decentralize the energy by producing small quantities in a localized way and reduce the strong constraints of performance and energy efficiency.

7.2 Proposed MPPT Control Algorithm

The description of the method consists in modeling the systems of production of wind energy with variable speed of the wind which can change at every moment, the determination of the speed of the generator piloted by MPPT algorithm to follow the small oscillations and points of maximum power while identifying to create a mathematical model. The wind parameters are mentioned in Table 7.1.

The wind turbine system is grounded on PMSG which consists of blades, creator, and control system and power electronic transformers. The PMSG is direct drive which makes it fluently controllable since there is thus no need for a gearbox reducing the complexity and size of the entire system as well as high necklace viscosity and veritably

Table 7.1 Wind parameters.

Parameters	Value
Blade radius	$R = 50$
Power coefficient	$C_{\text{pmax}} = 0.48$
Optimal relative wind speed	$\lambda_{\text{max}} = 0.014$
Mechanical speed multiplier	$G = 87.4$
Moment of inertia	$J = 0.0014 \text{ kg m}^2$
Damping coefficient	$F = 0.0050$
Density of air	$d = 1.2$
Rated rotational speed	$w_r = 314 \text{ rad/s}$

Table 7.2 PMSG parameters.

Parameters	Value
Rated power	$P_r = 1.5\,\text{MW}$
Rated stator voltage	$V_s = 76.2\,\text{V}$
Nominal frequency	$f_s = 50\,\text{Hz}$
Number of pole pair	$P = 6$
Stator resistor	$R_s = 0.0083\,\Omega$
Stator inductor	$L_s = 0.00331\,\text{H}$
Direct inductance	$L_d = 0.017\,\text{H}$
Quadrature inductance	$L_q = 0.017\,\text{H}$
Load torque	$C_t = 10\,\text{N. m}$

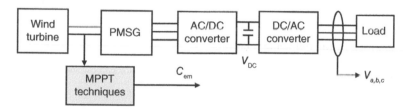

Figure 7.1 System block diagram.

low indolence, which makes it veritably effective. So the energy is transmitted directly to the network. The PMSG parameters are mentioned in Table 7.2.

The combination of the power converter with the aerodynamic and MPPT controllers is used to manage speed, torque, and generator side power under normal and grid fault conditions which it is very important to control, limit, and track the maximum power during higher wind speeds.

The diode-controlled method is used to convert the voltage obtained from the generator into DC voltage to charge the DC link capacitor, control the active and reactive power and control the voltage of the intermediate circuit [13] which supplies a three-phase inverter, connected to the network through a transformer. Figure 7.1 shows the PMSG-based wind turbine with its control system.

The proposed system consists in modeling a new MPPT control strategy developed in the literature [14, 15] to track the maximum power point with respect to the speed variation [16, 17] and control the PMSG speed. The implementation of fuzzy logic controller is to operate the turbine at its optimum value to achieve optimum generation speed with the deviation of the wind speed. The main advantage of fuzzy logic controller is quick, easier, and more efficient response. The brief description of the proposed controller is explained in the Section 7.4.

7.3 Wind Energy Conversion System

7.3.1 Wind Turbine Characteristics

The mechanical power, P_m, captured by the turbine is given by Eq. (7.1) [18]:

$$P_m = \frac{1}{2} \cdot C_p(\lambda, \beta) \cdot \rho \cdot \pi \cdot R^2 \cdot v^3 \tag{7.1}$$

where ρ, R, and v represent respectively the density of air, the blade radius, and the wind speed and C_p (λ, β) is the turbine power coefficient.

Additionally, C_p is made as a function of two important causes: the specific speed (λ) and the pitch angle (β). The relation between C_p, λ, and β is formulated in Eq. (7.2) [19].

$$C_p(\lambda, \beta) = (0.44 - 0.0167\beta) \sin \frac{\pi(\lambda - 2)}{13 - 0.3\beta} - 0.0018(\lambda - 2)\beta \tag{7.2}$$

$$\frac{1}{\lambda_i} = \frac{1}{\lambda + 0.08} - \frac{0.035}{\beta^3 + 1} \tag{7.3}$$

According to Betz's law:

$$C_p < 59\% \tag{7.4}$$

Although its maximum theoretical value is around 0.59, in practice it is between 0.4 and 0.45 [20].

If Ω is the angular velocity of the turbine, the reduced speed λ is defined:

$$\lambda = \frac{R\Omega}{v} \tag{7.5}$$

with R the shaft of pale of the wind speed.

The turbine torque for the generator is obtained by the following relation:

$$C_T = \frac{P_m}{\Omega} = \frac{1}{2} \cdot \rho \cdot \pi \cdot R^3 \cdot v^2 \cdot C_p(\lambda, \beta) \tag{7.6}$$

The mechanical torque is calculated:

$$C_m = C_T + C_{em} + f\Omega \tag{7.7}$$

where C_{em} is the electromagnetic torque and f is a viscous friction coefficient.

Moreover, if we ignore the fact of torque becoming viscous friction ($f\Omega_m = 0$) compared to the mechanical torque C_m, we can then write:

$$C_{em} = C_m \tag{7.8}$$

In this study, the given wind speed is always lesser than the scale wind speed, thus β is zero. If the speed ratio λ is kept at its optimal value λ_{opt} which corresponds to the maximum of the power coefficient C_{pmax}, the value of the TSR is maximum and set for all the points of maximum power.

The estimated electromagnetic torque is then expressed by:

$$C_{em}^* = \frac{C_{pmax} \cdot \rho \cdot \pi \cdot R^2 \cdot \Omega_{mec}^3}{2 \cdot \lambda_{opt}^3} \tag{7.9}$$

7.3.2 Model of PMSG

Unlimited attraction machines have been widely used and widely applied to small wind turbines because of their high efficiency and less conservation. The PMSG is based in dq reference form. The fine model of PMSG is defined by [21]:

$$v_d = L_d \frac{di_d}{dt} + R_s i_d - L_q \omega_r i_q \tag{7.10}$$

$$v_q = L_q \frac{di_q}{dt} + R_s i_q - p\omega_r i_d + \omega_r \lambda_m \tag{7.11}$$

where p is the number of pole pair, R_s the resistance of the stator winding, L_d the direct inductance of the stator winding, L_q inductance in squaring of the stator winding, L_s the inductance of stator winding, λ_m is the magnetic flux, and ω_r is the electrical angular frequency.

The electromagnetic torque can be expressed as a function of the magnetic flux and current:

$$C_{em} = \frac{3}{2}p[(L_d - L_q)i_d i_q + f i_q \lambda_m] \tag{7.12}$$

7.4 Fuzzy Logic Command for the MPPT of the PMSG

This section presents the fuzzy regulator applied for tracking the wind speed to acquire the MPPT and to attack the nonlinear features of the PMSG, when compared to tip speed ratio (TSR) especially in case of recurrently changing wind pets. Also, fuzzy logic controller has come popular in the artificial control operations field, for working control, estimation, and optimization problems [22], it has an advantage of fast confluence, squishy input and handling nonlinearity, fast response, and robustness. Fuzzy regulators attach to the class of education-grounded systems. Their capital thing is to apply mortal instruction in the form of a computer program.

Figure 7.2 shows the block diagram of the proposed MPPT control system, applying the fault between the reference speed and the actual rotor speed and the difference of this fault is as inputs. The output is the duty cycle D [23]. As shown, the controller has two input variables: the speed error (e) and the change of speed error (de) [24]. The fuzzy logic controller output is the variation of electromagnetic torque (C_{em}^*). The rotor speed error (e) is calculated by the following equation for every simpling time [25]:

$$e(t) = \omega_{rref}(t) - \omega_r(t) \tag{7.13}$$

$$de(t) = e(t) - e(t-1) \tag{7.14}$$

where

$$e(t-1) = \omega_{rref} - \omega_r(t-1) \tag{7.15}$$

$$de = \omega_r(t-1) - \omega_r(t) \tag{7.16}$$

The two inputs of fuzzy logic controller are collective by two scaling forms ($k1$) and ($k2$), independently. Both regulators have multiplied by another scaling factor ($k3$) and by carrier factors which play vital role for fuzzy logic controller which induce system stability, oscillations and system damping [26]. The structure of a complete fuzzy control system generally corresponds of three stages: fuzzification, rule base lookup table, and defuzzification as shown in Figure 7.3 [27].

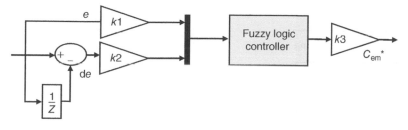

Figure 7.2 Scheme of the fuzzy logic controller.

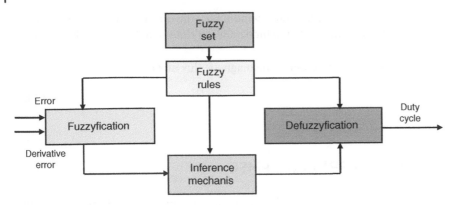

Figure 7.3 Basic structure of fuzzy logic controller.

7.4.1 Fuzzification

In this study, five fuzzy sets of NB, NS, ZE, PS, PB are chosen for (e), (de), and seven fuzzy sets of NB, NM, NS, ZE, PS, PM, PB are chosen for the controller output: Negative Big (NB), Negative Small (NS), Zero (ZE), Positive Big (PB), Positive Small (PS) are identified by the triangular membership function as advanced in Figure 7.4. The inputs universe is determined in −1 and 1, and labors of the fuzzy logic controller were adopted mid −1 and 1 with triangular class functions. Every fuzzy variable is a member of the endured with a degree of class μ varying in the middle of 0 and 1. The verbal rudiments used are the same as those used in utmost publications. The numerical system for the discrimination equations is of the Mamdani type. Aggregation and disaggregation styles are described as Minimum (min) and Maximum (maximum) independently.

Then the fuzzy set is processed in inference system where an appropriate fuzzy output is obtained using fuzzy rules.

7.4.2 Fuzzy Logic Rules

Based on authors' adept control instruction of the PMSG wind turbine system operation, these epicurean rules are asserted in fuzzy domain. Table 7.3 shows the fuzzy rules base with 49 rules, which the rules have been written in matrix form. The fuzzy logic controller was modeled using the MATLAB fuzzy-logic toolbox GUI. For example, if Error "e" is NB and Derivative "de" is NB, then convert in electromagnetic torque "dC_{em}" is NB.

7.4.3 Defuzzification

The defuzzification method uses centroid for processing. Then the fuzzy output is converted into the systematic crisp value as a form of duty cycle in defuzzification for monitoring the electromagnetic torque. The approach applied to the entire controller objective is the center of gravity method, which is given by Eqs. (7.13) and (7.14).

Table 7.3 Rule base of fuzzy logic controller.

e/de	NB	NS	ZE	PS	PB
NB	ZE	PB	ZE	NB	NS
NS	PS	ZE	ZE	NB	NS
ZE	ZE	ZE	ZE	ZE	ZE
PS	PS	PB	ZE	ZE	NS
PB	PS	PB	ZE	NB	ZE

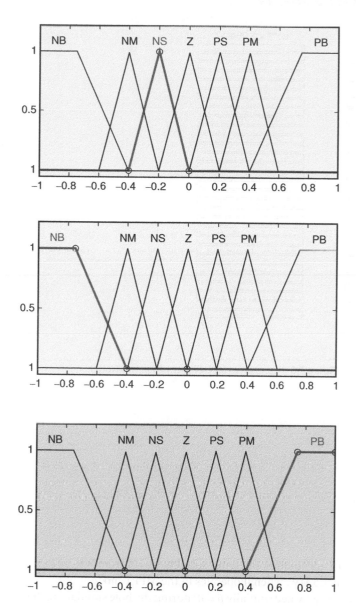

Figure 7.4 Structure membership functions of (*e*), membership functions of (*de*), and membership functions of (C^*_{em}).

The end output of the system is the weighted medium of all rules output, estimated as Figure 7.5.
Surface of fuzzy logic control is seen in Figure 7.6.

All forms of scaling, class mark function, fuzzification system and defuzzification system are preset and controlled constant during exploration apart from number of guides.

7.5 Results and Discussions

The system defined in Section 7.2 is executed in Matlab Simulink (Figure 7.1). The whole proposed fuzzy logic controller based MPPT algorithm control system consists of a model of the wind turbine, a fuzzy logic controller,

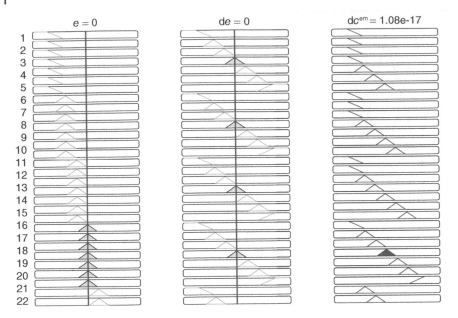

Figure 7.5 Defuzzification.

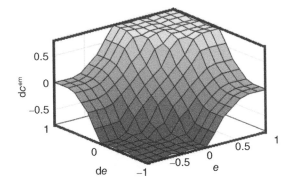

Figure 7.6 Control surface of fuzzy logic controller.

a PMSG model, to check the tracking ability of MPPT techniques, the wind speed is varied from 8 to 12 m/s as shown in Figure 7.7. The rated power of the turbine is 1.5 MW, when it operates in the MPPT mode.

Figures 7.8 and 7.9 show the pitch angle controller get activated (initially $\beta = 0$) during the hole simulation, then it decreases the C_p for protecting the wind turbine from damages. Initially C_p maintains its optimum ($C_p = 0.48$) for maximum power conversion and the λ is contributed and, hence, the $\lambda_{opt} = 0.014$ is imposed to bear at its optimal value. The value of the power coefficient does not reach the maximum theoretical value declared by Betz (0.59).

Figure 7.10 illustrates the mechanical power of wind turbine below the technique of MPPT. The characteristic of the power of a wind turbine is strongly nonlinear. It is clear that the mechanical power is most at distinct rotational speed for each wind speed.

Figure 7.11 presents the reel and reference rotor speeds which shows the performance of the fuzzy logic controller, which both describes make about consistent unstated. It is clear that the speed of the turbine is not adapted to that of the wind; however there is a good continuation of the reference value. All these quantities are of the same form because of the linear relationship that exists between them.

Figure 7.12 shows the response of the electromagnetic torque in the control case. The reversal of the direction of rotation makes it possible to deduce that the control is robust.

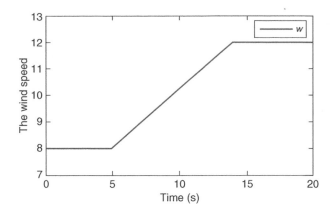

Figure 7.7 Wind speed profile.

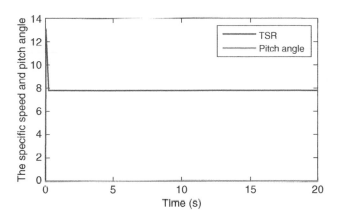

Figure 7.8 Tip speed ratio (λ).

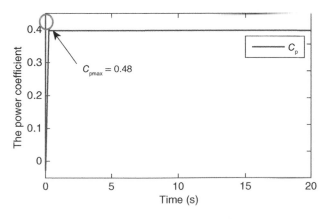

Figure 7.9 Power coefficient (C_p) of the turbine.

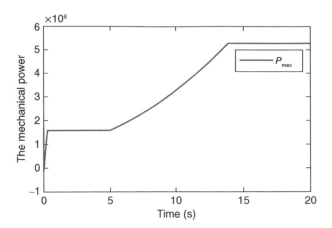

Figure 7.10 Mechanical power.

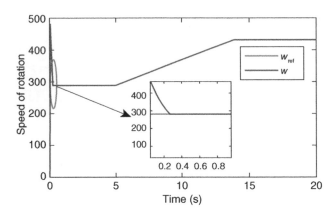

Figure 7.11 Reference and measured rotor speed.

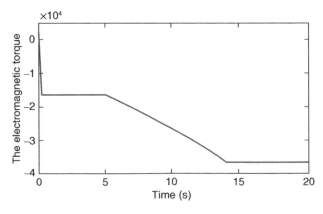

Figure 7.12 The electromagnetic torque.

The simulation results indicate that the advanced system has good dynamic and firm state performances throughout the MPPT process. The enhancement of fuzzy logic controller is assigned to its ability to convert the regulator parameters veritably snappily in answer to system dynamics particularly when the wind speed mutated abruptly. It is clear that the system response with fuzzy logic controller accords reform shadowing ability and the oscillation degree.

7.6 Conclusion

A new system fuzzy logic controller grounded MPPT for PMSG wind power system was designed and applied in this chapter. Proposed control system and modeling for wind energy under varying wind speed condition grounded on a 1.5 MW PMSG was bandied and successfully vindicated by both simulation results. Hence, fuzzy logic controller allows better shadowing ability and lesser oscillation rate contrasted with TSR which has lower response time and high delicacy without any fine computations. For that purpose, this system is more useful. The dynamic response of fuzzy regulator is much better semi-model free adaptive control under colorful operating conditions. The results obtained in the simulation are effective in terms of stability, fast tracking ability and oscillations, and it shows a good gesture of our regulators in terms of settling time, peak value and fall value to achieve these goals.

References

1 Sagbansua, L. and Balo, F. (2017). Decision making model development in increasing wind farm energy efficiency. *Journal of Renewable Energy* 109: 354–362.

2 Chen, J., Wu, H.B., and Sun, M. et al. (2012). Modeling and simulation of directly driven wind turbine with permanent magnet synchronous generator. In: *Proceedings of the 2012 IEEE Innovative Smart Grid Technologies, Asia (ISGT '12)* (May 2012), pp. 1–5.

3 Moutchou, R. and Abbou, A. (2019). MPPT and pitch angle control of a permanent magnet synchronous generator based wind emulator. *International Journal of Innovative Technology and Exploring Engineering (IJITEE)* ISSN: 2278-3075 8 (7).

4 Moutchou, R., Abbou, A., and vall Hemeyine, A. (2018). Control of the active and reactive powers of a permanent magnet synchronous generator decoupled by singular perturbations. In: *The 9th International Renewable Energy Congress (IREC 2018)*.

5 Bendib, B., Krim, F., Belmili, H. et al. (2014). Advanced fuzzy MPPT controller for a stand-alone PV system. *Energy Procedia* 50: 383–392.

6 Jeong, H.G., Seung, R.H., and Lee, K.B. (2012). An improved maximum power point tracking method for wind power systems. *Energies* 5: 1339–1354.

7 Chen, J., Jiang, L., Yao, W. (2013). A feedback linearization control strategy for maximum power point tracking of a PMSG based wind turbine. In: *2nd International Conference on Renewable Energy Research and Applications (ICRERA13)*, Madrid, Spain (October 2013), pp. 79–84.

8 Rhaili, S.E., Abbou, A., Marhraoui, S. et al. (2022). Optimal power generation control of 5-phase PMSG based WECS by using enhanced fuzzy fractional order SMC. *International Journal of Intelligent Engineering and Systems* 15 (2): 572–583.

9 Rahman, M.A., Vilathgamuwa, D.M., Uddin, M.N., and Tseng, K.J. (2003). Nonlinear control of interior permanent-magnet synchronous motor. *IEEE Transactions on Industry Applications* 39 (2): 408–416.

10 Kim, K.H., Jeung, Y.C., Lee, D.C., and Kim, H.J. (2012). LVRT scheme of PMSG wind power systems based on feedback linearization. *IEEE Transactions on Power Electronics* 27 (5): 2376–2384.

11 Shen, Y., Yao, W., Wen, J.Y. et al. (2018). Adaptive supplementary damping control of VSC-HVDC for interarea oscillation using GrHDP. *IEEE Transactions on Power Apparatus and Systems* 33 (2): 1777–1789.

12 Chen, J., Jiang, L., Yao, W., and Wu, Q.H. (2014). Perturbation estimation based nonlinear adaptive control of a full-rated converter wind-turbine for fault ride-through capability enhancement. *IEEE Transactions on Power Apparatus and Systems* 29 (6): 2733–2743.

13 Asghar, A.B. and Liu, X.D. (2017). Estimation of wind turbine power coefficient by adaptive neuro-fuzzy methodology. *Neurocomputing* 238: 227–233.

14 Stol, K.A. and Balas, M.J. (2003). Periodic disturbance accommodating control for blade load mitigation in wind turbines. *Journal of Solar Energy Engineering* 125: 379–385.

15 Ackermann, T. (2005). *Wind Power in Power System*. Wiley, Ltd.

16 Verma, Y.P. and Kumar, A. (2012). Dynamic contribution of variable speed wind energy conversion system in system frequency regulation. *Frontiers in Energy* 6 (2): 184–192.

17 Xia, Y., Ahmed, K.H., and Williams, B.W. (2011). A new maximum power point tracking technique for permanent magnet synchronous generator based wind energy conversion system. *IEE Transactions on Power Electronics* 26 (12): 3609–3620.

18 Rhaili, S.E., Abbou, A., El Hichami, N., and Marhraoui, S. (2021). Optimal power tracking through nonlinear backstepping strategy of a five-phase PMSG based wind power generation system. *International Journal on Technical and Physical Problems of Engineering* 13 (4): 115–122.

19 Moutchou, R. and Abbou, A. (2019). Comparative study of SMC and PI control of a permanent magnet synchronous generator decoupled by singular perturbations. In: *Proceedings of 2019 7th International Renewable and Sustainable Energy Conference, IRSEC 2019*, 9078310.

20 Moutchou, R., Abbou, A., and Rhaili, S.E. (2020). Super-twisting second-order sliding mode control of a wind turbine coupled to a permanent magnet synchronous generator. *International Journal of Intelligent Engineering and Systems* 14 (1): 484–495.

21 Kamel, T. and Abdelkader, D. (2014). Vector control of five phase permanent magnet synchronous motor drive. In: *International Conference on Electrical Engineering (ICEE)*, Boumerdes, Algeria (1–4 June 2014).

22 Moutchou, R. and Abbou, A. (2021). Control of grid side converter in wind power based PMSG with PLL method. *International Journal of Power Electronics and Drive Systems* 12 (4): 2191–2200.

23 Salaheddine, R., Abbou, A., and Marhraoui, S. et al. Vector control of five-phase permanent magnet synchronous generator based variable-speed wind turbine. In: *2017 International Conference on Wireless Technologies, Embedded and Intelligent Systems (WITS)*. doi: 10.1109/WITS.2017.7934647.

24 Adhikari, J., Prasanna, I.V., Ponraj, G., and Panda, S. (2016). Modelling, design, and implementation of a power conversion system for small-scale high altitude wind power generating system. *IEEE Transactions on Industry Applications* 26: 4388.

25 He, Y., Wang, Y., Wu, J. et al. (2010). A comparative study of space vector PWM strategy for dual three-phase permanent-magnet synchronous motor drives. In: *2010 Twenty-Fifth: Annual IEEE Conference on Applied Power Electronics Conference and Exposition (APEC)*, pp. 915–919.

26 Tiwari, R. and Babu, N.R. (2016). Fuzzy logic based MPPT for permanent magnet synchronous generator in wind energy conversion system. *IFAC Papers (online)* 49: 462–467.

27 Allouche, M., Abderrahim, S., Ben Zina, H., and Chaabane, M. (2019). A novel fuzzy control strategy for maximum power point tracking of wind energy conversion system. *International Journal of Smart Grid* 3 (3): 120–127.

8

A Novel Nearest Neighbor Searching-Based Fault Distance Location Method for HVDC Transmission Lines

Aleena Swetapadma[1], Shobha Agarwal[2], Satarupa Chakrabarti[1], and Soham Chakrabarti[1]

[1] *School of Computer Engineering, KIIT University, Bhubaneswar, India*
[2] *Department of Higher Technical Education and Skill Development, Jharkhand University, Ranchi, India*

8.1 Introduction

High voltage direct current transmission lines have maximum power transmission capability and hence are mostly implemented. Advantages of using HVDC transmission lines include long-distance power transmission, flexible, maximum power transmission, and losses. With all these advantages, there are certain drawbacks in HVDC transmission lines. Estimation of fault distance in transmission lines (HVDC) is a difficult task as the impedance is very small [1]. This aspect has been brought forward by various researchers and is highlighted below.

A technique has been suggested for locating fault in transmission line built around the natural frequency of distributed parameter line mode using one terminal current data in [2]. Dependence of fault distance and wave speed on the natural frequency of transmission (HVDC) line on the occurrence of the fault has been used. But it suffers from the drawback of selecting natural frequency because the loss of energy will be present in transient energy which is not considered. In [1], bipolar transmission system of 500 kV with 1000 km line length having frequency-dependent parameters line model had been studied. Post fault data obtained from the current and voltage measurements at both terminals based on the parameter line (distributed) model were required to locate faults. It has the disadvantage of high frequency and reach setting is only 90%. Drawbacks of the method are that it has the requirement of communication links and synchronized data.

In [3], traveling wave having different phase characteristics on different location has been studied using discrete Fourier analysis and characteristics of the phase frequency method. It was tested on different locations which has different phase distortion characteristics with frequency data of 40–60 Hz, sampling frequency of 4 kHz, and the phase angle ratio. The fault data are extracted for an interval of 10 km between two fault points to get the fittest least square nonlinear curve data. The transition resistance in three different cases is limited to 50 Ω. A fault-location method using the nearest neighbor principle with the post-fault voltage signals of a short time window has been proposed in [4]. It is based on the similarity of the existing pattern with captured signal measured at one of the line terminals. It has drawbacks of the high sampling frequency of 80 kHz to obtain absolute value below the threshold value of sampled voltage. Also, dependence on the buffered voltage signal of the faulted pole and pre-fault current range criteria for training of lines under different fault conditions is not defined.

In [5], a method has been proposed in which DC faults time variant source was changed because the Bergeron time-domain method is not suitable for the low-frequency components of fault. For better performance, a band-pass filter is required to obtain the appropriate frequency components. But the requirement of filter and tuning is costly and bulky for large power transmission. For the location of faults, a time-domain fault location algorithm [6, 7] has been proposed where a suitable frequency of transmitted electric energy was extracted. This

Artificial Intelligence-based Smart Power Systems, First Edition.
Edited by Sanjeevikumar Padmanaban, Sivaraman Palanisamy, Sharmeela Chenniappan, and Jens Bo Holm-Nielsen.

method has drawbacks of dependency on the frequency and the parameter of line. In [8], a genetic algorithm was studied for fault location avoiding the need for traveling wave theory, but it has the disadvantage of a large amount of data. A method to predict the location based on the natural frequency of traveling-wave has been proposed in [9]. In [10], an SVM-based approach has been proposed using DC voltage and current and AC RMS voltage. An algorithm for fault location for HVDC lines has been proposed in [5]. Here Bergeron model is adopted using unsynchronized two-end measurement. An electromagnetic time reversal-based method has been proposed in [11]. In [12], the authors used discrete wavelet transform and SVM for segmented DC lines that were based on the traveling wave.

The fault distance estimation approaches recommended for transmission lines (HVDC) are mostly based on the traveling wave. But the requirement of extra equipment at substations and high sampling rate poses to be the main drawback of this method. Traveling waves have demerits of wave-like sensitivities of wavefronts with transmission line parameters. A communication link is an essential part in a fault location method. To overcome the drawbacks of these methods, a novel fault distance estimation method for HVDC transmission lines has been proposed based on nearest neighbor searching (NNS). The input used in the NNS-based method is a current signal for locating the fault. The chapter is prepared as follows – NNS is described in Section 8.2, while the proposed method is highlighted in Section 8.3, results are summarized in Section 8.4 followed by a conclusion in Section 8.5.

8.2 Nearest Neighbor Searching

NNS technique works on the foundation of finding the nearest neighbor to unidentified data points depending on the known target [13]. The number of neighbors that would outline the output to which the data would belong determines how neighbor elements are calculated. The immediate neighbors are chosen via majority voting. The voting pattern can at times be weighted when the distance among the neighbors is accounted for. Also, different distance metrics are defined in order to measure distances between data and unidentified points [14]. Let the k-nearest neighbor set be defined as $X = \left\{ x_1^N, x_2^N, \dots x_k^N \right\}$ and the output corresponding to each of the neighbor as $C = \left\{ c_1^N, c_2^N, \dots c_k^N \right\}$. Each nearest neighbor weight can be calculated as:

$$W_i = \begin{cases} \dfrac{D_k^N - D_i^N}{D_k^N - D_1^N} & : D_k^N \neq D_1^N \\ 1 & : D_k^N = D_1^N \end{cases}, \quad i = 1,2,3, \dots k \tag{8.1}$$

where D_k^N represents the furthermost neighbor while D_1^N is the nearest neighbor point. Greater weights are usually assigned to the nearest neighbors.

Even though there are different ways of selection, the most common one is to use diverse values of k at each execution and comparing the results to select the best value for a particular problem. Bailey and Jain [15] modified the algorithm by introducing weight-based distance calculation. Guo et al. [16] mentioned the use of model-based approach where there would be a spontaneous selection of the value of k. NNS is centered on the instinctive assumption that nearby objects have high chances of similarity. Thus, in weight-based calculation of NNS, for each nearest neighbor, a weight is defined depending on the nearness of the point to the unidentified query point.

$$W(a, x_i) = \frac{\exp(-D(a, x_i))}{\sum_{i=1}^{k} \exp(-D(a, x_i))} \tag{8.2}$$

where $D(a, x_i)$ is the distance between the unidentified query point a and the ith data sample x. The notation to represent the problem for a dataset using NNS is based on the estimation of the value of different output variables (c) depending on the known and independent feature variables. The output can be given as

$$c_i = f(x_i, \text{training dataset}, \text{parameters}) \tag{8.3}$$

In the mathematical model of the nearest neighbor, it can be seen that the algorithm specifically uses local prior probabilities of data points. Let a query be considered as x_t, then the NNS algorithm will work in the following fashion,

$$y_t = \begin{matrix} \arg\max \\ c \in \{c_1, c_2, \dots c_m\} \end{matrix} \sum_{x_i \in N(x_t, k)} E(y_i, c) \tag{8.4}$$

where the predicted output of the query is given by y_t and the output present is given by m. The above equation can be expanded and written as follows,

$$y_t = \arg\max \left\{ \sum_{x_i \in N(x_t, k)} E(y_i, c_1), \sum_{x_i \in N(x_t, k)} E(y_i, c_2), \dots \sum_{x_i \in N(x_t, k)} E(y_i, c_m) \right\} \tag{8.5}$$

$$y_t = \arg\max \left\{ \sum_{x_i \in N(x_t, k)} \frac{E(y_i, c_1)}{k}, \sum_{x_i \in N(x_t, k)} \frac{E(y_i, c_2)}{k}, \dots \sum_{x_i \in N(x_t, k)} \frac{E(y_i, c_m)}{k} \right\} \tag{8.6}$$

Also, it is known,

$$p(c_j)_{(x_t, k)} = \sum_{x_i \in N(x_t, k)} \frac{E(y_i, c_j)}{k} \tag{8.7}$$

where $p(c_j)_{(x_t, k)}$ denotes the occurrence probability of jth class in the neighbor of x_t. Hence the equation can be re-written as:

$$y_t = \arg\max \{p(c_1)_{x_t, k}, p(c_2)_{x_t, k}, \dots p(c_m)_{x_t, k}\} \tag{8.8}$$

Therefore, it can be seen from the above set of mathematical equations that NNS applies prior probabilities for calculating the specific class for the given query, ignoring the distribution of the output around the point [17]. Another type is the weighted k-nearest neighbor algorithm as proposed in [17]. The algorithm is designed using the weight factor for the NNS algorithm. Three different outputs based weighted NNS algorithm could be accounted for. In the first design, the weight is based on the frequency of the output occurrence that can be expressed as follows:

$$W[c] = \frac{1}{\text{frequency}[c]} \tag{8.9}$$

where $W[c]$ is the weight factor that determines the weight of output c and frequency$[c]$ represents the amount of presence of output c in the whole dataset. Therefore, the NNS rule can be properly expressed as,

$$y(t) = \begin{matrix} \arg\max \\ c \in \{c_1, c_2, \dots c_m\} \end{matrix} \sum_{x_i \in N(x_t, k)} W[c] * E(y_i, c) \tag{8.10}$$

But this design comes with a drawback as it does not take into consideration the local distribution of the classes, so the modified design is expressed with respect to the above equation with an alpha coefficient factor while measuring the weight.

$$W[c] = \frac{1}{1 + \text{alpha} * \left(\frac{\text{frequency}[c]}{\sum_{i=1}^{m} \text{frequency}[c_j]} \right)} \tag{8.11}$$

This weight factor is bounded within a range of 1 to $\left(\dfrac{1}{1 + \text{alpha} * \left(\dfrac{\text{frequency}[c]}{\sum_{i=1}^{m} \text{frequency}[c_j]} \right)} \right)$. This is followed by another modification where the neighbor space around given unidentified data is considered instead of the whole

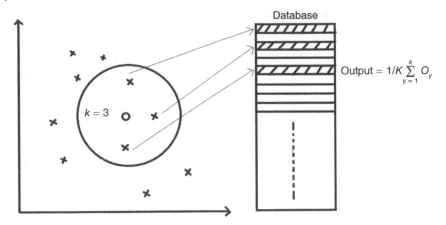

Figure 8.1 NNS method for prediction.

dataset. In a standard NNS algorithm depending on the value of k, the neighbors are used to classify the query data. In the modified approach, $(k + d)$ neighbors are considered for the distribution such that the distance is limited within a particular region. The mathematical expression of this approach is given as:

$$y(t) = \begin{array}{c} \arg\max \\ c \in \{c_1, c_2, \dots c_m\} \end{array} \sum_{x_i \in N(x_t, k)} W(c, x_t) * E(y_i, c) \tag{8.12}$$

where the weight factor is presented by $W(c, x_t)$ for output c and query x_t. The weight factor can be expressed as follows:

$$W(c, x_t) = \frac{1}{\sum\limits_{x_i \in N(x_t, k+d)} E(y_i, c)} \tag{8.13}$$

where d represents an input parameter, either given by the user or selected from the data. The term $\sum\limits_{x_i \in N(x_t, k+d)} E(y_i, c)$ denotes the occurrence of a particular output c in the $(k + d)$ neighborhood of the data query x_t [17]. Figure 8.1 shows the functioning NNS method with three nearest neighbors. In this work for calculating fault distance in bipolar LCC–HVDC transmission lines, NNS approach has been considered.

8.3 Proposed Method

The proposed NNS-based fault location estimation scheme has various steps as displayed in Figure 8.2. The following Sections 8.3.1 and 8.3.2 present a detailed explanation of the proposed approach.

8.3.1 Power System Network Under Study

A ±500 kV, 24 pulse linked transmission line (HVDC) of 1100 km on CIGRE, Benchmark system is studied. Figure 8.3 presents the line diagram of the bipolar transmission line. The ripples in DC current are smoothed with smoothing reactors per pole of 0.5 Henry. All the modeling and simulation work have been carried out using MATLAB [18]. In HVDC transmission lines, fault occurrence is a common phenomenon. Several types of faults occur in HVDC transmission lines such as pole-to-ground faults, pole-to-pole faults, and pole-to-pole ground.

After designing the HVDC power system, it is tested for various fault cases. The rectifier end current signals (DC) are recorded and 1 kHz sampling frequency is used for sampling the signals. To carry out the feature extraction

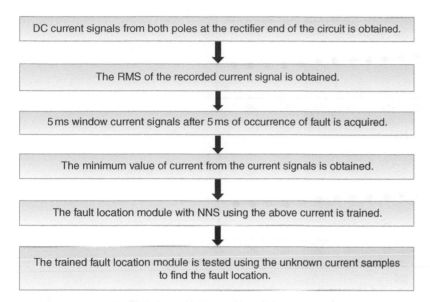

DC current signals from both poles at the rectifier end of the circuit is obtained.

↓

The RMS of the recorded current signal is obtained.

↓

5 ms window current signals after 5 ms of occurrence of fault is acquired.

↓

The minimum value of current from the current signals is obtained.

↓

The fault location module with NNS using the above current is trained.

↓

The trained fault location module is tested using the unknown current samples to find the fault location.

Figure 8.2 Flowchart of the planned fault location estimation approach.

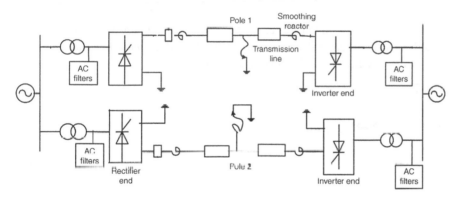

Figure 8.3 LCC–HVDC transmission lines.

process, first RMS value of current signals is obtained. Then 5 ms window of samples is taken after 5 ms of occurrence of fault. The minimum value of current is taken from the current signals. The input given to the fault location unit based on NNS is the current value that has been obtained. Figure 8.4 shows the input feature obtained after varying the fault location in the step of 50 km (50–1050 km) during P1G faults. Input current of pole 1 is shown in Figure 8.4a, while Figure 8.4b displays the input current of pole 2.

8.3.2 Proposed Fault Location Method

NNS-based fault location method first creates a database with known target values and testing with unknown samples using the nearest neighbor. The proposed NNS method uses the current of both poles obtained from different fault locations to create the database. The fault cases used to create the database involve P1G fault, P2G fault, P1P2G fault and P1P2 fault, fault resistance (0–100 Ω with a step size of 2.5 Ω), fault location (5–1195 km with a step size of 5 km), etc. The features obtained from the current act as an input parameter to the NNS method for the creation of a database to locate the faults. After obtaining the final database, it is verified with test samples

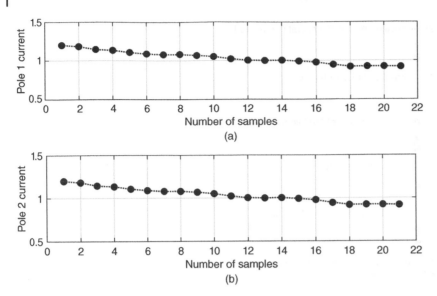

Figure 8.4 Input feature obtained during P1G fault varying different fault locations. (a) Pole 1 current feature during P1G fault varying different fault locations. (b) Pole 2 current feature during P1G fault varying different fault locations.

that are not familiar. The nearest neighbors are varied and tested, and the optimal number is found to be one. The NNS-based method is tested with varying different distance matrices such as Euclidean, Mahalanobis, and Minkowski as shown in Eqs. (8.14)–(8.17). The Euclidean distance is given in Eq. (8.14),

$$D_E = \sqrt[2]{\left((x_1 - x_2)^2 + (y_1 - y_2)^2\right)} \tag{8.14}$$

The Mahalanobis distance is given in Eq. (8.15),

$$D = \sqrt[2]{\sum_{i=1}^{n} \left((x_i - y_i)^2\right)} \tag{8.15}$$

The Minkowski distance is given in Eq. (8.16),

$$D = \left((x_1 - x_2)^2 + (y_1 - y_2)^2\right)^{\frac{1}{p}} \tag{8.16}$$

Euclidean distance is found to be the optimal distance to locate the fault. The efficacy of the proposed fault location approach has been discussed in the subsequent Section 8.4.

8.4 Results

The fault-location scheme based on NNS is verified with several test (fault) cases. These cases are generated varying nearest neighbor, distance matrices, fault locations, fault resistance, and fault types. The efficiency of the proposed method in locating the fault is checked in terms of percentage error, is denoted as E, and calculated using Eq. (8.17),

$$E = \left[\frac{|\text{Actual Location} - \text{Estimated Location}|}{\text{Line Length}}\right] * 100 \tag{8.17}$$

Table 8.1 Performance varying nearest neighbor.

Nearest neighbor	Actual fault location (km)	Estimated fault location (km)	E (%)
1	25	25.00	0.000
	50	50.00	0.000
	75	75.00	0.000
	100	100.0	0.000
2	125	117.50	0.681
	150	152.50	0.227
	175	165.00	0.909
	200	192.50	0.681
3	225	210.00	1.363
	250	200.00	4.545
	275	261.66	1.212
	300	298.33	0.151
4	325	310.00	1.363
	350	342.50	0.681
	375	365.00	0.909
	400	380.00	1.818
5	425	409.00	1.454
	450	431.00	1.727
	475	432.00	3.909
	500	469.00	2.818

8.4.1 Performance Varying Nearest Neighbor

In NNS method, it is very important to find the optimal number of nearest neighbors required to estimate the fault locations properly. In this work, using various nearest neighbors (1–10), proposed method is tested. Few test case instances are presented in Table 8.1 changing the nearest neighbor for P1G fault. It can be seen from Table 8.1 that for estimating location, the error is more in different nearest neighbor values except one. Figure 8.5 shows the percentage of fault cases with error values for different nearest neighbors. From Figure 8.5, it can be observed that one nearest neighbor has better performance than others. Hence in this work, the optimal value for the nearest neighbor is taken as one to estimate the fault location.

8.4.2 Performance Varying Distance Matrices

In the NNS method, distance matrices play an important part and it is necessary to find the optimal number of the nearest neighbor. Various distance matrices such as Euclidean, Mahalanobis, Minkowski, and cosine have been used to find an optimal distance matrix. Test results from varying distance matrices are given in Table 8.2 during the P2G fault. It can be seen from Table 8.2 that for estimating location, the error is less with Euclidean distance. Figure 8.6 shows the percentage of fault cases with error values for different distances. From Figure 8.6, it can be observed that Euclidean distance has better performance than others. Hence in this work, the optimal distance used is Euclidean distance to locate the DC faults.

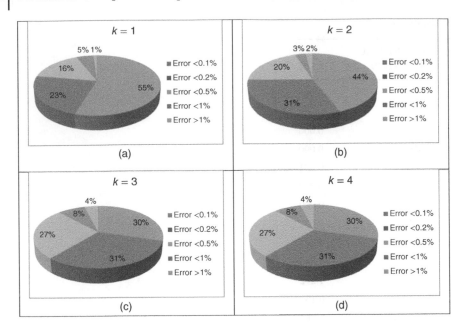

Figure 8.5 Error obtained varying the nearest neighbor. (a) Error for LG fault. (b) Error for LLG fault. (c) Error for LL fault. (d) Error for LLLG fault.

Table 8.2 Performance varying distance.

Distance matrices	Actual fault location (km)	Estimated fault location (km)	*E* (%)
Euclidean	517	515	0.181
	817	815	0.181
	917	915	0.181
	1017	1020	0.272
Mahalanobis	517	525	0.727
	817	815	0.181
	917	910	0.636
	1017	1025	0.727
Minkowski	517	515	0.272
	817	815	0.272
	917	915	0.181
	1017	1020	0.272

8.4.3 Near Boundary Faults

There is always the possibility of faults occurring near the boundary of transmission lines (LCC–HVDC). Most of the fault location estimation schemes proposed do not estimate the near boundary faults correctly. To design an approach that is robust, different near boundary faults have been tested within 0.1–20 km. Table 8.3 illustrates the test results of near boundary P1G faults. It can be deduced from the table that near boundary faults can be estimated correctly with very less errors using the proposed NNS-based approach. Hence the method can be implemented efficiently to trace the DC faults at near end of the line correctly.

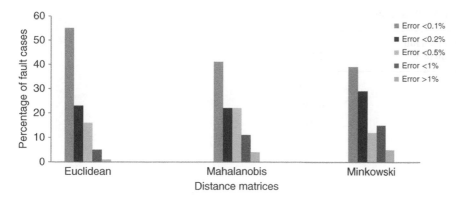

Figure 8.6 Error obtained varying distance matrix.

Table 8.3 Performance of near boundary P1G faults.

Actual fault location (km)	Estimated fault location (km)	E (%)
0.5	1.0	0.045
1.5	2.0	0.045
2.5	3.0	0.045
3.5	3.0	0.045
4.5	4.0	0.045
5.5	5.0	0.045
6.5	7.0	0.045
7.5	7.0	0.045
8.5	9.0	0.045
9.5	10.0	0.045

8.4.4 Far Boundary Faults

Faults may occur in the far boundary of the HVDC lines. Maximum estimation schemes for fault location do not estimate the far boundary faults as they have less reach setting. The suggested method based on NNS has been verified with different far-boundary faults from 1080 to 1099 km. Table 8.4 presents the test results of far boundary P2G faults. It can be visualized from the table that far boundary faults can be estimated correctly using the proposed NNS method.

8.4.5 Performance During High Resistance Faults

When a fault occurs through a high resistance medium, it is possible that the fault location method cannot detect it correctly. Hence the estimation method based on NNS to detect fault location has been verified with different resistance values. Table 8.5 shows some of the test results of faults in HVDC transmission lines with varying resistance. NNS-based method accurately estimates the location of fault with less error as shown in Table 8.5. It can be concluded that fault resistance does not affect the proposed NNS-based method.

Table 8.4 Performance of far boundary P2G faults.

Actual fault location (km)	Estimated fault location (km)	E (%)
1081.5	1080	0.136
1083.5	1085	0.136
1085.5	1085	0.045
1087.5	1090	0.227
1089.5	1090	0.045
1091.5	1090	0.136
1093.5	1095	0.136
1095.5	1095	0.045
1097.5	1095	0.227
1099.5	1095	0.409

Table 8.5 Performance during high resistance faults.

Fault resistance (Ω)	Actual fault location (km)	Estimated fault location (km)	E (%)
6	21	25	0.363
	121	125	0.363
	221	210	1.000
	321	325	0.363
36	721	725	0.363
	821	825	0.363
	921	925	0.363
	1921	1030	0.818
66	21	20	0.090
	121	120	0.090
	221	225	0.363
	321	325	0.363
96	721	720	0.090
	821	825	0.363
	921	920	0.090
	121	1005	1.454

8.4.6 Single Pole to Ground Faults

The fault location method based on NNS has been verified by varying fault locations during single pole to ground faults. Results of different tested fault cases tested with NNS-based method are presented in Table 8.6. Also, the percentage error in tracing the location of the fault lies within 1% for most test cases. Hence the NNS-based method estimates the location of the single pole to ground faults accurately in HVDC transmission lines.

Table 8.6 Performance during single pole to ground fault.

Fault type	Actual fault location (km)	Estimated fault location (km)	E (%)
P1G	31	30	0.090
	131	130	0.090
	431	430	0.090
	631	630	0.090
	831	830	0.090
	931	930	0.090
P2G	33	35	0.181
	133	135	0.181
	233	235	0.181
	333	335	0.181
	433	430	0.272
	633	630	0.272

Table 8.7 Performance during double pole to ground fault.

Actual fault location (km)	Estimated fault location (km)	E (%)
29	25	0.363
129	135	0.545
229	230	0.090
329	315	1.272
529	530	0.090
629	630	0.090
829	830	0.090

8.4.7 Performance During Double Pole to Ground Faults

The NNS-based method has been tested during double pole-to-ground faults. Some of the test results of the double pole-to-ground faults are shown in Table 8.7. Figure 8.7 shows the estimated fault location and error in location for the P1P2G fault. Figure 8.7a shows the actual fault location and estimated fault location during the P1P2G fault. Figure 8.7b shows the percentage error obtained for the above fault locations. Also, the percentage error in tracing the location of the fault lies within 1% for most test cases. Hence, the NNS-based fault location method estimates the location of P1P2G faults accurately.

8.4.8 Performance During Pole to Pole Faults

The NNS-based fault distance estimation method has been tested during pole-to-pole faults (P1P2). Table 8.8 presents the results of some of the test cases for the P1P2 fault. The percentage error in tracing the location of

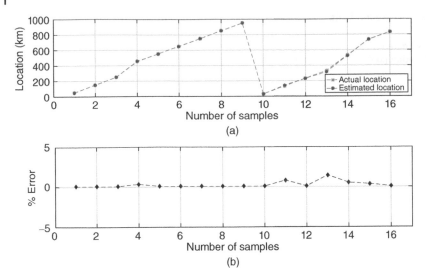

Figure 8.7 Performance during single pole to ground P1P2G fault. (a) Fault location (b) Error.

Table 8.8 Performance during P1P2 fault.

Actual fault location (km)	Estimated fault location (km)	E (%)
33	35	0.181
133	120	1.181
233	250	1.545
333	330	0.272
533	530	0.272
633	630	0.272
733	725	0.727
833	830	0.272
933	935	0.181

the fault lies within 1% for most test cases. Hence the proposed fault location approach is able to approximate the position of P1P2 faults accurately.

8.4.9 Error Analysis

NNS-based estimation method for tracing fault distance has been verified with all types of faults and fault location error has been calculated for various fault cases. The performance of the proposed method has been assessed in terms of percentage error. Figure 8.8 shows the error analysis of the proposed estimation method for fault distance based on NNS. Figure 8.8 presents the percentage error in various ranges and the percentage of fault cases belonging to that range. In almost all test fault cases, the percentage error lies within 1% with NNS-based fault distance estimation method. Hence the NNS-based fault distance estimation method estimates the locations correctly in bipolar LCC–HVDC transmission lines.

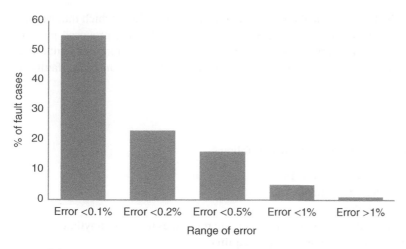

Figure 8.8 Error analysis.

8.4.10 Comparison with Other Schemes

Although HVDC transmission lines are implemented more recently, some improvements have been suggested by researchers. For tracing fault location, various estimate methods in HVDC transmission lines have been presented in recent years. A comparison of the proposed approach for fault location with previously suggested methods is presented in Table 8.9. Comparisons are made based on different parameters such as measurements used, sampling frequency, methods used, fault resistance, reach-setting accuracy, and the polarity of the lines. Some of the methods use traveling waves for fault location estimation. The drawback of the traveling wave method is that it involves a high sampling rate and installation of extra equipment at substations. Some of the methods used measurements from both sides of line, which will require a communication link. Most of the methods use high sampling frequency, which will be difficult to realize. Some of the approaches are incapable of locating faults with high fault resistance. Some of the methods overlooked the reach setting of the proposed method,

Table 8.9 Comparison of the proposed approach with previous methods.

Methods	Techniquesused	Measurement required	Sampling frequency (kHz)	Fault resistance	Reach setting (%)	Polarity of lines
He et al. [9]	Traveling-wave	Single end	100	Up to 50 Ω	90	Bipolar
Song et al. [2]	PRONY algorithm	Single end	50	100 Ω	90	Monopolar
Yuansheng et al. [5]	Traveling-wave theory with the Bergeron method	Both end	1000	500 Ω	–	Bipolar
Zhang et al. [8]	Electromagnetic time-reversal method	Both end	50	800 Ω	–	Monopolar
Li et al. [8]	Phase frequency	Single end	4	50 Ω	–	Bipolar
Johnson et al. [10]	SVM	Single end	1	Not mentioned	–	Bipolar
Livani et al. [12]	DWT and SVM	Single end	>100	70 Ω	–	Monopolar
Proposed scheme	NNS	Single end	1	100 Ω	99	Bipolar

which is also important. Some of the methods are designed for monopolar transmission lines, which may or may not work for bipolar transmission lines. As mentioned in Table 8.9, it can be seen that the proposed method uses only one end measurement, has a low sampling frequency, has more reach setting, and locates faults with high resistance. Hence proposed NNS-based method is more suitable for tracing the location of the faults in HVDC lines.

8.4.11 Advantages of the Scheme

The many advantages of the proposed scheme can be highlighted as follows:

1. %Error in locating the faults for the test cases lies within 1%.
2. It has a reach-setting of 99.9% of the line length (0–1099 km).
3. The sampling frequency used is very low (1 kHz).
4. As only one-end measurement has been used, coordination between the relay does not incur anytime delay.
5. Communication requirement is evaded as data from one end are required.

8.5 Conclusion

An NNS-based fault location method has been suggested in this work for bipolar LCC–HVDC lines. The fault location method based on NNS used rectifier end current signals. The NNS-based method for tracing fault location has been verified by the varying nearest neighbor, distance matrices, fault location, type, resistance, etc. The advantage of NNS-based method is that the sampling rate is less unlike methods using traveling-wave which has high sampling rate. Another advantage of the NNS-based method is it uses only one end data; hence, communication link is not needed. Yet another advantage of the proposed method is it has reach setting up to 99% of the line length. Also, for test cases, the error in tracing fault location lies within 1%.

Acknowledgment

This work has been funded by SERB-DST, India, under an early career research award (ECR scheme) project titled "Intelligent Fault Distance Estimation Scheme for High Voltage AC and High Voltage DC Transmission Lines: A Comparative Study of Various Artificial Intelligent Techniques to Explore a Suitable Scheme" with sanction order no. ECR/2017/000013.

References

1 Suonan, J., Gao, S., Song, G. et al. (2010). A novel fault-location method for HVDC transmission lines. *IEEE Transactions on Power Delivery* 25: 1203–1210.
2 Song, G., Chu, X., Cai, X. et al. (2014). A fault-location method for VSC-HVDC transmission lines based on natural frequency of current. *International Journal of Electrical Power and Energy Systems* 63: 347–352.
3 Li, C. and He, P. (2018). Fault-location method for HVDC transmission lines based on phase frequency characteristics. *IET Generation, Transmission and Distribution* 12: 912–916.
4 Farshad, M. and Sadeh, J. (2013). A novel fault-location method for HVDC transmission lines based on similarity measure of voltage signals. *IEEE Transactions on Power Delivery* 28: 2483–2490.

5 Liang, Y., Wang, G., and Li, H. (2015). Time-domain fault-location method on HVDC transmission lines under unsynchronized two-end measurement and uncertain line parameters. *IEEE Transactions on Power Delivery* 30: 1031–1038.

6 Song, G., Suonan, J., and Xu, Q. (2005). Parallel transmission lines fault location algorithm based on differential component net. *IEEE Transactions on Power Delivery* 20: 2396–2406.

7 Song, G., Suonan, J., and Xu, Q. (2006). Time-domain fault location for parallel transmission lines using unsynchronized currents. *International Journal of Electrical Power and Energy Systems* 28: 253–260.

8 Li, Y., Zhang, S., and Li, H. (2012). A fault location method based on genetic algorithm for high-voltage direct current transmission line. *European Transactions on Electrical Power* 22: 866–878.

9 He, Z., Liao, K., Li, X. et al. (2014). Natural frequency-based line fault location in HVDC lines. *IEEE Transactions on Power Delivery* 29: 851–859.

10 Johnson, J. and Yadav, A. (2017). Complete protection scheme for fault detection, classification and location estimation in HVDC transmission lines using support vector machines. *IET Science, Measurement and Technology* 11: 279–287.

11 Zhang, X., Tai, N., Wang, Y., and Liu, J. (2017). EMTR-based fault location for DC line in VSC-MTDC system using high-frequency currents. *IET Generation, Transmission and Distribution* 11 (10): 2499–2507.

12 Livani, H. and Evrenosoglu, C.Y. (2014). A single-ended fault location method for segmented HVDC transmission line. *Electric Power Systems Research* 107: 190–198.

13 Sutton, O. (2012). Introduction to *k* nearest neighbour classification and condensed nearest neighbour data reduction – the *k* nearest neighbours algorithm. University Lectures, University of Leicester, pp. 1–10.

14 Chatzigeorgakidis, G., Karagiorgou, S., and Athanasiou, S. (2018). FML-kNN: scalable machine learning on big data using k-nearest neighbor joins. *Journal of Big Data* 5: 1–27.

15 Bailey, T. and Jain, A. (1978). A note on distance – weighted *k*-nearest neighbor rules. *IEEE Transaction on Systems, Man, and Cybernetics* 8: 311–313.

16 Guo, G., Wang, H., Bell, D. et al. (2003). KNN model-based approach in classification. *Lecture Notes in Computer Science* 2888: 986–996.

17 Dubey, H. (2013). Efficient and accurate kNN based classification and regression. A Master Thesis Presented to the Center for Data Engineering. International Institute of Information Technology, 32.

18 (2016). MATLAB user's guide, R2016a Documentation. MathWorks Inc, Natick, MA.

9

Comparative Analysis of Machine Learning Approaches in Enhancing Power System Stability

Md. I. H. Pathan[1], Mohammad S. Shahriar[2], Mohammad M. Rahman[3], Md. Sanwar Hossain[4], Nadia Awatif[5], and Md. Shafiullah[6]

[1] Department of Electrical and Electronic Engineering, Hajee Mohammad Danesh Science and Technology University, Dinajpur, Bangladesh
[2] Department of Electrical Engineering, University of Hafr Al-Batin, Hafr Al Batin, Saudi Arabia
[3] Information and Computing Technology Division, Hamad Bin Khalifa University, College of Science and Engineering, Doha, Qatar
[4] Department of Electrical and Electronic Engineering, Bangladesh University of Business and Technology, Dhaka, Bangladesh
[5] Department of Electrical, Electronic and Communication Engineering, Military Institute of Science and Technology, Dhaka, Bangladesh
[6] King Fahd University of Petroleum & Minerals, Interdisciplinary Research Center for Renewable Energy and Power Systems, Dhahran, Saudi Arabia

9.1 Introduction

The modern-day power systems are usually operated at the highest capacity to meet the higher energy demand. This energy demand is increasing due to the development of different kinds of technological infrastructures and their usage by the growing population. Therefore, operating the power systems economically functions at maximum capacity to make them profitable. However, consistent operation of the power systems at the highest capacity may cause the violation of operational constraints. Consequently, the systems become oscillatory due to adding constraint's violation disturbances, which eventually cause the system's instability. Moreover, the recent trends toward renewable energy (RE) to integrate into the power systems are declining the dependency on fossil fuels and making the overall systems environmentally friendly, increasing the intermittency of operational power conditions simultaneously. This intermittent nature of renewable energy sources, which is ultimately reflected in the overall power supply system, introduces the system's low-frequency oscillations (LFOs). The power networks connected through weak tie lines are undesirably affected by LFOs, which is a type of electronic frequency causing a rhythmic pulse in the system and the range of that frequency is between 0.1 and 2.5 Hz [1]. Therefore, the LFOs may result in the complete blackout of the grid due to the lag of appropriate damping action in time, as this situation brings the system toward dynamic instability [2]. Intending to protect the system from such cases, proper damping action is necessary at the right time.

Although an ample number of researchers have tried to handle such situations in the last few decades, still opportunities are there to contribute to this area. Generator excitation control based automatic voltage regulator (AVR) is one of the possible techniques to overcome the LFO issues proposed by the researchers. In contrast, the LFOs are increased due to rotors' inadequate damping torque, which are declined in high-gain AVR-based synchronous machines [3]. In this situation, the power system stabilizer (PSS) can be employed to improve the stability profile of the system by suppressing the LFOs [4]. The crucial factor in achieving the appropriate action from PSS in time is the proper tuning of key parameters of PSS. This critical task is addressed by the researchers in [5] using different approaches. However, the usages of flexible alternating current transmission system (FACTS) devices in power networks are remarkably increased due to the development of power electronics devices. The steady-state performance of power systems is investigated through integrating FACTS devices in the systems, and noticeable improvement is observed [6–8]. It is reported in [9] that the enhancement of dynamic parameters, such as phase angle, voltage, current, and impedance, of power networks can be achieved by incorporating the FACTS devices in the systems. Generally, overall performance is enhanced by incorporating the FACTS devices in the systems. In references [10–13], the stability of electric power networks is evaluated by embedding the FACTS devices into the systems and shows a significant enhancement in terms of optimal power supply, maintaining desired voltage levels, optimally dispatching reactive power to the system, and minimizing inter-area oscillations, respectively. In this case, artificial intelligence (AI) techniques were employed to analyze these critical power system parameters. There are mainly three types of connections available for FACTS devices: series, shunt, and combination of series and shunt. Hence, the unified power flow controller (UPFC) is considered as one of the most intelligible and multifunctional FACTS devices that can combined function the series and shunt operations [14]. The FACTS devices can be used in tuning the power system's parameters, such as bus voltages, phase angles of bus voltages, line reactance, etc. Therefore, the steady-state power flow profile among the transmission lines can be enhanced with the aid of FACTS devices. In addition, the minimization of power loss, transient stability, voltage regulation, and LFOs can be improved through the implementation of FACTS devices [15–17]. The coordination between PSS and thyristor-controlled series capacitor (TCSC) is constantly synchronized using the dolphin echolocation optimization (DEO) method in case of improving the stability profile of the power network [18]. The UPFC-PSS-based LFO damping mainly depends on the strategical approach of maintaining the control parameters. In case of mitigating the LFOs from the system, the appropriate synchronization between PSS and UPFC must be confirmed to ensure the stability of the system [19]. To enhance the stability of power networks, the investigation was conducted with the aid of computational intelligence techniques, including water cycle algorithm (WCA) [19], backtracking search algorithm (BSA) [20], genetic algorithm (GA) [21], particle swarm optimization (PSO) [22], and differential evolution (DE) [23] for properly optimizing of PSS parameters. However, most methods evaluated the system's stability through offline tuning of PSS parameters following a particular loading condition (LC).

On the other hand, the loading conditions of power networks are continuously varied due to the volatile nature of utility customers. Consequently, the investigated approaches for tuning the PSS parameters are not feasible for real-time implementation in power systems. On the contrary, the performance of real-time/online tuning of PSS parameters was also evaluated employing some AI approaches, including extreme learning machine (ELM), genetic programming (GP), artificial neural network (ANN), and support vector machine (SVM) [24–27], respectively. Furthermore, the neuro-fuzzy technique, a data-driven hybrid approach of neural networks (NNs) and fuzzy systems, was developed to mitigate LFOs and enhance the overall system's performance by appropriately tuning PSS parameters instead of using conventional PSS [28]. In the case of controlling the real-time dynamics in power networks, the feasibility of application was conducted using the neuro-fuzzy intelligent system in [29].

This chapter evaluated a comparative analysis of the applicability, feasibility, and robustness of AI techniques to enhance the power systems' stability by suppressing LFOs through real-time tuning of PSS key parameters for arbitrarily chosen loading conditions. The considered AI techniques for this analysis were group method of data handling (GMDH), neurogenetic (NG), ELM, and multi-gene genetic programming (MGGP). After that, the AI-based results were compared with the conventionally tuned PSS-based results that demonstrated the superiority of AI in enhancing the power system stability. In this case, two power system models were selected to show the efficacy of AI techniques. The first power network consists of a single machine infinite bus (SMIB) system with integrated PSS and the second is the SMIB test system with UPFC coordinated PSS. Two performance indicators, namely the minimum damping ratio (MDR) and eigenvalue, were considered to investigate the efficacy in case of damping out of LFOs through online adjustment of PSS key parameters—time constant (T_1) and gain (K). Then the impact of tuned PSS parameters was measured and depicted in time domain frame to stabilize the oscillations.

The rest of the chapter is structured as follows: Section 9.2 interprets the system modeling of the selected two electric power networks. Section 9.3 explains the AI techniques illustrated on power system stability improvement through the PSS parameter's online adjustment. Section 9.4 illustrates the development procedures of the proposed

AI techniques alongside the data generation and processing procedures. Section 9.5 demonstrates the comprehensive analysis of the results obtained from the employed machine learning (ML) models. Lastly, Section 9.6 concludes the article with findings and a few research directions.

9.2 Power System Models

This chapter considered two power system networks to investigate the feasibility and applicability of the AI techniques in improving the stability of power systems by mitigating the LFOs through appropriately tuning key parameters of the PSS. First, this section briefly describes the constructional overview of the selected networks.

9.2.1 PSS Integrated Single Machine Infinite Bus Power Network

A PSS-integrated SMIB power network is sketched in Figure 9.1a. The PSS is connected to the terminal of the synchronous generator, whereas the infinite bus is connected to the other terminal through a transmission line of specific reactance. A single-stage lead-lag PSS is associated with the first network to evaluate the performance and applicability of AI techniques in tuning the PSS parameters for damping out of LFOs. The structural sketch of the stabilizer is depicted in Figure 9.1b. To manage the input and output signals of the stabilizer, a washout filter and control signal limiter are integrated with the main part of the stabilizer, respectively.

The eigenvalues of the state matrix corresponding to a power network represent the modes of that power network in case of operating within a stable area or not. Therefore, a linearized model following a specific operating point of the first electric network is considered to form the state matrix and state equations of the whole network. Hence, the six-order state-space model of the overall first electric network aggregates the fourth-order SMIB system and second-order PSS model as reported in [30–33]. The overall state matrix (A_{c1}) of first network in state space can be described by Eq. (9.1) as:

$$X_1 = A_{c1}\Delta X_1 \tag{9.1}$$

In representing the operational modes of power networks, eigenvalues of the state matrix of the corresponding network are evaluated. Negative real parts of all eigenvalues can ensure the stability of the power network. In contrast, any eigenvalues with a positive real part or zero real value will make the system unstable. Therefore, it can be said that the placement of all eigenvalues of the state matrix within the negative side of the complex plane will confirm the system's stability. Again, any minor disturbance in the network is reflected in the state of the network, and hence the change of eigenvalues of the state matrix of the corresponding network. Owing to changing the values of eigenvalues, the network starts to oscillate, which is known as LFOs. This oscillation may gradually shift the eigenvalues to the positive side of the imaginary axis $(j\omega)$ if there is no proper action is

Figure 9.1 Structure of first electric networks – (a) PSS integrated SMIB test system and (b) structure of single-stage lead-lag PSS.

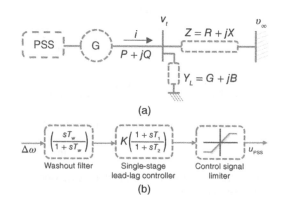

taken in time through adjusting the key PSS parameters, such as time constant (T_1) and gain constant (K) values. Therefore, any minor disturbance can ultimately cause the system's instability by introducing LFOs. Hence, the PSS plays a key role by tuning its parameters for mitigating the LFOs within the quickest possible time and placing the eigenvalues appropriately within the negative area of the complex plane. Therefore, the superior operation of power system largely depends on the appropriate tuning of crucial PSS parameters (T_1 and K) subjected to any undesired condition.

9.2.2 PSS-UPFC Integrated Single Machine Infinite Bus Power Network

Regarding investigating the impacts of applying the AI techniques in enhancing the stability profile of power networks, a double-stage lead-lag PSS embedded to the SMIB network equipped with a UPFC is considered as the second electric network. The general structural view of the UPFC-PSS coordinated SMIB test system is depicted in Figure 9.2a,b. In addition, the washout filter and control signal limiter were also used for the same purpose, like the first network.

Unlike the connecting approach of UPFC with SMIB test network, the procedure of UPFC with the network, the integrating procedure of PSS to the terminal of synchronous machine alongside the infinite bus connection for the second network is exactly same as the first network. However, boosting transformer (BT) and excitation transformer (ET) were employed to couple the UPFC with the SMIB test network. Hence, the terminals of a DC-link capacitor were coupled to the ET and BT ends through two voltage source converters (VSCs), which are the VSC-1 and VSC-2, respectively. The UPFC is usually operated based on four controlling parameters: the phase angles and amplitude modulation ratios of both transformers. In accordance with the specifications mentioned in Figure 9.2a,

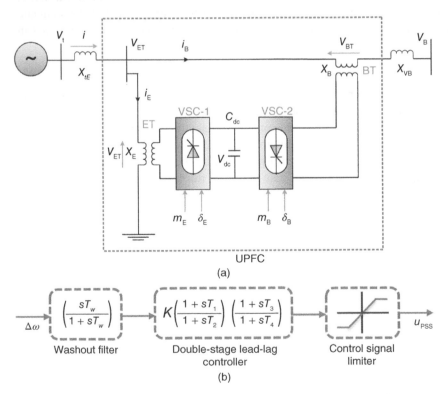

Figure 9.2 Structure of second electric networks – (a) UPFC-PSS integrated SMIB test system and (b) structure of double stage lead-lag PSS.

δ_E and m_E are the phase angle and amplitude modulation ratio for ET, and δ_B and m_B are the phase angle and amplitude ratio for booster transformer, respectively. These phase angles and amplitude ratios mainly control the operation of excitation and booster transformers. A thorough explanation of these parameter settings and the selected system can be found in [26].

As alluded earlier, the eigenvalues of the state matrix corresponding to a network indicate the operational modes within a stable area or not for that network. Therefore, the linearized model of the SMIB network following a particular operating point and the linearized PSS and UPFC models are considered to construct the state matrix and state equation of the complete second network. In this case of the second network, the complete model aggregating the fourth-order SMIB system, second-order PSS, and third-order UPFC comprises a nine-order state matrix in the state space. The overall state matrix (A_{c2}) of the second electric network in state space can be formulated by Eq. (9.2) as reported in [33]:

$$X_2 = A_{c2} \Delta X_2 \tag{9.2}$$

Finally, the complete second electric network consists of nine eigenvalues for the state matrix (A_{c2}) of the state space model after including the states of the UPFC. In representing the operational modes of power networks, eigenvalues of the state matrix of the corresponding networks are evaluated. In case of having negative real parts of all eigenvalues can ensure the stability of the power network. In contrast, any eigenvalues with a positive real part or zero real value will bring the system into unstable mode. Therefore, it can be said that the placement of all eigenvalues of the state matrix within the negative side of the complex plane will confirm the system's stability. Again, any minor disturbance in the network is reflected in the state of the network, and hence the change of eigenvalues of state matrix of the corresponding network. Owing to changing the values of eigenvalues, the network starts to oscillate, which is known as LFOs. This oscillation may gradually shift the eigenvalues to the positive side of the imaginary axis ($j\omega$) if there is no proper action is taken in time through adjusting the key PSS parameters, such as time constant (T_1) and gain constant (K) values. Therefore, any minor disturbance can ultimately cause the system's instability because of introducing LFOs. Hence, the PSS plays a key role by tuning its parameters for mitigating the LFOs within the quickest possible time and placing the eigenvalues appropriately within the negative area of the complex plane. Therefore, the superior operation of the power system largely depends on the appropriate tuning of crucial PSS parameters (T_1 and K) subjected to any undesired condition.

9.3 Methods

In this chapter, four ML algorithms were considered to investigate the feasibility and applicability of AI in improving power systems' stability. These methods are illustrated shortly from Sections 9.3.1 to 9.3.4.

9.3.1 Group Method Data Handling Model

The multiparametric complex, nonlinear, and unstructured systems are modeled using an inductive mathematical regression algorithm's GMDH. The Russian Mathematician and Cyberneticist Ivakhnenko introduced the naturally inspired GMDH as referred to in [34]. In selecting the number of hidden layers and neurons to adjust the actual target, the technique usually follows the automatically structural and parametric optimization. In the case of mapping the actual nonlinear functional response, the GMDH algorithm usually considers the quadratic polynomials of different possible combinations of two parameters from all the input parameters as reported in [35, 36]. In each layer of the GMDH network, some of the less efficient neurons are excluded concerning better neurons by evaluating the polynomials for minimizing the mapping errors. In this way, the mapping error is minimized in each step regarding input features and target response and reconfigured the network. The least-square regression approach is considered in the base structure of the GMDH model [34–36]. The estimation procedure of nonlinear functional output from a set of the multiparametric datasets can be explained as follows: for a provided input parameter set $X = (x_1, x_2, x_3, \ldots x_n)$, in lieu of using actual polynomial "g," the second order polynomial "$\hat{g}$" is hypothesized in such a way that the estimated target response "$\hat{y}$" becomes as nearest as possible to the original target "y." The input-target relation can be expressed for a dataset of multi-input and single output with "i" observations as:

$$y_i = g(x_{i1}, x_{i2}, x_{i3}, \ldots x_{in}), \quad (i = 1, 2, 3, \ldots M) \tag{9.3}$$

where, $i \ and \ n$ are the observation and input numbers, respectively. The proposed GMDH model is trained following multiparametric input data with second-order transfer functions to hypothesize the target response ($\hat{y}$) as:

$$\hat{y}_i = \hat{g}(x_{i1}, x_{i2}, x_{i3}, \ldots x_{in}), \quad (i = 1, 2, 3, \ldots M) \tag{9.4}$$

To develop the GMDH algorithm, Eq. (9.5) is followed during the training period to minimize the error between the actual and estimated targets.

$$\sum_{i=1}^{M} [\hat{g}(x_{i1}, x_{i2}, x_{i3}, \ldots x_{in}) - y_i]^2 \to \min \tag{9.5}$$

Volterra functional series is used to construct the second-order polynomials from a multi-parametric input dataset during the training period of the GMDH algorithm, which is expressed by Eq. (9.6) as:

$$y = b_0 + \sum_{i=1}^{n} b_i x_i + \sum_{i=1}^{n} \sum_{j=1}^{n} b_{ij} x_i x_j + \sum_{i=1}^{n} \sum_{j=1}^{n} \sum_{k=1}^{n} b_{ijk} x_i x_j x_k + \ldots \tag{9.6}$$

where $b_0, b_i,$ and b_{ij} are the polynomial coefficients. Hence, the second-order polynomial can be approximated by Eq. (9.7) to map the layers and neurons of the GMDH algorithm.

$$\hat{y} = \hat{g}(i, j) = b_0 + b_1 x_i + b_2 x_j + b_3 x_i x_j + b_4 x_i^2 + b_5 x_j^2 \tag{9.7}$$

Different statistical indices are considered regarding adjusting the weights of the second-order polynomials and corresponding neurons of the GMDH network. Hence, better adjustment of weights assists the estimated response to be closer as much as possible toward the actual targeted response.

9.3.2 Extreme Learning Machine Model

ELM is a ML model capable of efficiently handling different model-based analysis using the single-layer feed-forward neural network (SLFN) concept. The over-fitting issues and poor normal neural network training rate can be overcome using the ELM algorithm as reported in [37]. Gradient-based algorithm with backpropagation is not followed for functioning the ELM algorithm. For evaluating the weights of the ELM network, Moore–Penrose generalized inverse procedure is employed. In general, the ELM algorithm is first trained and then tested with two data sets. In the testing phase, the algorithm is used to conjecture the target to be proved the robustness of the trained algorithm. Some influencing variables and corresponding outcomes are considered to enter the ELM algorithm during the training stage, and the learning processes are completed through several iterations. After the training, the trained ELM model can predict according to the given inputs and memory stored following the training. The architectural view of the ELM model is sketched in Figure 9.3.

9.3.3 Neurogenetic Model

An ANN, which can perform different tasks from examples, categorize things, and develop relationships, is assembled with a GA to develop a hybrid version of a NG AI algorithm [33]. The flow diagram of the NG algorithm

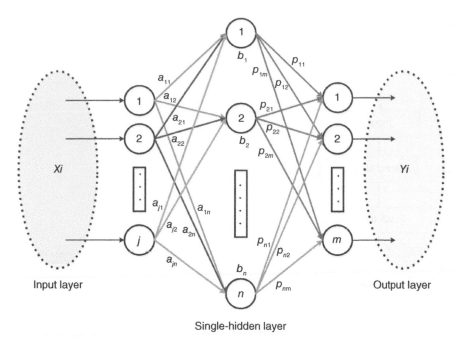

Figure 9.3 Single hidden-layer feedforward architecture for extreme learning machine algorithm.

is depicted in Figure 9.4. ANN is suitable to handle highly nonlinear, nondifferential, and complex systems. For example, to solve the problems of power networks, which is highly nonlinear, the multilayer perceptron feedforward neural network (NN) is one of the most efficient algorithms among many NN algorithms.

Usually, an ANN network is constructed using three layers (input, hidden, and output) connected through some bias and weight coefficients. The input layer multiplies necessary weights with the input vector and forwards them to the hidden layer, whereas the hidden layer processes the data using suitable activation functions. After that, the hidden layer forwards the processed information to the output layer by adding and multiplying necessary bias and weight coefficients, respectively. Finally, the output layer processed the received information by adding biases and multiplying weights required to execute them as target values at the output terminal. This procedure continues until the desired outputs are achieved by changing the values of bias and weight coefficients. Among many training algorithms, Levenberg–Marquardt algorithm is considered in this case for training the NN algorithm. However, the neural network is a local optimization approach that can trap in the local minima position. Therefore, a GA is deployed in the neural network structure to evaluate the biases and weights of the network and to overcome the trapping of local optima of the network during the training phase. In the NG model, the GA is employed to execute the essential optimization tasks to optimize the biases and weights of the layers. In this way, the GA can improve the performance of the original ANN model. During the optimization procedure, GA performs the crossover and mutation operations to tune the biases and weights to enhance the overall NG model's performance by overcoming local optimum trappings. The details of GA and ANN can be found in [38–41].

9.3.4 Multigene Genetic Programming Model

MGGP, an approach based on the Darwinian natural selection principle, was introduced by Koza as reported in [42]. MGGP model can construct estimation equations without any previous form of existing equations compared to other statistical ML algorithms. For a set of input parameters a_i, $i = 1, 2, 3, \ldots, n$, a mathematical equation is formed defining the relation between target Y and input variables concerning each structure of the tree that is

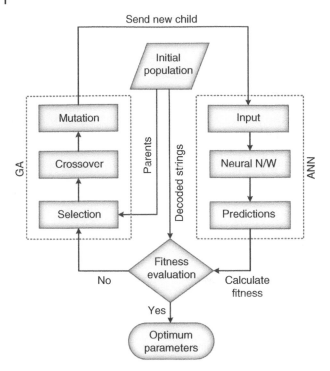

Figure 9.4   The functional flow diagram for neurogenetic system.

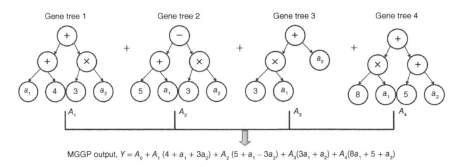

Figure 9.5   Structure of trees for MGGP algorithm to predict target Y with respect to input variables a_1 and a_2.

represented for an individual of population in the case of implementing GP algorithm [43]. Indeed, two categories of GP algorithms are available in the literature [27, 44, 45], and where the single-gene GP category is the base of the multi-gene model. Therefore, the GP-based MGGP network defines every one of the total populations set as a structure of the tree, each of which is known as a gene of the MGGP algorithm, to forecast the target Y by receiving the input variables a_i, as depicted in Figure 9.5.

The mathematical structure for an MGGP is approximated through the selection and crossover of the parent individuals, where the selection process is like the standard single-gene genetic programming (SGGP). On the other hand, the MGGP algorithm follows two types of crossovers [27]: the single point low level and double points high-level crossovers. Through these crossover processes, new individuals are generated from the parent individuals. Unlike the crossover process, the mutation, replication, up-gradation of the best solution for the next iteration, and convergence criteria checking are the same for SGGP, and MGGP approaches. In the process of MGGP algorithm development, $+, -, \cos, \sin, \log, \exp, \ldots$, etc. are considered as a functional set, and a_i is the terminal set, which is considered as real power (P_e), reactive power (Q_e), and terminal voltage (V_t) for the MGGP model development in this chapter.

9.4 Data Preparation and Model Development

The data production and processing for developing the ML models and the corresponding model's development procedures are elucidated in the Sections 9.4.1 and 9.4.2.

9.4.1 Data Production and Processing

To compare the ML models, namely the GMDH, NG, MGGP, and ELM, performance in suppressing the LFOs, 1000 different loading conditions were generated, which were the combinations of real power (P_e), reactive power (Q_e), and terminal voltage (V_t) of the synchronous machine within the operating limits for the selected power system networks. The ranges of operating conditions of the networks are given in Table 9.1 [46].

After that, the two sets of 1000 operating circumstances were used to obtain the corresponding two sets of optimized values of crucial PSS parameters $(T_1$ and $K)$ for respective networks with the aid of offline simulation. In this regard, a maximization problem was formulated for the first network to optimize the MDR [20]. In contrast, two decision-making parameters, e.g. damping ratio and damping factor, were aggregated to formulate multiobjective minimization for the second power system network [47]. Detailed parameter optimization techniques for the first and second networks can be found in [20, 47], respectively. It is worth mentioning that all the PSS parameters are interrelated, and any change impacts all the parameters. Therefore, all other PSS parameters of respective networks except T_1 and K were kept constant throughout the optimization process.

9.4.2 Machine Learning Model Development

To develop the ML models, the operating situations $(P_e, Q_e,$ and $V_t)$ of the power system networks were considered as the inputs and the optimized PSS parameters $(T_1$ and $K)$ obtained offline were considered as the outputs. Besides, 70% of the available data were employed for the training, and the remaining were used for testing pur poses. In the case of GMDH, the systematic trial and error processes were followed to adjust the model specifications for the training and the testing datasets. It is worth noting that minimizing the mean squared error (MSE) between the predicted and offline-optimized datasets was set as the objective function. Likewise, the other ML models, ELM, NG, and MGGP, were developed.

Table 9.1 Operating ranges in per unit (pu) for both electric networks.

Variables	PSS-SMIB		UPFC-PSS-SMIB	
	Max	Min	Max	Min
Real power (P_e)	1.15	0.40	1.30	0.60
Reactive power (Q_e)	0.40	0.01	0.40	−0.16
Terminal voltage (V_t)	1.10	0.90	1.06	0.98

9.5 Results and Discussions

In the Sections 9.5.1 and 9.5.2, the developed GMDH machine-learning algorithm is investigated with the system's eigenvalues, MDR, and time-domain results in real-time conjecturing of PSS crucial parameters to mitigate the LFOs for improving the network stability. Besides, the developed model is also compared with the other three ML models, namely the NG, ELM, and MGGP, and the conventional approach for online PSS parameter prediction. Finally, sustainability to handle the external disturbances is also assessed through time-domain simulated results after initiating extra mechanical torque to the synchronous machine deploying the GMDH model in the electric networks. Two electric power networks are considered to investigate the applicability of the proposed approach in enhancing the online stability of power systems.

9.5.1 Eigenvalues and Minimum Damping Ratio Comparison

In regard to analyzing eigenvalues of a particular system, it can be alluded that any eigenvalue in the right half plane of the complex plane will persuade the system into instability. In this case, a higher number of eigenvalues in positive real parts reflect the tendency to become unstable. Moreover, more positiveness of eigenvalues will have a greater impact on the instability of the system. Therefore, it can be said that the placement of eigenvalues within the negative area of the complex plane will ensure higher stability and vice versa. Intending to investigate the applicability of AI techniques in damping out of LFOs through the system's eigenvalue analysis concerning different loading circumstances, this chapter considered three arbitrary loading conditions (combinations of P_e, Q_e, and V_t of the synchronous machine) for each electric network. The considered loading conditions for both networks are accumulated in Table 9.2. The loading conditions were produced so that it does not transgress the machine's voltage constraint and thermal limits. Besides, the MDR for the selected loading conditions of both networks was also evaluated to ensure the feasibility of AI-based real-time conjecture of PSS parameters in improving the overall system stability. Furthermore, the conventional approach-based eigenvalues and MDR values were also evaluated for both electric networks following the same loading conditions to compare and explore the suitability of using the AI models on the addressed power systems.

The eigenvalues corresponding to the randomly selected three loading circumstances, tabulated in Table 9.2, of the first network are summarized in Tables 9.3–9.5 after adopting the AI methodologies (GMDH, NG, ELM, and MGGP models). With a view of investigating the applicability of ML models on power systems, the ML model-based eigenvalues were compared with the eigenvalues of conventional approach in the same tables. Tables 9.6–9.8 arrange the eigenvalues for light, nominal, and heavy loading conditions (HLCs) of second electric network after employing the four different ML models and conventional approach in online prediction of PSS parameters (T_1 and K).

Table 9.2 Loading conditions (LC) of two electric networks.

	Items	P_e(pu)	Q_e(pu)	V_t(pu)
First electric network	Loading condition (LC) #1	0.60	0.01	0.98
	Loading condition (LC) #2	0.95	0.28	1.03
	Loading condition (LC) #3	0.43	0.21	1.09
Second electric network	Light loading condition (LLC)	0.60	0.01	0.98
	Nominal loading condition (NLC)	0.98	−0.16	1.00
	Heavy loading condition (HLC)	1.30	0.40	1.06

Table 9.3 Eigenvalues of first electric network for LC #1.

Parameters	GMDH	NG	ELM	MGGP	Conventional
Eigenvalues	−0.349	−0.349	−0.349	−0.349	−0.337
	−17.547	−17.499	−17.492	−17.490	−17.516
	$-3.387 \pm 5.641i$	$-3.338 \pm 5.374i$	$-3.242 \pm 5.313i$	$-3.305 \pm 5.293i$	$-5.452 \pm 6.835i$
	$-2.929 \pm 4.737i$	$-3.002 \pm 4.963i$	$-3.101 \pm 5.016i$	$-3.039 \pm 5.036i$	$-0.886 \pm 4.179i$
Minimum damping ratio	0.515	0.518	0.521	0.517	0.207

Table 9.4 Eigenvalues of first electric network for LC #2.

Parameters	GMDH	NG	ELM	MGGP	Conventional
Eigenvalues	−0.358	−0.358	−0.358	−0.358	−0.338
	−17.676	−17.697	−17.703	−17.706	−18.379
	$-3.036 \pm 5.074i$	$-3.267 \pm 5.125i$	$-3.241 \pm 5.164i$	$-3.273 \pm 5.176i$	$-5.284 \pm 7.414i$
	$-3.211 \pm 4.551i$	$-2.970 \pm 4.530i$	$-2.994 \pm 4.493i$	$-2.960 \pm 4.486i$	$-0.621 \pm 3.595i$
Minimum damping	0.513	0.538	0.532	0.534	0.170

Table 9.5 Eigenvalues of first electric network for LC #3.

Parameters	GMDH	NG	ELM	MGGP	Conventional
Eigenvalues	−0.349	−0.349	−0.349	−0.349	−0.336
	−17.353	−17.226	−17.291	−17.288	−16.398
	-3.283 ± 5.854	$-2.991 \pm 5.344i$	$-3.391 \pm 5.570i$	$-3.510 \pm 5.554i$	$-6.226 \pm 6.181i$
	$-3.130 \pm 4.530i$	$-3.486 \pm 4.956i$	$-3.053 \pm 4.790i$	$-2.936 \pm 4.816i$	$-0.672 \pm 4.371i$
Minimum damping	0.489	0.488	0.520	0.521	0.152

Table 9.6 Eigenvalues of second electric network for LLC.

Parameters	GMDH	NG	ELM	MGGP	Conventional
Eigenvalues	−0.390	−0.391	−0.391	−0.391	−0.400
	−1.367	−1.434	−1.386	−1.400	−6.593
	−83.399	−83.492	−83.490	−83.490	−87.562
	−127.447	−126.877	−126.891	−126.889	−110.031
	−977.231	−977.806	−977.792	−977.793	−993.512
	$-1.367 \pm 0.422i$	$-1.305 \pm 0.067i$	$-1.386 \pm 0.093i$	$-1.400 \pm 0.118i$	$-0.718 \pm 0.295i$
	$-4.223 \pm 2.942i$	$-4.077 \pm 2.849i$	$-4.081 \pm 2.851i$	$-4.081 \pm 2.851i$	$-0.615 \pm 3.968i$
Minimum damping	0.821	0.820	0.820	0.820	0.153

Table 9.7 Eigenvalues of second electric network for NLC.

Parameters	GMDH	NG	ELM	MGGP	Conventional
Eigenvalues	−0.199	−0.199	−0.199	−0.199	−0.206
	−1.537	−1.269	−1.254	−1.224	−6.695
	−80.778	−80.7697	−80.766	−80.759	−86.497
	−125.221	−125.248	−125.260	−125.282	−110.705
	−982.247	−982.221	−982.210	−982.190	−994.471
	$−1.301 \pm 0.110i$	$−1.429 \pm 0.127i$	$−1.434 \pm 0.154i$	$−1.444 \pm 0.199i$	$−0.676 \pm 0.320i$
	$−4.090 \pm 3.688i$	$−4.099 \pm 3.692i$	$−4.103 \pm 3.694i$	$−4.111 \pm 3.698i$	$−0.419 \pm 4.610i$
Minimum damping	0.743	0.743	0.743	0.744	0.090

Table 9.8 Eigenvalues of second electric network for HLC.

Parameters	GMDH	NG	ELM	MGGP	Conventional
Eigenvalues	−0.142	−0.143	−0.143	−0.143	−0.147
	−1.090	−1.296	−1.888	−2.092	−7.269
	−82.216	−82.454	−82.523	−82.579	−87.048
	−139.392	−137.287	−136.717	−136.241	−112.996
	−964.966	−967.168	−967.762	−968.259	−991.096
	$−1.234 \pm 0.622i$	$−1.362 \pm 0.157i$	$−1.150 \pm 0.190i$	$−1.122 \pm 0.161i$	$−0.677 \pm 0.274i$
	$−5.244 \pm 3.453i$	$−4.846 \pm 3.227i$	$−4.71 \pm 3.160i$	$−4.602 \pm 3.100i$	$−0.427 \pm 4.800i$
Minimum damping ratio	0.835	0.832	0.831	0.829	0.089

It is condign to allude again that all other correlated PSS parameters were set as fixed values during the optimization phase for corresponding ML models and electric networks. Furthermore, MDR values were also measured for the randomly selected loading conditions (Table 9.2) of first and second electric networks for four ML algorithms alongside the conventional method and presented in Figure 9.6. It is worth mentioning that the higher MDR values stabilize the power networks through mitigation of LFOs subjected to any disturbance. The real-time conjectured PSS key parameters using four ML models along with the conventional approach for the arbitrarily selected loading conditions of both networks are obtained and summarized in Table 9.9.

The eigenvalues accumulated in Tables 9.3–9.8 for the selected operating circumstances, as presented in Table 9.3, of first and second electric power networks, respectively, lie in the negative area of the complex plane. As stated before, the eigenvalues with negative real parts firmly ensure the system's stable operation. However, eigenvalues for all the considered loading attempts of both networks in the case of adopting all ML models locate at a far distance from the imaginary axis with better optimal positions than the conventional approach. Again, in the case of MDR value investigation, the MDR values for each of the considered loading circumstances are almost identical when all the considered ML models are employed in online prediction of key PSS parameters for both networks, but much higher compared to conventionally evaluated MDR value. This excellence in MDR value evaluation is secured for all the randomly chosen operating conditions of both networks, as it is sketched

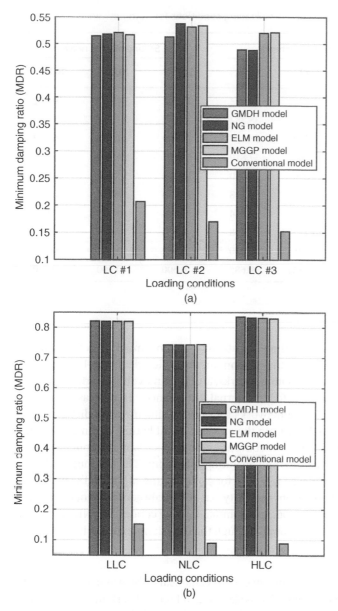

Figure 9.6 MDR value comparison for − (a) the first electric network and (b) second electric network for arbitrarily produced three operating conditions of each network in case of four ML models and conventional approach.

in Figure 9.6a for the first network and Figure 9.6b for the second network evaluation. According to the notes explained earlier in case of eigenvalue and MDR value impact on power system's stability improvement, the ML model incorporated power networks maintain distant and better optimal locations for the eigenvalues, and higher MDR values for stability improvement of power systems within quickest possible time as compared to the conventional approach. Therefore, it can be concluded that the ML algorithms/AI (e.g. GMDH, NG, ELM, and MGGP, etc.) are suitable for application in power systems for enhancing the overall system's stability through appropriately online prediction of PSS parameters to damp out of LFOs.

Table 9.9 Online estimated PSS parameters for randomly selected three loading conditions for both networks.

Items		First electric network			Second electric network		
		LC #1	LC #2	LC #3	LLC	NLC	HLC
Gain constant (K)	GMDH	26.512	25.841	39.093	26.202	24.127	29.012
	NG	26.506	25.689	38.382	25.596	24.152	27.267
	ELM	26.598	25.740	38.204	25.609	24.167	26.822
	MGGP	26.541	25.706	37.819	25.601	24.189	26.354
	Conventional	7.090	7.090	7.090	15.000	15.000	15.000
Time constant (T_1)	GMDH	0.227	0.193	0.218	0.985	0.984	0.987
	NG	0.224	0.195	0.214	0.984	0.984	0.986
	ELM	0.223	0.195	0.218	0.984	0.984	0.985
	MGGP	0.223	0.195	0.220	0.984	0.984	0.500
	Conventional	0.685	0.685	0.685	0.500	0.500	0.500

9.5.2 Time-Domain Simulation Results Comparison

To analyze the applicability and compatibility of ML algorithms for stability enhancement of power systems through damping out of undesired LFOs, oscillations of three major components of electric networks, which are the change of rotor angles and rotor angular frequencies for both networks, and DC-link voltage variations of the second electric network, are considered to make stable subjected to external disturbances. In this case, three ML models (e.g. GMDH, NG, and ELM) are implemented in the networks for online adjustment/prediction of PSS parameters to damp out LFOs and make the systems stable within the quickest possible time and measured the component's response in the time-domain frame. Additionally, the conventional approach was also applied to measure similar time domain response following the same features and operating conditions. The ML model adopted responses were then compared to the conventionally achieved response for exploring the ML model's superiority of implication on the power system in case of enhancing system's stability. An arbitrarily generated operating circumstance, e.g. $P_e(pu) = 1.15$, $Q_e(pu) = 0.40$, and $V_t(pu) = 1.06$, and nominal loading conditions (NLCs) for first and second networks, respectively, were commonly followed for all through the time domain frame investigation.

9.5.2.1 Rotor Angle Variation Under Disturbance

In the case of observing the sustainability of rotor angle variation, an extra mechanical pulse was assimilated with the normal response of rotor angle at one second for both networks, as sketched in Figure 9.7. Hence, the mechanical torque was continued as an external disturbance for four cycles to the first and second electric networks under the mentioned operating circumstances.

Owing to assimilating the disturbances on the networks, the rotor of the synchronous machine started to vibrate, which was revealed as the oscillations of rotor angles for both networks. To make the oscillations stable by suppressing undesired LFOs, new impacts of PSS were imposed on both networks through instantly predicting and deploying the key PSS parameters (T_1 and K) by separately employing three different ML models (GMDH, NG, and ELM) and a conventional approach, and measured the responses accordingly. It can be noticed from Figure 9.7a,b for first and second electric networks, respectively, that the responses obtained using all the developed ML models were almost overlapped each other, and become stable by nearly 3.2 seconds. In contrast, although the conventional approach confirmed the first network to be stable taking marginally less than the double time of ML model-based time, which was almost 6.0 seconds, the approach could not ensure the mitigation of oscillations for

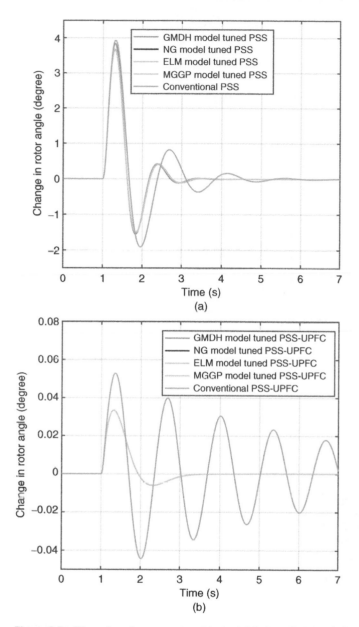

Figure 9.7 Time-domain responses with the initiation of external disturbances – (a) rotor angle variation for the first electric network and (b) second electric network.

the second network even within 7.0 seconds. Therefore, the applications of AI are meaningful in the area of power systems for enhancing the system's stability through appropriately impacting the PSS on the networks.

9.5.2.2 Rotor Angular Frequency Variation Under Disturbance

In evaluating the angular frequency variation of the rotor of synchronous machine subjected to external disturbances, extra mechanical torque was conjoined with normal angular frequency responses of both networks. The disturbances were incited to both networks at one second and prolonged to four cycles. Due to having disturbances

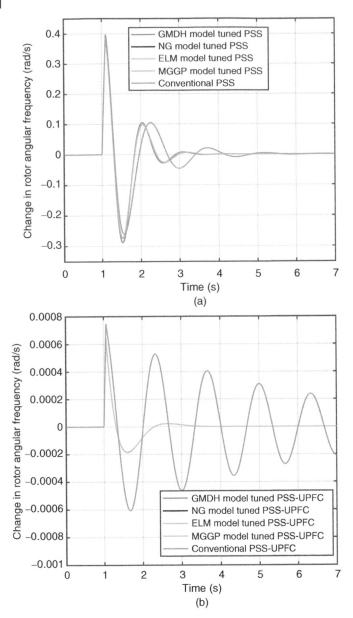

Figure 9.8 Time-domain responses with incitation of external disturbances – rotor angular frequency variation for (a) first electric network and (b) second electric network.

of the networks, the angular frequencies of the rotor of both machines started to oscillate immediately. In this situation, new impacts of PSS were imposed on both networks through estimating in real time and deploying the key PSS parameters (T_1 and K) by separately implementing three different ML models (GMDH, NG, and ELM) and conventional approach, and evaluating the responses accordingly. All the responses are plotted in Figure 9.8a for the first network and Figure 9.8b for the second network.

Figure 9.8 demonstrates that the responses achieved employing the considered three ML models are very close to each other and become stable by approximately 3.2 seconds. On the other hand, although the conventional approach ensured the first network to be stable, requiring about the double time of ML model-based consumed

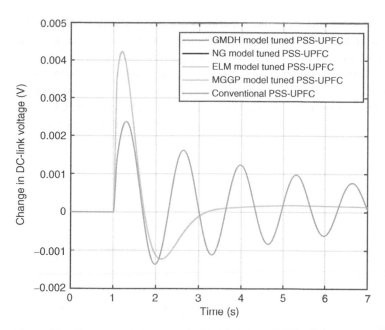

Figure 9.9 Time-domain responses with incitation of external disturbances – DC-link voltage variation for the second electric network.

time, which was nearly 5.5 seconds, the approach could not ensure the suppression of oscillations (e.g. LFOs) for the second network even within 7.0 seconds. Hence, it can be concluded based on the evaluation of angular frequency responses following external disturbances that the ML algorithms are applicable in enhancing power systems stability due to their appropriately quicker response in online conjecturing of PSS parameters to damp out LFOs as compared to the conventional way.

9.5.2.3 DC-Link Voltage Variation Under Disturbance

Similar to the time domain analysis of the other two components of the synchronous machine, the DC-link voltage variation was also analyzed for the second network using three different ML models and conventionally measured response. All the responses are evaluated in a time domain frame and depicted in Figure 9.9. Similar stability demonstrations were also noticed, like the rotor angles and angular frequencies. Therefore, it can be concluded that the applications of AI are meaningful in the area of power systems for enhancing the system's stability through appropriate online tuning of PSS parameters (T_1 and K) by damping out the undesired LFOs subjected to disturbances.

9.6 Conclusions

This chapter has proposed and summarized the feasibility of applying AI techniques in electric power networks for mitigating unwanted LFOs and enhancing the overall stability profile. Four ML models (e.g. GMDH, ELM, NG, and MGGP) were developed to analyze the networks' real-time operating performance. The data sets were generated considering two distinct electric networks, SMIB with PSS only, and UPFC coordinated SMIB with PSS. The real-time performance of the networks was investigated based on three stability measurement indices (e.g. MDR, eigenvalues, and time-domain simulations) for several operating conditions. In this regard, the developed ML tools were implemented for online adjustment of PSS parameters through appropriate prediction so that the undesired

LFOs are damped out in the shortest possible time. Furthermore, the superiority and applicability of the AI techniques in enhancing the network's stability were tested by comparing them with conventional approaches. The proposed meta-heuristic optimization-based ML models executed better effectivity, efficiency, and robustness in all operating situations by accurately predicting the PSS parameters in real time. Therefore, it can be concluded that the AI techniques are suitable and effective in electric power networks to improve the network's stability by mitigating the undesired LFOs via real-time fine-tuning of PSS parameters. However, the applicability and potentiality of the AI techniques in enhancing the power grid's stability can further be investigated in deeper insight by implementing the proposed ML models in multimachine power systems and then, the small-scale real power systems.

References

1 Kundur, P. (1994). *Power System Stability and Control*. New York: McGraw-Hill, Inc.

2 Bhukya, J. and Mahajan, V. (2019). Optimization of damping controller for PSS and SSSC to improve stability of interconnected system with DFIG based wind farm. *International Journal of Electrical Power & Energy Systems* 108: 314–335. https://doi.org/10.1016/J.IJEPES.2019.01.017.

3 Sambariya, D.K. and Prasad, R. (2013). Design of PSS for SMIB system using robust fast output sampling feedback technique. In: *Proceedings of 7th International Conference on Intelligent Systems and Control (ISCO 2013)*, 166–171. Tamilnadu, India: IEEE.

4 Jolfaei, M.G., Sharaf, A.M., Shariatmadar, S.M., and Poudeh, M.B. (2016). A hybrid PSS–SSSC GA-stabilization scheme for damping power system small signal oscillations. *International Journal of Electrical Power & Energy Systems* 75: 337–344. https://doi.org/10.1016/j.ijepes.2015.08.024.

5 Assi Obaid, Z., Cipcigan, L.M., and Muhssin, M.T. (2017). Power system oscillations and control: classifications and PSSs' design methods: a review. *Renewable and Sustainable Energy Reviews* 79: 839–849. https://doi.org/10.1016/J.RSER.2017.05.103.

6 Alam, M.S., Razzak, M.A., Shafiullah, M., and Chowdhury, A.H. (2012). Application of TCSC and SVC in damping oscillations in Bangladesh power system. In: *2012 7th International Conference on Electrical and Computer Engineering*, 571–574. IEEE https://doi.org/10.1109/ICECE.2012.6471614.

7 Alam, M.S., Shafiullah, M., Hossain, M.I., and Hasan, M.N. (2015). Enhancement of power system damping employing TCSC with genetic algorithm based controller design. In: *International Conference on Electrical Engineering and Information Communication Tech, (ICEEICT)*, 1–5. IEEE https://doi.org/10.1109/ICEEICT.2015.7307353.

8 Eslami, M., Shareef, H., and Mohamed, A. (2010). Application of PSS and FACTS devices for intensification of power system stability. *International Review of Electrical Engineering* 5 (2): 552–570.

9 Siddiqui, A.S., Khan, M.T., and Iqbal, F. (2017). Determination of optimal location of TCSC and STATCOM for congestion management in deregulated power system. *International Journal of Systems Assurance Engineering and Management* 8 (1): 110–117. https://doi.org/10.1007/s13198-014-0332-4.

10 Prasad, D. and Mukherjee, V. (2016). A novel symbiotic organisms search algorithm for optimal power flow of power system with FACTS devices. *Engineering Science and Technology, an International Journal* 19 (1): 79–89. https://doi.org/10.1016/j.jestch.2015.06.005.

11 Inkollu, S.R. and Kota, V.R. (2016). Optimal setting of FACTS devices for voltage stability improvement using PSO adaptive GSA hybrid algorithm. *Engineering Science and Technology, an International Journal* 19 (3): 1166–1176. https://doi.org/10.1016/j.jestch.2016.01.011.

12 Mukherjee, A. and Mukherjee, V. (2016). Chaotic krill herd algorithm for optimal reactive power dispatch considering FACTS devices. *Applied Soft Computing* 44: 163–190. https://doi.org/10.1016/j.asoc.2016.03.008.

13 Alizadeh, M. and Tofighi, M. (2013). Full-adaptive THEN-part equipped fuzzy wavelet neural controller design of FACTS devices to suppress inter-area oscillations. *Neurocomputing* 118: 157–170. https://doi.org/10.1016/j .neucom.2013.03.001.

14 Khan, M.T. and Siddiqui, A.S. (2016). FACTS device control strategy using PMU. *Perspectives in Science* 8: 730–732. https://doi.org/10.1016/j.pisc.2016.06.072.

15 Made Wartana, I., Singh, J.G., and Ongsakul, W. et al. (2011). Optimal placement of UPFC for maximizing system loadability and minimize active power losses by NSGA-II. In: *Proceedings of the 2011 International Conference and Utility Exhibition on Power and Energy Systems: Issues and Prospects for Asia, ICUE 2011*, IEEE, Pattaya, Thailand, pp. 1–8. https://doi.org/10.1109/ICUEPES.2011.6497710.

16 Institute of Electrical and Electronics Engineers., Thailand. Kānfaifā Sūan Phūmiphāk., and Asian Institute of Technology., "Proceedings of the 2011 International Conference & Utility Exhibition on Power and Energy Systems: Issues and Prospects for Asia (ICUE) : Amari Orchid Pattaya Hotel, Pattaya City, Thailand, 28–30 September 2011," 2008.

17 Elgamal, M.E., Lotfy, A., and Ali, G.E.M. (2012). Voltage profile enhancement by fuzzy controlled MLI UPFC. *International Journal of Electrical Power & Energy Systems* 34 (1): 10–18. https://doi.org/10.1016/j .ijepes.2011.08.001.

18 Hussain, A.N. and Hamdan Shri, S. (2018). Damping improvement by using optimal coordinated design based on PSS and TCSC device. In: *2018 3rd Scientific Conference of Electrical Engineering, SCEE 2018* (July 2018), IEEE, Baghdad, Iraq, pp. 116–121. https://doi.org/10.1109/SCEE.2018.8684209.

19 Khodabakhshian, A., Esmaili, M.R., and Bornapour, M. (2016). Optimal coordinated design of UPFC and PSS for improving power system performance by using multi-objective water cycle algorithm. *International Journal of Electrical Power & Energy Systems* 83: 124–133. https://doi.org/10.1016/j.ijepes.2016.03.052.

20 Shafiullah, M., Rana, M.J., and Coelho, L.S. et al. (2017). Power system stability enhancement by designing optimal PSS employing backtracking search algorithm. In: *2017 6th International Conference on Clean Electrical Power (ICCEP)*, Santa Margherita Ligure, Italy, pp. 712–719. https://doi.org/10.1109/ICCEP.2017.8004769.

21 Hassan, L.H., Moghavvemi, M., Almurib, H.A.F., and Muttaqi, K.M. (2014). A coordinated design of PSSs and UPFC-based stabilizer using genetic algorithm. *IEEE Transactions on Industry Applications* 50 (5): 2957–2966. https://doi.org/10.1109/TIA.2014.2305797.

22 Abido, M.A., Al-Awami, A.T., and Abdel-Magid, Y.L. (2006). Analysis and design of UPFC damping stabilizers for power system stability enhancement. In: *IEEE International Symposium on Industrial Electronics*, IEEE, Montreal, QC, Canada, vol. 3, pp. 2040–2045. https://doi.org/10.1109/ISIE.2006.295887.

23 Vanitila, R. and Sudhakaran, M. (2012). Differential evolution algorithm based weighted additive FGA approach for optimal power flow using muti-type FACTS devices. In: *2012 International Conference on Emerging Trends in Electrical Engineering and Energy Management (ICETEEEM)*, 198–204. IEEE https://doi .org/10.1109/ICETEEEM.2012.6494459.

24 Rana, M.J., Shahriar, M.S., and Shafiullah, M. (2019). Levenberg–Marquardt neural network to estimate UPFC-coordinated PSS parameters to enhance power system stability. *Neural Computing and Applications* 31 (4): 1237–1248. https://doi.org/10.1007/s00521-017-3156-8.

25 Shafiullah, M., Rana, M.J., Shahriar, M.S. et al. (2021). Extreme learning machine for real-time damping of LFO in power system networks. *Electrical Engineering* 103 (1): 279–292. https://doi.org/10.1007/s00202-020-01075-7.

26 Shahriar, M.S., Shafiullah, M., and Rana, M.J. (2017). Stability enhancement of PSS-UPFC installed power system by support vector regression. *Electrical Engineering* 100: 1–12. https://doi.org/10.1007/s00202-017-0638-8.

27 Shafiullah, M., Rana, M.J., Shahriar, M.S., and Zahir, M.H. (2019). Low-frequency oscillation damping in the electric network through the optimal design of UPFC coordinated PSS employing MGGP. *Measurement* 138: 118–131. https://doi.org/10.1016/J.MEASUREMENT.2019.02.026.

28 Sabo, A., Wahab, N.I.A., Othman, M.L. et al. (2020). Application of neuro-fuzzy controller to replace SMIB and interconnected multi-machine power system stabilizers. *Sustainability* 12 (22): 1–42. https://doi.org/10.3390/su12229591.

29 Açikgöz, H., Keçecioğlu, Ö.F., and Şekkeli, M. (2019). Real-time implementation of electronic power transformer based on intelligent controller. *Turkish Journal of Electrical Engineering and Computer Sciences* 27 (4): 2866–2880. https://doi.org/10.3906/elk-1807-315.

30 Yu, F.T.S. (1983). *Electric Power System Dynamics*. Academic Press Inc.

31 Machowski, J., Lubosny, Z., Bialek, J.W., and Bumby, J.R. (2008). *Power System Dynamics: Stability and Control*, 2e, 855. Wiley.

32 Shafiullah, M., Juel Rana, M., Shafiul Alam, M., and Abido, M.A. (2018). Online tuning of power system stabilizer employing genetic programming for stability enhancement. *Journal of Electrical Systems and Information Technology* 5 (3): 287–299. https://doi.org/10.1016/j.jesit.2018.03.007.

33 Shahriar, M.S., Shafiullah, M., Rana, M.J. et al. (2020). Neurogenetic approach for real-time damping of low-frequency oscillations in electric networks. *Computers and Electrical Engineering* 83: 1–14. https://doi.org/10.1016/j.compeleceng.2020.106600.

34 Farlow, S.J. (1981). The GMDH algorithm of Ivakhnenko. *The American Statistician* 35 (4): 210–215. https://doi.org/10.1080/00031305.1981.10479358.

35 Amiri, M. and Soleimani, S. (2021). ML-based group method of data handling: an improvement on the conventional GMDH. *Complex & Intelligent Systems* 7 (6): 2949–2960. https://doi.org/10.1007/s40747-021-00480-0.

36 Najafzadeh, M., Barani, G.A., and Hessami-Kermani, M.R. (2015). Evaluation of GMDH networks for prediction of local scour depth at bridge abutments in coarse sediments with thinly armored beds. *Ocean Engineering* 104: 387–396. https://doi.org/10.1016/j.oceaneng.2015.05.016.

37 Huang, G.-B., Zhou, H., Ding, X., and Zhang, R. (2012). Extreme learning machine for regression and multiclass classification. *IEEE Transactions on Systems, Man, and Cybernetics Part B: Cybernetics* 42 (2): 513–529. https://doi.org/10.1109/TSMCB.2011.2168604.

38 Katoch, S., Chauhan, S.S., and Kumar, V. (2021). A review on genetic algorithm: past, present, and future. *Multimedia Tools and Applications* 80 (5): 8091–8126.

39 Guo, G., Xie, Z., Zhang, Y. et al. (2020). The ferromagnetic and half-metal properties of hydrogen adatoms, fluorine adatoms and boron adatoms adsorbed at edges of zigzag silicene nanoribbon. *Physica E: Low-dimensional Systems and Nanostructures* 116 (July 2019): 113733. https://doi.org/10.1016/j.physe.2019.113733.

40 Sharif, A.A. and Hosseinzadeh Aghdam, M. (2019). A novel hybrid genetic algorithm to reduce the peak-to-average power ratio of OFDM signals. *Computers and Electrical Engineering* 80: 1–13. https://doi.org/10.1016/j.compeleceng.2019.106498.

41 Zou, J., Han, Y., and So, S.S. (2008). Overview of artificial neural networks. *Methods in Molecular Biology* 458: 15–23. https://doi.org/10.1007/978-1-60327-101-1_2.

42 Koza, J.R. (1994). Genetic programming as a means for programming computers by natural selection. *Statistics and Computing* 4 (2): 87–112. https://doi.org/10.1007/BF00175355.

43 Koza, J.R. (2003). *Genetic Programming: On the Programming of Computers By Means of Natural Selection Complex Adaptive Systems*. Cambridge, Massachusetts London, England: A Bradford Book, The MIT Press.

44 Alavi, A.H. and Gandomi, A.H. (2011). A robust data mining approach for formulation of geotechnical engineering systems. *Engineering Computations (Swansea, Wales)* 28 (3): 242–274. https://doi.org/10.1108/02644401111118132.

45 Walker, J.A. and Miller, J.F. (2008). The automatic acquisition, evolution and reuse of modules in Cartesian genetic programming. *IEEE Transactions on Evolutionary Computation* 12 (4): 397–417. https://doi.org/10.1109/TEVC.2007.903549.

46 Ilius Hasan Pathan, M., Juel Rana, M., Shoaib Shahriar, M. et al. (2020). Real-time LFO damping enhancement in electric networks employing PSO optimized ANFIS. *Inventions* 5 (4): 61. https://doi.org/10.3390/inventions5040061.

47 Shahriar, M.S., Shafiullah, M., and Asif, M.A. et al. (2016). Design of multi-objective UPFC employing back-tracking search algorithm for enhancement of power system stability. In: *2015 18th International Conference on Computer and Information Technology, ICCIT 2015* (June 2016), IEEE, Dhaka, Bangladesh, pp. 323–328. https://doi.org/10.1109/ICCITechn.2015.7488090.

10

Augmentation of PV-Wind Hybrid Technology with Adroit Neural Network, ANFIS, and PI Controllers Indeed Precocious DVR System

Jyoti Shukla[1], Basanta K. Panigrahi[2], and Monika Vardia[1]

[1]*Department of Electrical Engineering, Poornima College of Engineering, RTU, Jaipur, India*
[2]*Department of Electrical Engineering, Institute of Technical Education & Research, SOA University, Bhubaneswar, India*

10.1 Introduction

The researchers always concentrated on the issue of generating power which is environment friendly and pecuniary. The persistent use of conventional sources of energy gives rise to many causes of uncertainties in the environment such as the release of toxic substances and leading unbalance in the atmosphere. So, it is vitally important that the energy we are consuming gives its best to output with less investment and encourages on generating power eco-friendly. And that will also exaggerate the system reliability and maintain the consistency with less dependency on sources. As recognizing the present need of every type of consumers that also fulfills the essential requirements of sustainably developed society, authors presented the intelligent photovoltaic (PV)-Wind Hybrid Power System with the dynamic voltage restorer (DVR) which is controlled by novel techniques. The relevance of using the Neural Network (NN) controllers, adaptive neuro-fuzzy inference system (ANFIS) controllers, and PI controller in the system is for improving the dynamics of the system and feasibility of network. In [1], the aspect of renewable energy in playing the role of preserving the environment with its various sources and concluded that the applications of renewable sources are increasing rapidly as it can supply two-third of power of the global energy need with high level of efficiency. In [2], the review is presented on the use of various components of Hybrid energy generation, and the challenges faced for increasing the efficiency of the system are elaborated. Various types of comparisons on the techniques are made in [3] on the concerned issue of maximizing the power of maximum power point tracking (MPPT) of PV system for making it compatible with consumers and operators. The presented analysis also focuses on the advanced Artificial Neural Network (NN) Techniques which have shown better performance than the other listed techniques.

In [4], the Wind energy conversion system based on PMSG is represented with its controlling theory under grid fault conditions for maintaining a unity power factor by three control strategies as DC-Link voltage control, pitch angle control, and rotor speed control. Further in this chapter, authors have used the pitch angle controlling technique with different controllers and obtained better results. For the system effectiveness, the DVR is conferred in [5], as it depends on the capability of voltage injection and energy storage system, so a *D-Q* controlling strategy is addressed so far, despite this chapter does not include any upcoming strategies for future reference. In [6], the Hybrid Power generation system with an Incremental Conductance technique is discussed and the improvement in low voltage ride-through capability is presented. For enhancing the system stability there can be more progressive techniques discussed and inclusion of applications could be done.

Considering all the above literature, the techniques applied to the system are not much astute and perceptive in response. If the renewable energy power generation comes which is not yet familiar with every consumer,

Artificial Intelligence-based Smart Power Systems, First Edition.
Edited by Sanjeevikumar Padmanaban, Sivaraman Palanisamy, Sharmeela Chenniappan, and Jens Bo Holm-Nielsen.

there is an essential requirement to make it more productive and transmits power efficiently due to uncertainties in the generation of power from PV and Wind energy sources. Therefore, the profitable generation of power from PV and Wind energy sources in the uneven conditions must also be undertaken by using radical controllable techniques.

As concerning the above inscription, in this chapter the advance controlling techniques are proposed which are completely based on NNs and Discrete PI Controllers. Form the best of observation and recognition of the authors, this approach as yet has not been addressed in the literature.

The primary motive of this chapter is to make the system stronger and more fortified by the use of ingenious and innovative controllers so as to make the power from renewable energy sources limited not only to urban areas only but also to remote locations. The DVR is a high-tech gadget that manages the sags and swells that occur in the path of generation. It is a member of the flexible alternating current transmission system (FACTS) devices family. In the realm of electronics, the FACTS devices have set a standard for preserving the efficiency and consistency of the power system [7]. The DVR offers various capabilities for upgrading the performance of the system by making the system much feasible and more reliable. In the proposed configuration the Wind Turbine taken is fully based on PMSG-Based Wind Energy Conversion System generation and for reducing fluctuations in the system among all the other FACTS devices the DVR is taken for constant output voltage as proven from its performance.

This chapter is structured as follows: In Section 10.2 the proposed topologies of the system are discussed and their controlling techniques are also explained further. Section 10.3 demonstrates and validates the presented techniques applied to the system as from their results obtained. The results obtained are also compared with and without the use of the DVR in the system with the constant voltage obtained from the proposed DVR which is supplying the power to grid. In Section 10.4 this chapter is concluded and summarized.

10.2 PV-Wind Hybrid Power Generation Configuration

The proposed configuration has emerged as the best alternative source of energy than conventional sources of energy. The Hybrid system connections are made as to the PV system which is controlled by NN Predictive and ANFIS Controller is connected to the step-up transformer of 30 kV and then with the buses to the load and DVR [8]. Simultaneously, the PMSG-Based Wind Turbine System is controlled by PI, Fuzzy Logic and NN-based NARMA-L2 Controllers for pitch angle control purpose which is also connected to the step-up transformer and

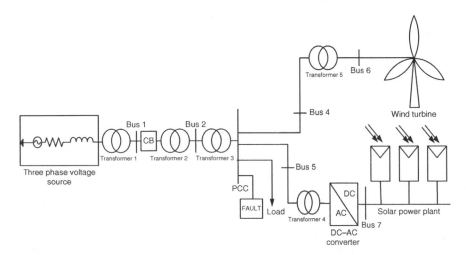

Figure 10.1 Single-line diagram of the proposed hybrid system.

then in parallel to load and DVR of the system with buses for supplying the power to the grid of the Hybrid system. The main merits of the proposed system are that it gives the fast recovery of voltage, reduction in power oscillations overshoot, controllable rotor torque and speed and DVR is preventing the system from the overvoltage and therefore in the system applied controllers enhance the stability of power system with LVRT requirements.

The suggested model is a basic depiction of a grid-connected wind energy power plant and solar energy power plant. A wind turbine, solar panel, three-phase source, transformers, circuit breakers, and a fault block connected to the load are all included in the proposed model, which is simulated in MATLAB and SIMULINK. The single line diagram of the suggested hybrid model is depicted in Figure 10.1.

10.3 Proposed Systems Topologies

10.3.1 Structure of PV System

The PV power generation system consists of the number of cells connected within it. Such that the current and voltage generation through these cells depends on the processing system applied to them for generation, as the accurate choice gives better and proficient results for system. So, for generating productive results from PV system, the double-diode structure is applied. The array type implemented for system is as SunPower SPR-X21-345-COM. And the configuration of the parallel strings connected is 109, whereas series connected are 15 modules per string. In Figure 10.2, double-diode equivalent circuit for a solar cell is as shown respectively. The total current equation obtained from the double-diode structure in Figure 10.2 is as described below [9]:

$$I = I_{\text{ph}} - I_{\text{sat1}} \left(e^{q \frac{V + Ir_s}{\eta' kT}} - 1 \right) - I_{\text{sat2}} \left(e^{q \frac{V + Ir_s}{\eta'' 2kT}} - 1 \right) - \frac{V + Ir_s}{r_p} \tag{10.1}$$

The description of equation variable is as I denotes output current of the PV module, V indicates the terminal voltage, K defines as Boltzmann constant (1.38×10^{-23} J/K), q is as electric charge (1.6×10^{-9} C) of the cell, whereas T denotes the cells temperature (K) and the η' is an ideality factor.

Also taking into consideration that all the solar cells are in proportion and they are working in cognate operating conditions. Likewise, all the voltages and currents of the cell multiplied by the series and parallel connected string of cell then subsequently series resistance r_s and parallel resistance r_p enhanced by a factor of N_s and consequently divide by N_p. The V_t is thermal voltage at the terminals of cell and I_{phase} is the generation of charge carriers produced by the PV cell.

$$I_{\text{phase,field}} = N_p \cdot I_{\text{phase,cell}} \tag{10.2}$$

$$V_{\text{t,field}} = N_s \cdot V_{\text{t,cell}} \tag{10.3}$$

$$r_{\text{p,field}} = \frac{N_s}{N_p} \cdot r_{\text{p,cell}} \tag{10.4}$$

$$I_{\text{sat1,field}} = N_p \cdot I_{\text{sat2,cell}} \tag{10.5}$$

Figure 10.2 Proposed solar *PV* cell equivalent circuit.

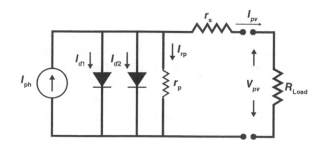

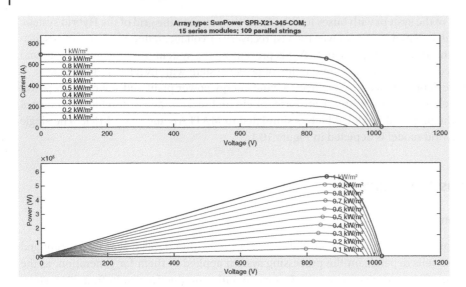

Figure 10.3 Electrical characteristics of the *PV* array.

Table 10.1 Obtained parameters from *PV* array.

Parameters	Realized values
Maximum power	344.94 W
Open circuit voltage (V_{oc})	68.2 V
Cells per module (N_{cell})	96
Short-circuit current (I_{sc})	6.39 A
Light generated current (IL)	6.392 A
Voltage at maximum power point (V_{mp})	57.3 V
Current at maximum power point (I_{mp})	6.02 A

The *I-V* and *P-V* characteristics of the *PV* array system are expressed in Figure 10.3 based on different values taken for solar irradiance for obtaining the required results from *PV* array as mentioned below. And Table 10.1 of the PV system describes the specifications taken for obtaining the uniform results from the system.

In the PV system, the Boost Converter plays a significant role in intensifying and mollifying the power generated by the system and it consequently elevates the generation and efficiency of the system which is necessarily required for any consumer. It is a step-up converter and belongs to the family of switched-mode power supply (SMPS) electronic equipment that contains within it distinctive type of components as inductor, capacitor, and transistor for mitigating the output voltage ripples effectively. In the proposed configuration the Boost Converter circuit is set accordingly as shown in Figure 10.4 which further converts 730 V of DC voltage into 440 V of AC voltage and incorporating the 3-level IGBT-inverter within it. The inverter is then connected to inductive grid filter and then to a low-frequency transformer for stepping up the desired voltage as 440 V–30 kV and then to the grid. In the grid, the PV system is connected to the proposed DVR system parallelly as well as to the load of the system for accomplishing a well-managed Hybrid system.

Figure 10.4 Simplified presentation of boost converter circuit.

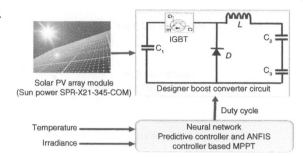

Solar PV array module
(Sun power SPR-X21-345-COM)

Designer boost converter circuit

Duty cycle

Temperature ⟶
Irradiance ⟶

Neural network
Predictive controller and ANFIS
controller based MPPT

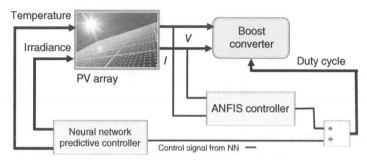

Figure 10.5 Schematic circuit of applied technique to MPPT.

10.3.2 The MPPTs Technique

In concept of utilizing the renewable power, the PV system is taken as primary alternate to generation [10]. The PV system runs with a profitable and consistent supply given MPPT. There are various types of techniques available for imposing into MPPT of the PV system that are discussed in detail by Khan and Mathew [11]. Each technique has its own set of advantages and disadvantages, but there is an urgent need to use a system-beneficial technique such as MPPT. After detailed analysis on the various methods we had implemented, the novel NN Predictive Controller technique with potent ANFIS Controller technique in the PV system is selected as MPPT shown in Figure 10.5.

10.3.3 NN Predictive Controller Technique

For maintaining the new interconnections of the PV system it is important that the system becomes more advanced in technology as well as generates output power and its response should be dynamic in nature. The use of Predictive Controller offers distinctive merits to the PV system as making of system robust in nature, effective measurement of training data with high efficiency, maintaining systems accuracy and gives fast prediction of plants performance which increases its credibility and feasibility. Further, the performance of the system is improved by parallel and distributed processing of learning that is advantageous [12].

Figure 10.6 the NN Predictive Controllers performance is presented as identification step of system as step 1 and in Figure 10.7 the controlling operation NN controller is performed which is based on prediction and it is the step 2 of controlling, it includes two steps for proper implementation of the proposed controller.

As stated in Figures 10.6 and 10.7 about the performance of controller that signifies the step-wise role of Predictive Controller in controlling the MPPT of the PV system. The controller's performance obtained is as described below:

$$J = \sum_{j=n_1}^{n_2}(r(t+j) - y_{\mathrm{mp}}(t+j))^2 + \rho\sum_{j=1}^{n_u}(u'(t+j-1) - u'(t+j-2))^2 \tag{10.6}$$

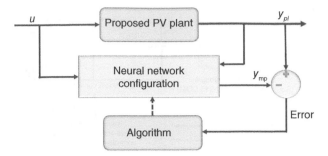

Figure 10.6 NN predictive controllers identification of system.

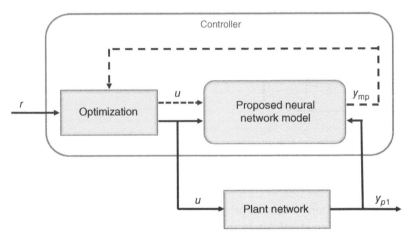

Figure 10.7 Predictive control operational representation.

In the above equation, the n_1, n_2, and n_u specifies as horizons used for tracking of error in the first step of controlling as identification, u' implies the signal used for controlling, ρ defines as that on the performance index the sum of squares has incremental control on operation and y_{mp} and r is the NN model response and it is desired response of the system. In addition to it, the u' use to minimize J in turn gives the u as input to plant. In Figure 10.8 the proposed NN controller's performance is obtained after training it with the input data and with the 200 epochs.

10.3.4 ANFIS Technique

The ANFIS technique is qualitative in use and favorable in all types of conditions where it is to be implemented. This technique is applied with the NN Predictive Controller for enhancing and strengthening the system with the performance. Despite using a simple NN Controller as MPPT, we have used here the combination of potent and useful controllers together for elevating the required amount of power generation with fewer efforts and also making the system more advanced due to the dependency of PV power on weather conditions [13]. It offers fewer distortions in steady-state condition and gives a fast dynamic response to the system. In the proposed controller, the FIS system is based on the back-propagation algorithm which is obtained while training the Takagi–Sugeno-based Fuzzy Inference System.

The suggested ANFIS system rules are detailed below, and they are based on the assumption of two input variables, x and y, with one output, f_1, as illustrated in Figure 10.9.

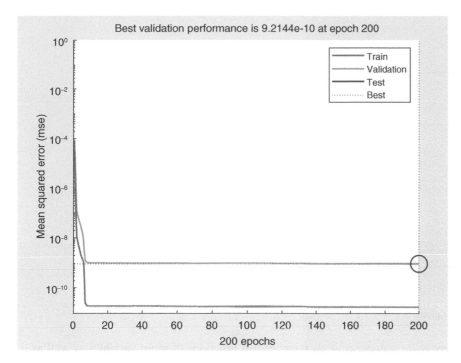

Figure 10.8 NN predictive controllers performance obtained.

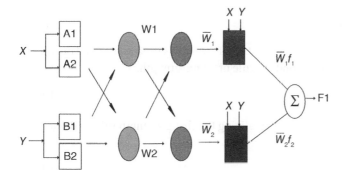

Figure 10.9 Equivalent proposed architecture of ANFIS.

This Fuzzy Inference System is based on Takagi–Sugeno system with as A_1, B_1, A_2, and B_2 are the following membership functions use as input to the system. The rule set for Fuzzy if-then rules are described as:

$$\text{if } x \text{ is } A_1 \text{ and } y \text{ is } B_1, \text{ THEN } f_1 = p_1 x + q_1 y + r_1 \tag{10.7}$$

$$\text{if } x \text{ is } A_2 \text{ and } y \text{ is } B_2, \text{ THEN } f_2 = p_2 x + q_2 y + r_2 \tag{10.8}$$

The use of such architecture is shown in Figure 10.9, and it also represents the layer 5 of the Fuzzy Inference System. The layers of the system are as described:

$$\text{Layer } 1 : O_{1,i} = \mu A_i(x) \quad \text{for } i = 1,2$$

$$= \mu B_{i-2}(y) \quad \text{for } i = 3,4 \tag{10.9}$$

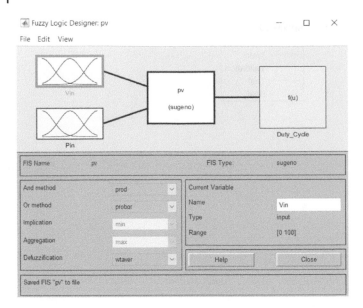

Figure 10.10 Sugeno type FIS for MPPT.

The layer 1 describes that every node is dependent on membership functions of input variables as well as it is an adaptive node of system where output of ith node in layer is as $O_{1,i}$, μ can be of any function as Gaussian, trapezoidal, and triangle. The generalized membership functions are given as $\mu_A(x) = \dfrac{1}{1+\left|\frac{x-c_i}{a_i}\right|^{2b_i}}$.

$$\text{Layer 2}: O_{2,i} = w_i = \mu_{Ai}(x)\mu_{Bi}(y), \quad \text{for } i = 1,2 \tag{10.10}$$

In this layer, it accepts the values from layer 1 as in the form of output and further it is used as for AND operation for these values. It is also known as fixed node, and it corresponds to rule of firing strength.

$$\text{Layer 3}: O_{3,i} = \overline{w}_i = \frac{w_i}{w_1 + w_2} \tag{10.11}$$

This layer is a fixed node that performs key functions such as normalizing the firing strength rule by dividing it with the sum of all firing strength rules.

$$\text{Layer 4}: O_{4,i} = \overline{w}_i f_i = \overline{w}_i(p_i x + q_i y + r_i) \tag{10.12}$$

It is referred as adaptive node of layers and $\overline{w}_i$ represents the normalized firing strength and the parameters p_i, q_i, r_i are used to optimize within the training process.

$$\text{Layer 5}: \sum_i \overline{w}_i f_i = \frac{\sum_i w_i f_i}{\sum_i w_i} \tag{10.13}$$

Layer 5 is the last layer and sums up all the signals of the input to get the desired output. In Figure 10.10, the implemented ANFIS system output is shown.

10.3.5 Training Data

The training data used for generating the duty cycle for Boost Converter is described here. Since the NN Controllers need the more data for calculating the output, we have taken more amount of data as input and the results obtained by training the data are highly efficient in performance as shown in Table 10.2.

Table 10.2 Data used for training the controller.

Irradiance (W/m²)	Temperature (°C)	Duty cycle
1000	20	0.9999
900	35	0.9998
800	20	0.9998
700	30	0.9998
600	40	0.9997
500	35	0.9997
400	30	0.9996
300	45	0.9996
200	40	0.9995
100	45	0.9995

In the proposed MPPT technique, the combination of NN Predictive Controller and ANFIS Controller is highly justified as it improves the performance of the PV system, and it accomplishes the needs of humans for giving a good consistent supply. The combination of two potent controllers gives new heights to the system by making it more intelligent.

10.4 Wind Power Generation Plant

For upgrading the Wind Power System, we have used the PMSG-Based Wind Turbine Power Plant for increasing and assisting the performance of the system [14]. The The Wind power availability in the environment is utmost same as for the PV system and that counts the reason behind for making of PV-Wind Hybrid Power generation system [15]. The PMSG-Based Wind system possesses various advantages of making the system more qualitative as it is economically viable, and it has a gearless system that reduces the maintenance charges.

By controlling the PMSG Wind system will give better outcomes for increasing and intensifying the power generation successfully [16]. The PMSG-Based Wind Turbine system stator voltage equations are described below as *d-q* or Park's transformation of rotating reference frame as:

$$\frac{\mathrm{d}}{\mathrm{d}t}id_{st} = \frac{1}{L_{dst}}V_{dst} - \frac{R_{st}}{L_{dst}}i_{dst} + \frac{L_{qst}}{L_{dst}}p\omega_r i_{qst} \tag{10.14}$$

$$\frac{\mathrm{d}}{\mathrm{d}t}iq_{st} = \frac{1}{L_{qst}}V_{qst} - \frac{R_{st}}{L_{qst}}i_{qst} + \frac{L_{dst}}{L_{qst}}p\omega_r i_{dst-\lambda_a}\rho\frac{\omega_r}{L_{qst}} \tag{10.15}$$

The equations of PMSG rotor are as:

$$\frac{\mathrm{d}}{\mathrm{d}t}id_{rt} = \frac{1}{L_{drt}}V_{drt} - \frac{R_r}{L_{drt}}i_{drt} + \frac{L_{qrt}}{L_{drt}}p\omega_s i_{qrt} \tag{10.16}$$

$$\frac{\mathrm{d}}{\mathrm{d}t}iq_{rt} = \frac{1}{L_{qrt}}V_{qrt} - \frac{R_{rt}}{L_{qrt}}i_{qrt} + \frac{L_{drt}}{L_{qrt}}p\omega_s i_{drt-\lambda_a}\rho\frac{\omega_{st}}{L_{qrt}} \tag{10.17}$$

The stator flux linkage equations are as:

$$\Psi_{dst} = L_{dst}i_{dst} - \Psi_m \tag{10.18}$$

$$\Psi_{qst} = L_{qst}i_{qst} \tag{10.19}$$

The Electromagnetic Torque (T_e) is described as:

$$T_e = \frac{3}{2} P(\Psi_m iq_{st} + (L_{dst} - L_{qst})id_{st}iq_{st}) \tag{10.20}$$

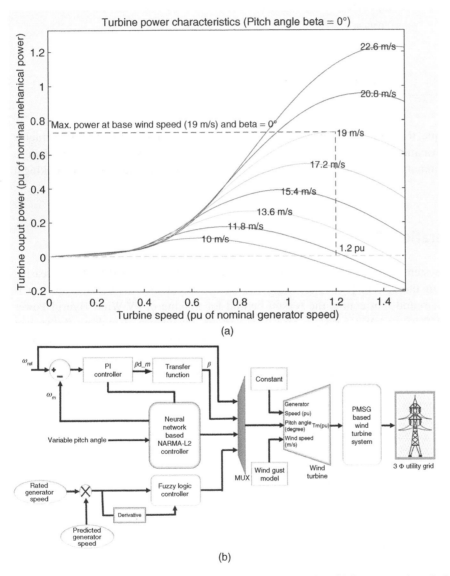

(a)

(b)

Figure 10.11 (a) The characteristics of PMSG-based wind turbine. (b) Representation of pitch angle control techniques applied to the PMSG based wind turbine system.

Table 10.3 Ratings of the proposed wind turbine system.

Parameters	Values
Base wind speed	19 m/s
Output power	700 W
Maximum power at base wind speed	0.73
Gust start time	5 s

The active and reactive power equations of the PMSG System are as:

$$P_s = \frac{3}{2} \left(V_{dst}\, i_{dst} + V_{qst} i_{qst} \right) \tag{10.21}$$

$$Q_s = \frac{3}{2} \left(V_{qst}\, i_{dst} - V_{dst} i_{qst} \right) \tag{10.22}$$

where R and L are the inductance and resistance of the system, and V, I, Ψ are the voltage, current, and flux linkage of the Turbine System. In Figure 10.11, the characteristics of PMSG-Based Wind Turbine System are represented and plot in between of output power and speed of the Turbine System. As in Table 10.3, the ratings of the proposed system are formulated.

10.5 Pitch Angle Control Techniques

There are several methods for controlling the system, but the most magnified and skillful approach is based on pitch angle control of Wind Turbine profitably [17, 18]. The pitch angle control technique implemented is based on a combination of three main dexterously controllers as PI, Fuzzy, and NN-based NARMA-L2 Controllers. The use of the three controllers is to magnify the harvesting of wind energy of the system and smoothing the unwanted variations of the system. In Figure 10.11, the complete representation of three virtuous controllers as pitch angle controlling technique for the system is shown. In the system, variable pitch angles taken for analyzing the controllers performance are 0°, 5°, 10°, …40°, 45°, and from the respective three controllers, performance obtained is highly admirable and efficient as 0.9999–0.9998.

10.5.1 PI Controller

The pitch angle technique presented in this chapter is the combination of the three controllers and using of PI Controller is one of them. The PI Controllers makes the system easier and smoother [19]. The pitch angle control is achieved feasible by one of the three, the PI Controller, in order to keep a constant output power. The plant's dynamic behavior is characterized by the relationship between pitch demand and pitch angle. The following is the desired pitch angle for the proposed Wind Turbine configuration:

$$\beta_{d\_m} = K_p e + K_i \int e\, dt \tag{10.23}$$

$$\text{as } e = \omega_{\text{ref}} - \omega_m \tag{10.24}$$

where the output of controller is denoted by β_{d_m}, ω_{ref} is the generator reference speed, and ω_m is the generator speed. In Figure 10.12, the tuned response of the controller is generated with $K_p = 1.49$ and $K_i = 3.129$.

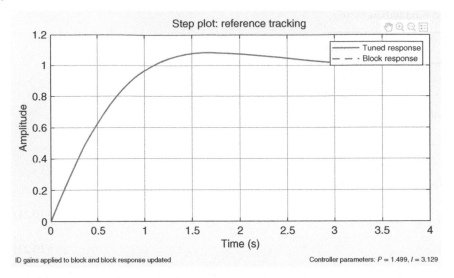

Figure 10.12 Response plot obtained after tuning the controller.

10.5.2 NARMA-L2 Controller

The NARMA-L2 Controller is also known as Feedback Linearization Controller, and it is based on the nonlinear autoregressive moving average algorithm [20, 21]. The concept of using it behind is that of its fast action which makes the vigorous in actions. It also has the capability that learning from experience and consideration of input with its relevance and then realize the theory from previous results. It does so by converting the system's nonlinear dynamics to linear dynamics and minimizing the system's nonlinearities. This technique is used with the other two controllers as PI and Fuzzy Logic Controller. The controller operational process depends firstly on to identifying the system which has to be controlled. Then training of the NN is done by the input given to the controller as variable pitch angles. The general model representing the nonlinear discrete-time systems of the controller is as follows:

$$y(k + d) = N[y(k), y(k - 1), \dots, y(k - n + 1), u(k), u(k - 1), \dots, u(k - n + 1)] \tag{10.25}$$

where $u(k)$ is input to the system and $y(k)$ is output of the system. The merit of the controller is also that the controlling action which is to be done can follow the reference for the input of the system as to follow the output

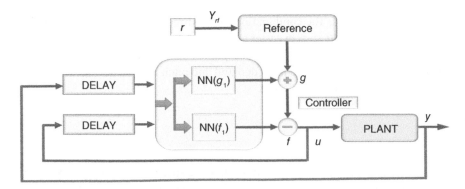

Figure 10.13 Schematic of NARMA-L2 controller.

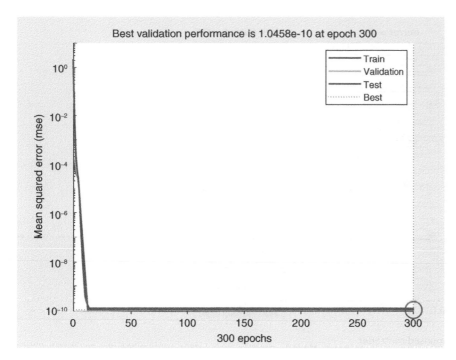

Figure 10.14   NN controllers training performance.

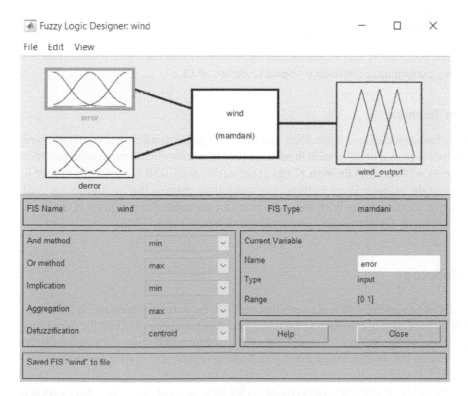

Figure 10.15   The proposed fuzzy logic controller designer view.

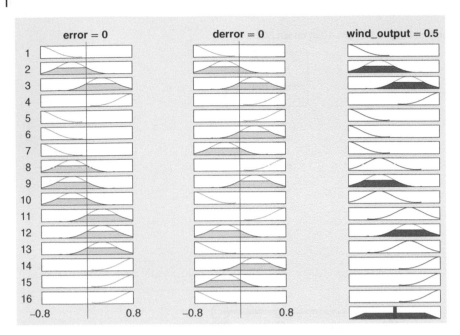

Figure 10.16   Rules formation of proposed controller.

as $y(k+d) = y_r(k+d)$. The resulting performance of the controller is as:

$$u(k) = \frac{y_{rf}(k+d) - f'[y(k), y(k-1), \dots, y(k-n+1), u(k-1), \dots, u(k-n+1)]}{g'[y(k), y(k-1), \dots, y(k-n+1), u(k-1), \dots, u(k-n+1)]} \qquad (10.26)$$

In Figure 10.13, for the pitch angle control of Wind turbine System the NARMA-L2 schematic controlling process is shown. The Controller's training performance obtained is shown in Figure 10.14.

10.5.3   Fuzzy Logic Controller Technique

The pitch angle control strategy is based on the three controllers, and Fuzzy Logic Controller is the very finest type of controller which upgrades the performance of the system in advance [22]. The controller in the system provides the rated power of the generator as well as an evaluation of the generator's power, and the system's output is utilized to control the pitch angle of the planned Wind Turbine. For controlling purpose the two inputs are used which generate the respective output for the system. With 16 rules formed for the controller as shown in the block diagram of Figure 10.15, rules from Fuzzy Logic Designer generated are shown in Figure 10.16.

10.6   Proposed DVRs Topology

The DVR is a high-tech gadget that manages the sags and swells that occur in the path of generation. It is a member of the FACTS devices family. In the realm of electronics, the FACTS devices have set a standard for preserving the efficiency and consistency of the power system [7]. The DVR offers various capabilities for upgrading the performance of the system by making the system much feasible and more reliable [23, 24]. It plays an essential and imperative role in making the system distortion less. Its applications are wide enough. As system faces many discrepancies while running and for overcoming such situations we have realized the DVR in the proposed PV-Wind Hybrid System. It is further controlled by a decisive and dexterous controller known as ANFIS with the

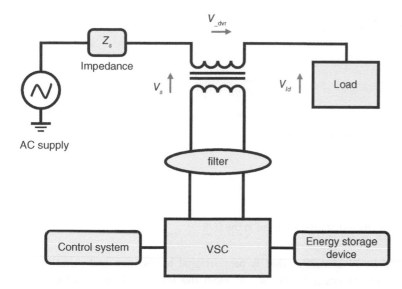

Figure 10.17 General structure of the proposed DVR to the system.

PI Controller in the DVR of the system. The structure of the DVR includes various beneficiary components such as an energy storage device, VSC, injection transformer, control system used for controlling it, and harmonic filter escorted in it as shown in Figure 10.17. The voltage equation of the DVR is as follows:

$$V_{dvr} = V_{ld} + Z_{th}I_{ld} - V_{th} \tag{10.27}$$

where V_{ld} is the voltage of the load, Z_{th} denotes the impedance of the load, V_{th} indicates the systems voltage during fault, and I_{ld} represents the current of the load. Taking into consideration V_{ld} as the reference voltage the system then the voltage equation can be as follows:

$$V_{dvr} = V_{ld}^{\angle 0} + z_{th}^{\angle all-\theta} - V_{th}^{\angle \delta} \tag{10.28}$$

$$\text{As, } \theta = \tan^{-1}\left(\frac{\theta_l}{p_l}\right) \tag{10.29}$$

10.7 Proposed Controlling Technique of DVR

For restoring the voltage of the load the system uses the skilled DVR system for maintaining the system's active power during voltage sag conditions. There are various controlling strategies proposed by Ogunboyo et al. [25]. The very beneficial and new controlling strategy is proposed in this chapter as the system is to be controlled by the ANFIS and PI Controller for increasing the response of the system and output of the Hybrid Power generation. The DVR is a controlled voltage source of the system and installed between the supply and sensitive and critical loads [26, 27]. The proposed controllers satisfied the performance of the system during the faulty conditions it sustains the voltage of the load unchanged. The controlling technique proposed in the system is shown in Figure 10.18.

10.7.1 ANFIS and PI Controlling Technique

The PI Controller proposed in the system is Discrete PI Controller which is a feedback controller. The system is controlled by the weighted total of the errors and the summation of that value. The error signal and integral

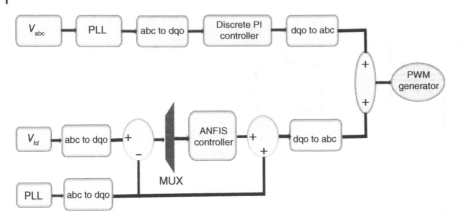

Figure 10.18   Complete representation of controlling techniques proposed in the DVR.

gain inputs are used to create a discrete-time PI controller that is proportional to the size and duration of the error. The mistake is then multiplied by the integral gain, K_i, to yield the integrated offset, which has already been corrected. As the PI Controller gives the efficient output for the system, but its applications are reduced until the operation is not of a wide range. So we have proposed the ANFIS Controller with it for widening the area and results of the system. It is based on the feed-forward network and hybrid learning rules. It greatly optimizes the Fuzzy Inference System and handles conditions of the network in which it is implemented. The learning algorithm in it has Sugeno-type Fuzzy Inference System used for training the controller [28, 29]. In this, the controller utilizes the nonlinear functions of the network in online series of diagnosing the components. The FIS has five layers of operation which are based on Back-Propagation NN structure are as follows:

$$\text{Layer } 1 : O_i^1 = L_i(x_i), \text{ for } i = 1, 2, \ldots, p \tag{10.30}$$

In this layer, the use of the membership functions as $L_1, L_2, \ldots, L_q$ are equal to node functions that are used in regular fuzzy systems.

$$\text{Layer } 2 : O_i^2 = L_i(x_i) \text{ AND } L_j(x_j) \tag{10.31}$$

This layer contains the sum of all incoming nodes as well as each node's output.

$$\text{Layer } 3 : O_i^3 = \frac{O_i^2}{\sum\limits_i O_i^2} \tag{10.32}$$

This layer gives a connection between all the normalized firing strengths and the total of all firing values.

$$\text{Layer } 4 : O_i^4 = O_i^3 \sum_{j=1}^{p} P_j x_j + c_j \tag{10.33}$$

The DVRs structure is represented by Layer 4. With every IF-THEN rule, it displays the system's variables as input and output.

$$\text{Layer } 5 : O_i^5 = \sum_i O_i^4 \tag{10.34}$$

The output of the system's total equation is described in the last layer. The Sugeno type Fuzzy Inference System, also known as the ANFIS, is depicted in Figure 10.19.

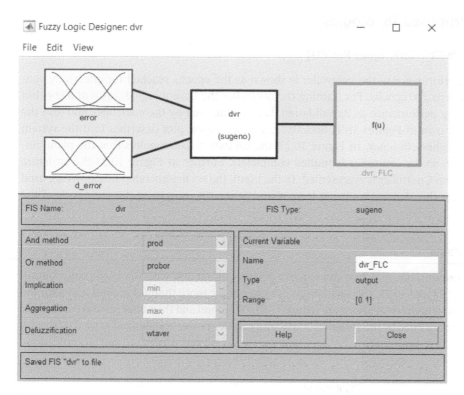

Figure 10.19 FIS of the ANFIS controller.

Figure 10.20 After training the network of NN Predictive Controller obtained performance.

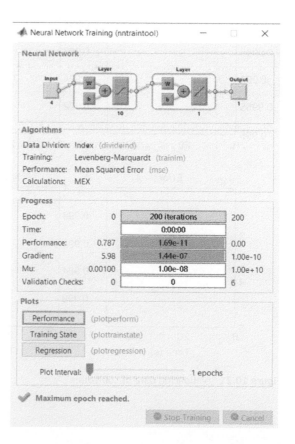

10.8   Results of the Proposed Topologies

10.8.1   PV System Outputs (MPPT Techniques Results)

In Figure 10.20, the training performance of the controller is shown as the epochs reached for effectively controlling the MPPT of the system as 200 epochs. For training the controller the Levenberg–Marquardt algorithm is used. As in result, the training performance as its validating data is generated by the controller and also the regression plot is obtained as shown in Figures 10.21 and 10.22. The regression plot describes that the system is trained completely with its higher efficiency. In Figure 10.21, the validation data denotes that the plant output and NN output is the same, so the controller is trained completely. Further in Figure 10.23 the structure plot obtained from Takagi–Sugeno Controller is represented. In the input, the six membership functions are used and with a combination of these the 36 rules are being formed and generating output as the duty cycle of the controller.

10.8.2   Main PV System outputs

In Figures 10.24–10.27, the representation of active and reactive power without and with the DVRs contribution is shown. As in both Figures 10.24 and 10.25, the oscillations are present for more than 0.1 second and with the contribution of the DVR in the system the oscillations are reduced till 0.07 second and then after several oscillations it is stabilized. The active power generated by the system is 2 MW, and reactive power is stabilized at $t = 0.07$ second in Figures 10.26 and 10.27.

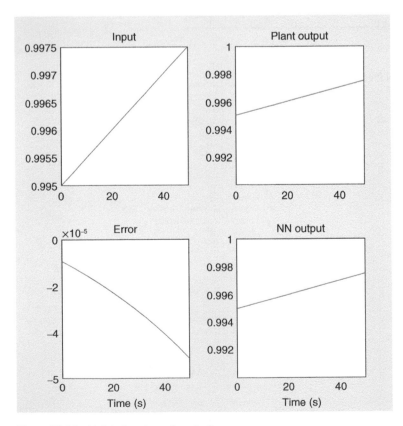

Figure 10.21   Validating data of controller.

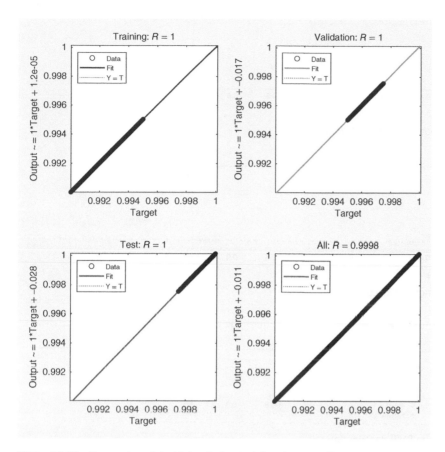

Figure 10.22   Regression plot obtained after training the controller.

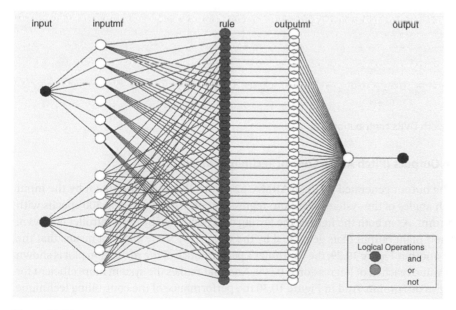

Figure 10.23   ANFIS structure of MPPT technique.

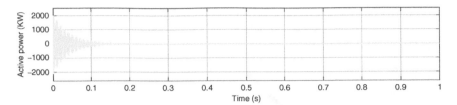

Figure 10.24   The Active Power without the DVRs contribution.

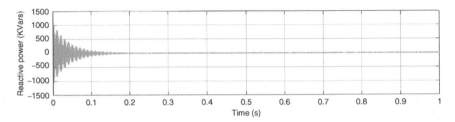

Figure 10.25   The Reactive Power without DVRs contribution.

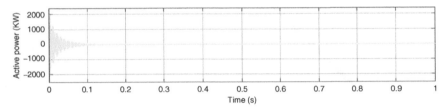

Figure 10.26   The Active Power with the DVRs contribution.

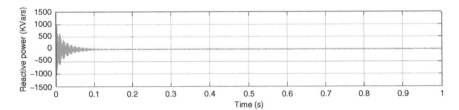

Figure 10.27   The Reactive Power with DVRs contribution.

10.8.3   Wind Turbine System Outputs (Pitch Angle Control Technique Result)

Figures 10.28 and 10.29 show the output generated by the NARMA-L2 Controller while training it by the input variables for controlling the pitch angles of the system. The Auto-regressive controller trained at 300 epochs with the Levenberg–Marquardt algorithm. As in both the figures, the testing data generated by the controller is shown. In Figure 10.28 the NN output and plant output plots generated by the system are same and that means that the system is trained successfully. In another Figure 10.29, the controller's performance of the regression plot is shown which represents the maximum value reached of Regression = 0.999. And that makes the system more efficient for controlling the desired pitch angles of Turbine. And in Figure 10.30 the performance of the controlling technique

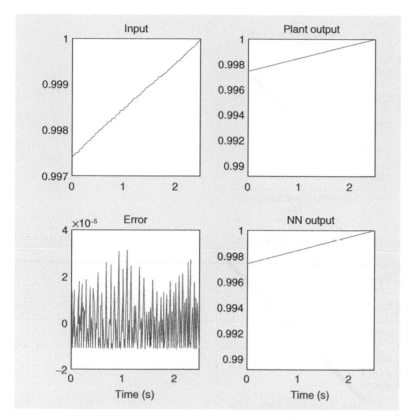

Figure 10.28 Testing data obtained after training the controller.

applied to the system and the output generated by this controlling technique as Fuzzy Logic, PI, and NARMA-L2 are represented.

10.8.4 Proposed PMSG Wind Turbine System Output

In Figure 10.31 the Electromagnetic Torque, T_e, generated by the proposed PMSG-Based Wind Turbine system is shown which is stabilized after 0.08 second with the DVRs contribution after reaching its maximum value. Further in Figure 10.32 the three-phase voltage obtained with the DVRs contribution is represented. As for starting cycles of 0.01–0.02 second, the transient period comes and then after at 0.03 second the system again regains and supplies a constant supply of voltage with reaching its maximum voltage.

In Figures 10.33 and 10.34, the active and reactive power generated by the PMSG-Based Wind Turbine system without the DVRs contribution is shown. In these figures, the oscillation is continued till 0.1 second, and the maximum power reaches by the system is 1 MW where it becomes constant and reactive power reaches till negative values. Then in Figures 10.35 and 10.36 the active and reactive power with the DVRs contribution is presented, and the oscillation is damped till 0.07 second only. In these figures the active primarily reaches its maximum value of 4 MW and then stabilizes at 1 MW with this the reactive with damping of oscillations at early stage stabilizes soon.

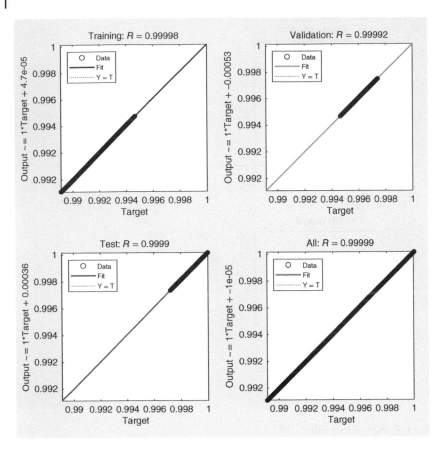

Figure 10.29 Regression plot obtained after training the controller.

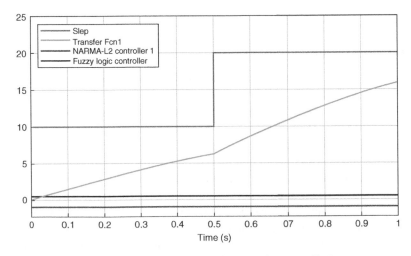

Figure 10.30 Performance of the controlling techniques applied to system.

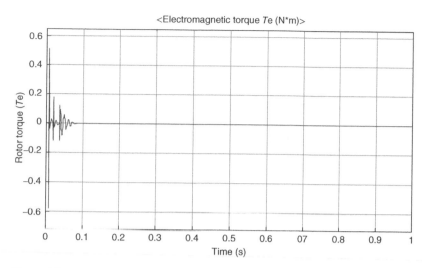

Figure 10.31 The electromagnetic torque obtained of turbine.

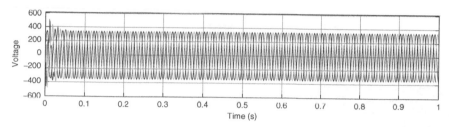

Figure 10.32 The three-phase voltage obtained of the system.

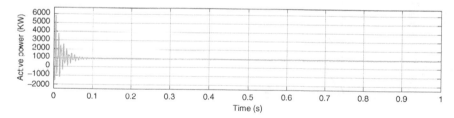

Figure 10.33 The active power of the system without the DVRs contribution.

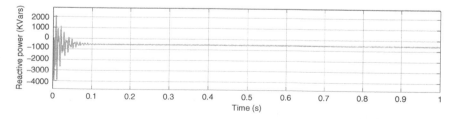

Figure 10.34 The reactive power of the system without DVRs contribution.

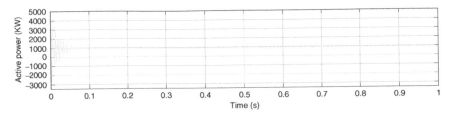

Figure 10.35 The active power obtained with the DVR contribution.

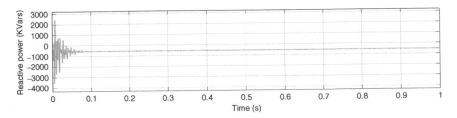

Figure 10.36 The reactive power obtained with the DVRs contribution.

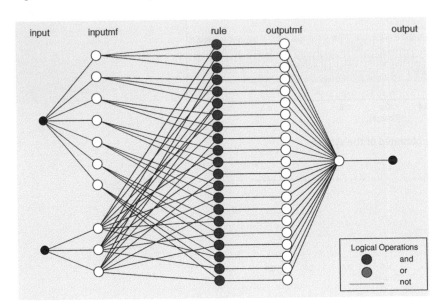

Figure 10.37 ANFIS structure obtained of DVR.

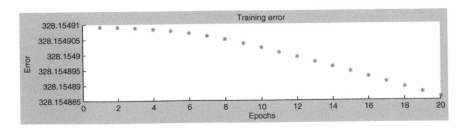

Figure 10.38 The training performance from ANFIS controller.

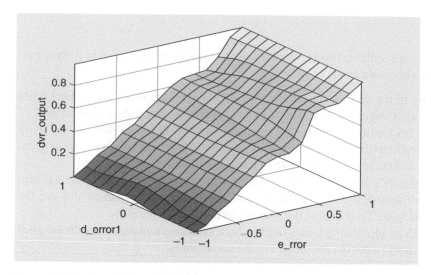

Figure 10.39 Surface viewer of the proposed ANFIS controller in system.

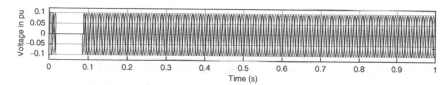

Figure 10.40 Voltage to the system without DVR.

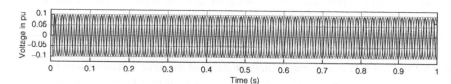

Figure 10.41 Voltage to the system with DVR.

10.8.5 Performance of DVR (Controlling Technique Results)

In Figures 10.37 and 10.38 the controlling technique applied to the DVR system and the output generated by training the controller with maximum epochs 20 is presented, and the ANFIS structure generated by the Takagi–Sugeno type Fuzzy Logic Controller is shown with its 7 as input 1 and in another 3 as input 2 as membership functions are presented with making of 21 rules for controlling of the DVR and obtaining the Total Harmonic Distortion (THD) level below 5% that has been obtained also. In Figure 10.39 the surface viewer generated by the ANFIS Controller is shown within input error and derivative of error is shown in the output, and the controlling of the DVR as output is presented.

10.8.6 DVRs Performance

In Figure 10.40 the output voltage generated by the Hybrid system is presented which is without the DVRs contribution and has voltage sag as generated by the system. As by the contribution of the DVR in the PV-Wind Hybrid system, the continuous voltage, three-phase voltage is obtained within minimum-1 and maximum-1 values and that means that the system is ought to maintain the unity power factor for a stable generation as shown in Figure 10.41.

10.9 Conclusion

The proposed PV-Wind Hybrid system has validated its performance with maintaining the efficiency of the network as the following conclusions can be drawn from the system:

(i) By implementing the techniques in the system, the results that are obtained are better than many other techniques applied to Hybrid Power Generation System as it is showing the efficiency of network as 98%.

(ii) By applying in the PV system the NN Predictive controlling technique with ANFIS and in PMSG-Based Wind system the NN NARMA-L2 Controller with PI and Fuzzy Logic Controller, the system has increased its accuracy in running and paramount, and the results obtained from these controllers are well justified.

(iii) The optimistic performance of the DVR controller with its ANFIS with PI controlling technique has also intensified the efficiency and outcomes of the individually PV and Wind system as explained and also the Hybrid generation with maintaining the THD level of the system less than 5% significantly.

(iv) The main purpose of making such an adorable system is fulfilled here which is more familiar with the need of present power generation status and meet the consumer's demands skillfully.

References

1 Gielen, D., Boshell, F., Saygin, D. et al. (2019). The role of renewable energy in the global energy transformation. *Energy Strategy Reviews* 24: 38–50.

2 Anoune, K., Bouya, M., Astito, A., and Abdellah, A.B. (2018). Sizing methods and optimization techniques for PV-wind based hybrid renewable energy system: a review. *Renewable and Sustainable Energy Reviews* 93: 652–673.

3 Podder, A.K., Roy, N.K., and Pota, H.R. (2019). MPPT methods for solar PV systems: a critical review based on tracking nature. *IET Renewable Power Generation* 13: 1615–1632.

4 Lyu, X., Zhao, J., Jia, Y. et al. (2019). Coordinated control strategies of PMSG-based wind turbine for smoothing power fluctuations. *IEEE Transactions on Power Systems* 34 (1): 391–401.

5 Pradhan, M. and Mishra, M.K. (2019). Dual P-Q theory based energy-optimized dynamic voltage restorer for power quality improvement in a distribution system. *IEEE Transactions on Industrial Electronics* 66 (4): 2946–2955.

6 Benali, A., Khiat, M., Allaoui, T., and Denai, M. (2018). Power quality improvement and low voltage ride through capability in hybrid wind-PV farms grid-connected using dynamic voltage restorer. *IEEE Access* 6: 68634–68648.

7 Gandoman, F.H., Ahmadi, A., Sharaf, A.M. et al. (2018). Review of FACTS technologies and applications for power quality in smart grids with renewable energy systems. *Renewable and Sustainable Energy Reviews* 82: 502–514.

8 Panigrahi, B.K., Ray, P.K., Rout, P.K. et al. (2018). Detection and classification of faults in a microgrid using wavelet neural network. *Journal of Information and Optimization Sciences* 39 (1): 327–335.

9 Chaibi, Y., Allouhi, A., Malvoni, M. et al. (2019). Solar irradiance and temperature influence on the photovoltaic cell equivalent-circuit models. *Solar Energy* 188: 1102–1110.

10 Dashtdar, M. and Dashtdar, M. (2020). Voltage and frequency control of islanded micro-grid based on battery and MPPT coordinated control. *Mapta Journal of Electrical and Computer Engineering (MJECE)* 2 (1): 1–19.

11 Khan, M.J. and Mathew, L. (2019). Comparative study of maximum power point tracking techniques for hybrid renewable energy system. *International Journal of Electronics* 106 (8): 1216–1228.

12 Mishra, R. and Saha, T.K. (2020). Performance analysis of model predictive technique based combined control for PMSG-based distributed generation unit. *IEEE Transactions on Industrial Electronics* 67 (10): 8991–9000.

13 Panigrahi, B.K., Rout, P.K., Ray, P.K., and Kiran, A. (2018). Fault detection and classification using wavelet transform and neuro fuzzy system. In: *2018 International Conference on Current Trends towards Converging Technologies (ICCTCT)*, 1–5. IEEE.

14 Peng, X., Liu, Z., and Jiang, D. (2021). A review of multiphase energy conversion in wind power generation. *Renewable and Sustainable Energy Reviews* 147: 111172.

15 Jain, A., Shankar, S., and Vanitha, V. (2018). Power generation using permanent magnet synchronous generator (PMSG) based variable speed wind energy conversion system (WECS): An overview. *Journal of Green Engineering* 74 (4): 477–504.

16 Muñoz, C.Q.G. and García Márquez, F.P. (2018). Wind energy power prospective. In: *Renewable Energies* (ed. F. García Márquez, A. Karyotakis and M. Papaelias). Cham: Springer https://doi.org/10.1007/978-3-319-45364-4_6.

17 Ahmadi, M., Lotfy, M.E., Howlader, A.M. et al. (2019). Centralised multi-objective integration of wind farm and battery energy storage system in real-distribution network considering environmental, technical and economic perspective. *IET Generation, Transmission & Distribution* 13 (22): 5207–5217.

18 Abir, A., Mehdi, D., and Lassaad, S. (2016). Pitch angle control of the variable speed wind turbine. In: *2016 17th International Conference on Sciences and Techniques of Automatic Control and Computer Engineering (STA)*, 582–587. IEEE.

19 Krommydas, K.F. and Alexandridis, A.T. (2015). Modular control design and stability analysis of isolated PV-source/battery-storage distributed generation systems. *IEEE Journal on Emerging and Selected Topics in Circuits and Systems* 5 (3): 372–382.

20 Mokri, S.S. and Shafie, A.A. (2008). Real time implementation of NARMA L2 feedback linearization and smoothed NARMA L2 controls of a single link manipulator. In: *2008 International Conference on Computer and Communication Engineering*, 691–697. IEEE.

21 John, S. and Pedro, J.O. (2013). Neural network-based adaptive feedback linearization control of antilock braking system. *International Journal of Artificial Intelligence* 10 (S13): 21–40.

22 Singh, K. (2020). Load frequency regulation by de-loaded tidal turbine power plant units using fractional fuzzy based PID droop controller. *Applied Soft Computing* 92: 106338.

23 Swain, P., Panigrahi, B.K., Dalai, R.P. et al. (2019). Mitigation of voltage sag and voltage swell by dynamic voltage restorer. In: *2019 International Conference on Intelligent Computing and Control Systems (ICCS)*, 485–490. IEEE.

24 Anees, A.S. (2012). Grid integration of renewable energy sources: challenges, issues and possible solutions. In: *2012 IEEE 5th India International Conference on Power Electronics (IICPE)*, 1–6. IEEE.

25 Ogunboyo, P.T., Tiako, R., and Davidson, I.E. (2018). Effectiveness of dynamic voltage restorer for unbalance voltage mitigation and voltage profile improvement in secondary distribution system. *Canadian Journal of Electrical and Computer Engineering* 41 (2): 105–115.

26 Istanbuly, M., Halloum, M., and Panigrahi, B.K. (2021). Dynamic stability improvement of power system utilizing fuzzy logic and compare it with conventional stabilizer. In: *Advances in Intelligent Computing and Communication*, Lecture Notes in Networks and Systems, vol. 202. https://doi.org/10.1007/978-981-16-0695-3_16 (ed. S. Das and M.N. Mohanty), 153–161. Singapore: Springer.

27 Guo, Y. and Mohamed, M.E.A. (2020). Speed control of direct current motor using ANFIS based hybrid PID configuration controller. *IEEE Access* 8: 125638–125647.

28 Gharajeh, M.S. and Jond, H.B. (2020). Hybrid global positioning system-adaptive neuro-fuzzy inference system based autonomous mobile robot navigation. *Robotics and Autonomous Systems* 134: 103669.

29 Güler, I. and Übeyli, E.D. (2004). Application of adaptive neuro-fuzzy inference system for detection of electrocardiographic changes in patients with partial epilepsy using feature extraction. *Expert Systems with Applications* 27 (3): 323–330.

11

Deep Reinforcement Learning and Energy Price Prediction

Deepak Yadav[1], Saad Mekhilef[1,2,3], Brijesh Singh[4], and Muhyaddin Rawa[3,5]

[1] *Power Electronics and Renewable Energy Research Laboratory, Department of Electrical Engineering, University of Malaya, Kuala Lumpur, Malaysia*
[2] *School of Science, Computing and Engineering Technologies, Swinburne University of Technology, Hawthorn, Vic, Australia*
[3] *Smart Grids Research Group, Center of Research Excellence in Renewable Energy and Power Systems, King Abdulaziz University, Jeddah, Saudi Arabia*
[4] *Department of Electrical and Electronics Engineering, KIET Group of Institutions, Ghaziabad, India*
[5] *Department of Electrical and Computer Engineering, Faculty of Engineering, K.A. CARE Energy Research and Innovation Center, King Abdulaziz University, Jeddah, Saudi Arabia*

Abbreviations

ARIMA	autoregressive integrated moving average
LMP	locational marginal pricing
RSI	relative strength index
PV	solar photo-voltaic sources
IP	interior point method
OPF	optimal power flow
MPP	maximum power point
FERC	Federal Energy Regulatory Commission
ISO	independent system operator
RTO	regional transmission operator
i	node indices, $i = 1, 2, 3, \ldots$
k	node indices, $k = 1, 2, 3, \ldots; i \neq k$
u	network index, $z = 1, 2, 3, \ldots, m, n, \ldots U$
T_l	tie-lines index
G	total generators' number
D	total demands' number
C_i	cost coefficient of generators
C_s	cost coefficient of slack generator
P_{g_i}	active power output of generators
$\underline{P_{g_i}}$	lower limit of active power generation of conventional generators
$\overline{P_{g_i}}$	upper limit of active power generation of conventional generator
$P_{g_i}{}^{MPP}$	PVs' active power maximum power point
P_{d_i}	active power load at bus i

Artificial Intelligence-based Smart Power Systems, First Edition.
Edited by Sanjeevikumar Padmanaban, Sivaraman Palanisamy, Sharmeela Chenniappan, and Jens Bo Holm-Nielsen.
© 2023 The Institute of Electrical and Electronics Engineers, Inc. Published 2023 by John Wiley & Sons, Inc.

P_{loss}	losses (MW)
$\underline{Q_{g_i}}$	lower limit of conventional generators' reactive power
$\overline{Q_{g_i}}$	upper limit of conventional generators' reactive power
S_{ik}	apparent power (MVA) in line between nodes i and k of Network U
$\overline{S_{ik}}$	total capacity (MVA) in line between nodes $- i$ and k of Network U
S_{T_l}	tie-line's power flow (MVA) between networks U
$\overline{S_{T_l}}$	tie-line's total capacity (MVA) between networks U
V_i	voltage at i
$\underline{V_i}$	lower voltage boundary at node i
$\overline{V_i}$	upper voltage boundary at node i
$\mathcal{H}$	inequality constraints' vector
μ	constraints Lagrange multiplier

11.1 Introduction

Increased investment in clean and affordable energy resources is being driven by rapid rise in demand and rising global temperatures. Solar PV (PV) renewable resources have been encouraged, in addition to their ease of installation. The infrastructures of electricity and renewable energy resources have been changed because of economic, environmental, technological, and political incentives. The expanding power and energy markets are luring investors looking to make money in the real-time and future energy markets [1]. Higher PV and other renewable power injections in the power grid, however, pose some significant challenges. PV units raise voltage levels in the local region over permissible levels [2–4]. The increased penetration of PV units (on a big scale) may result in surplus power generation during peak hours, while demand is also lower during the middle of the day (peak hour of PV units). Without action, this excess power could cause harm to electrical devices including rotating machines connected to a network. The network's load bus can be penetrated up to 120–250% of the minimum daytime demand of the associated bus' maximum power point. With the help of modern converters and no over-voltage issues, this degree of PV penetration is desirable [5–7].

The market's liberalization, active customer engagement, operations, and developing technology of large-interconnected power networks are making it more complicated and dynamic. Now to resolve these new challenges, effective methods are required for planning and operating the future smart grids (SG). The smart grid uses bi-directional flow of power and information among all participants in the market, including producers, transmission and distribution (T&D), consumers, and demand response (DR) aggregators [8]. These under way power systems transformations increasing the uncertainties and complexities in business transactions as well as in physical power flow [9]. Therefore, the future SG should be able to monitor, predict, learn, schedule, and make decisions to provide more efficient and intelligent solutions in real-time [10–12]. Further, deep learning in combination with reinforcement learning (subsets of artificial intelligence [AI]) forms a new technology called deep reinforcement learning [13] which significantly helps robotics [14, 15], finance and business management [16–18], and natural language processing [19, 20]. Many power system challenges can be reduced to sequential decision-making activities. Convex optimization methods, programming methods, and other traditional methods are the most common as well as heuristic methods. By comparing qualitatively, the benefits and drawbacks of these strategies, according to deep reinforcement learning (DRL), are as follows:

A classical mathematical method, such as the Lyapunov optimization algorithm [21], is the first. The benefit of this strategy is that the mathematics is rigorous, allowing for real-time management. This type of technique, on the other hand, is based on explicit objective functional expressions, which are difficult to abstract from many real-world optimization decision circumstances. Furthermore, in sophisticated, high-dimensional circumstances, the Lyapunov condition (which is essential for the Lyapunov optimization procedure) cannot be ensured. The second factor to consider is the programming approach, which includes mixed integer programming [22, 23], dynamic programming [24, 25], and stochastic programming [26]. These methods can be used to tackle a wide range of optimization issues, including sequence optimization. This type of technique, however, requires each iteration to be recalculated from the beginning. Furthermore, in some cases, the calculation cost is too high to allow for real-time decision-making. Some programming techniques rely on precise renewable energy generation and load predictions, which are challenging to achieve in real-world circumstances.

Heuristic approaches, such as genetic algorithms (GAs) [27, 28], ant colony optimization (ACO) [29, 30], or particle swarm optimization (PSO) [31, 32], are another type of heuristic method. A heuristic approach can reach the local optimal solution with a certain probability for optimization issues, notably non-convex optimization problems, which is advantageous to solving the problem of big data scale and difficult scenarios. These methods, on the other hand, are less reliable and cannot be rationally verified using mathematics. In contrast to convex optimization approaches, DRL does not require an accurate objective function. DRL, on the other hand, evaluates decision behavior using the reward function. In comparison to convex optimization approaches, DRL can handle larger dimensional data. DRL, in contrast to programming methodologies, makes decisions based on the present situation, making real-time and online decisions. DRL is more resilient than heuristic approaches, with solid convergence findings and is better suited for decision-making situations.

Further to study a case using statistical and predictive techniques, this work extended the [33]; PV systems generate incredibly low-cost electricity, according to the explanation. As a result, making long-term decisions on how to operate conventional generators profitably has become more difficult. As a result, independent system operators (ISOs) and transmission system operators (TSOs) have indeed suggested inter-regional/market-to-market power exchange norms. The Federal Energy Regulatory Commission (FERC) of the United States has issued instructions in order no. 1000, even though ISO/regional transmission operators (RTO) interconnections are based on financial commitments. Pennsylvania-New Jersey-Maryland Interconnection (PJM) operators have also suggested that the system be tested for the ongoing projects in analyzing the technical aspects of the systems [34, 35]. In the article [36], the authors have found that the direct interconnections of identified PV units to identified load buses would provide optimal solutions for the M2M.

In this chapter, locational marginal prices (LMPs) have been calculated and forecasted using autoregressive integrated moving average (ARIMA) and have been implied with relative strength index (RSI) which provides signal to purchase electricity from wholesale market. With growing digital technologies, consumers and retailers would have access to electricity prices. Though to understand best prices in the market, it is hard to identify best time to purchase or bid for the electricity. ARIMA is a statistical method to forecast prices in market. The ARIMA stands for the autoregressive integrated moving average, which calculates the differenced time series data using lagged errors [37]. Further, the RSI stands for relative strength index. This RSI calculates index defined in range of 0–100. The lower value of RSI suggests the fall in energy prices (LMPs). RSI uses last 14 period values of LMPs at any certain node, then for the next time slot it excludes the first value of LMP. In more general way, RSI computes the 0–100 value using average value of increased price and average value of decreased price in last 14 periods [38]. In this way, the retailers and consumer not having market knowledge could choose their best time to enter the market as well as to exit the market. Therefore, in an M2M market the consumers would have more options to trade electricity with volatile prices.

The test has considered the LMPs as the energy price at nodes of the network. The unit-commitments for the generators have been allotted using security-constrained-based optimal flow (SC-OPF). Further, the interior-point (IP) optimization technique based on Karush–Kuhn–Tucker (KKT) optimality conditions has been used to obtain the optimal solutions. The optimal schedules of the generators have been obtained using IP-SCOPF.

11.2 Deep and Reinforcement Learning for Decision-Making Problems in Smart Power Systems

Deep reinforcement learning combines the sensing function of deep learning with the decision-making capabilities of reinforcement learning. Because it is an artificial intelligence technology that is closer to human thinking, it is termed real AI. Figure 11.1 depicts the DRL's core framework. Deep learning obtains targeted observation data from the surroundings and offers current environment state information. The reinforcement learning algorithm then translates the current state to the appropriate action and assesses values using the expected return [39, 40]. Decision-making behavior becomes a step-by-step procedure because of a continuous interaction process. The next section introduces the principles and algorithms for reinforcement learning.

11.2.1 Reinforcement Learning

A behavior strategy, or a policy that minimizes a satisfaction criterion, is calculated using reinforcement learning. Meanwhile, interacting with a particular environment through trial and error yields a long-term aggregate of rewards. A reinforcement learning framework consists of a decision-maker, termed the agent, functioning in an environment characterized by state variables to accomplish these functions. As a function of the current state s_t, the agent is capable of doing specific actions. As shown in Figure 11.1, after choosing an action at time t, the agent receives a scalar reward r_{t+1} and enters a new state s_{t+1} that is dependent on the current state and the chosen action. The following sections cover the mathematical underpinnings and ideas of reinforcement learning.

11.2.1.1 Markov Decision Process (MDP)

A Markov decision process, as illustrated in Figure 11.2, is a basic formulation of reinforcement learning that satisfies a Markov property. A Markov property is one in which future of process depends solely on its current state, and the agent is uninterested in the entire history. It can be summarized as follows:

$$P_s(s_{t+1}|s_0, a_0, \dots, s_t, a_t) = P_s(s_{t+1}|s_t, a_t) \tag{11.1}$$

where P_s is probability of state transition.

At each epoch, the agent performs an action that alters its environment's state and offers a reward. Value functions and an optimal policy are proposed to further process the reward value.

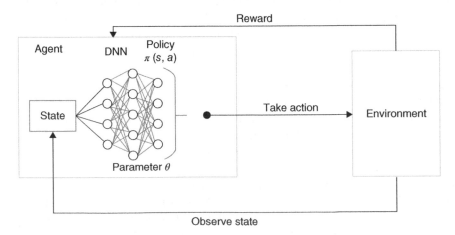

Figure 11.1 Deep reinforcement learning (DRL) structure.

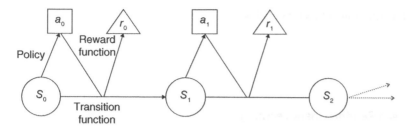

Figure 11.2 Illustration of Markov decision process (MDP).

Real World Examples That Fit MDP Markov decision processes may be used to model a wide range of problems in nature, business, and even our personal lives. Here's a sample of examples/problems:

- Self-driving car (speed/steering actions to maximize safety/time).
- Chess is a board game (actions are moves made by the pieces to improve their chances of winning the game).
- Actions indicate inventory moves in order to maximize throughput/time in complex logistical activities, such as those in a warehouse.
- Developing a humanoid robot capable of walking and running on a rough surface (actions are walking movements to optimize time to destination).
- Portfolio management for investing (actions are trades to optimize long-term investment gains).
- During a football game, making the greatest decisions possible (actions are strategic game calls to optimize chances of winning the game).
- A successful electoral strategy (actions constitute political decisions to optimize chances of winning the election).

11.2.1.2 Value Function and Optimal Policy

The return R_t must maximize the long-term cumulative reward after the current time t in the situation of a finite time horizon that ends at time T.

$$R_t = r_{t+1} + \alpha r_{t+2} + \alpha^2 r_{t+3} + \ldots = \sum_{i=0}^{\infty} \alpha^i r_{t+i+1} \tag{11.2}$$

where the discount factor $\alpha \varepsilon [0, 1]$, and α can take 1 only in episodic Markov decision process (MDPs).

Some methods use the value function $V(s)$, which indicates how beneficial it is for the agent to attain a particular state s, to discover the best policy. Such a function is dependent on the agent's actual policy:

$$V^\pi(s_t) = \mathbb{E}[R_t | s_t = s] = \mathbb{E}\left[\sum_{i=0}^{\infty} \alpha^i r_{t+i+1} | s_t = s\right] \tag{11.3}$$

Similarly, the value of taking "a" action in state "s" under a policy π is represented by the action-value function Q as:

$$Q^\pi(s_t, a_t) = \mathbb{E}[R_t | s_t = s, A_t = a] = \mathbb{E}\left[\sum_{i=0}^{\infty} \alpha^i r_{t+i+1} | s_t = s, A_t = a\right] \tag{11.4}$$

The Bellman equation can be used to express the Q function in an iterative manner in a Q-learning algorithm [41]:

$$Q^\pi(s_t, a_t) = \mathbb{E}[r_{t+1} + \alpha Q^\pi(s_{t+1}, a_{t+1}) | s_t, a_t] \tag{11.5}$$

In the long run, an optimal policy π^* is one that yields the greatest cumulative reward:

$$\pi^* = \frac{\arg \max}{\pi} V^\pi(s) \tag{11.6}$$

The best value function and action-value function at this time will be:

$$V^*(s) = \frac{\max}{\pi} V^\pi(s) \tag{11.7}$$

$$Q^*(s, a) = \frac{\max}{\pi} Q^\pi(s, a) \tag{11.8}$$

11.2.2 Reinforcement Learnings to Deep Reinforcement Learnings

The transition from RL to DRL has taken a long time to complete. State and action spaces in traditional tabular RL, such as Q-learning, are small enough for approximate value functions to be expressed as arrays or tables. The approaches can often discover the exact best value functions and policies in this instance [42]. However, when it comes to real-world implementations, these earlier solutions have a difficult design difficulty. Instead of being expressed as a table, the approximate value functions are represented as a parameterized functional form with a vector of weights (alike deep neural networks). DRL's capacity to acquire degrees of abstraction from data allows it to execute complex tasks with less prior knowledge [43, 44]. Further explanations about DL and DRL have been discussed below.

1. Spaces that are three-dimensional and continuous. Despite the fact that high dimensional and continuous state spaces or action spaces are generated by a range of real-world challenges, they cannot be stored in a table or function. The "curse of dimensionality" is the name given to this occurrence. Function approximation is used to extract features from models, value functions, or policies, and then attempts to generalize from them to produce an approximation of the full function using supervised learning techniques such as deep neural networks [45, 46].
2. The issue of exploration versus exploitation. When an agent begins collecting data about the environment, it must choose between learning more about the environment (exploration) and adopting the most promising strategy with the data obtained (exploitation). Uncertainty about the reward function and transition probabilities can be expressed as confidence intervals or posteriors of environment parameters in tabular RL. Different settings are used in DRL. One is that the agent only investigates when the learning chances are sufficiently valuable that it can perform well without a separate training phase. Another possibility is that the agent follows a training policy during the initial phase of interactions with the environment in order to collect training data and learn a test policy [39, 42].
3. To ensure convergence in RL, only tables and linearly parameterized approximators can be utilized. When there is no prior knowledge to guide the selection of basic functions, a large number of basic functions must be constructed to uniformly cover the state-action space, which is impracticable in high-dimensional issues. Non-linear approximators, such as convolutional neural networks (CNN), have been used with replay buffer and target networks to retrieve features of select regions of states [13, 44, 45].

11.2.3 Deep Reinforcement Learning Algorithms

Optimization, planning, management, and control challenges are all examples of DRL problems. As shown in Figure 11.3, solution strategies might be model-free or model-based, value-based, or policy-based. Control theory has a big influence on model-based DRL, and it's commonly presented in terms of several disciplines. Model-free DRL, on the other hand, ignores the model and is less concerned with its inner workings. The advantage of model-based DRL is that it is easy and efficient. If it is appropriate to estimate the space as linear, for example, learning the model will require far fewer samples. Model-based methods, on the other hand, are several orders of magnitude more sophisticated than model-free methods.

Value-based approaches learn from any trajectory taken from the same environment by iteratively increasing the value function until it converges. The iteration process of the Q function in tabular RL, such as Q-learning, is

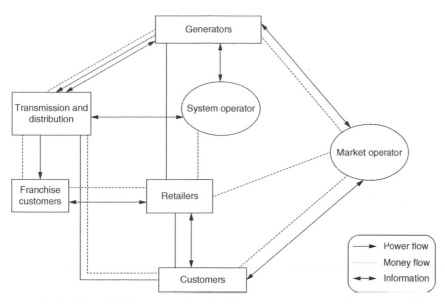

Figure 11.3 Electricity Market Participants.

represented in Eq. (11.9); however in DRL, it will update as shown in Eq. (11.10). The objective function can be defined as Eq. (11.11):

$$Q(s_t, a_t) \leftarrow Q(s_t, a_t) + \beta \left[r_{t+1} + \alpha \frac{\max}{a} Q(s_{t+1}, a) - Q(s_t, a_t) \right] \tag{11.9}$$

$$\theta_{t+1} = \theta_t + \beta \left(r_{t+1} + \alpha \frac{\max}{a} Q(s_{t+1}, a, \theta) - Q(s_t, a_t, \theta) \right) \nabla_\theta Q(s_t, a_t, \theta) \tag{11.10}$$

$$J(\theta) = \mathbb{E} \left[\left(r_{t+1} + \alpha \frac{\max}{a} Q(s_{t+1}, a_{t+1}, \theta) - Q(s_t, a_t, \theta) \right)^2 \right] \tag{11.11}$$

where β is the learning rate and θ is the set of function approximator parameters [47, 48].

Policy-based techniques directly optimize the quantity of interest while remaining stable under function approximations by redefining the policy at each step and computing the value according to this new policy until the policy converges. The gradient of the objective function is first established as policy parameters, as shown in Eq. (11.11) before the weight matrix is updated in Eq. (11.12).

$$\nabla_\theta J(\theta) = \mathbb{E} \left[\sum_{t=0}^{T} \nabla_\theta \log \pi_\theta(a_t | s_t) \sum_{t=0}^{T} r(s_t, a_t) \right] \quad \theta \leftarrow \theta + \nabla_\theta J(\theta) \tag{11.12}$$

11.3 Applications in Power Systems

These applications encompass a wide range of power system decision, control, and optimization challenges, including energy management, demand response, electricity market, operational control, and many more. This section goes over some of the most common application fields.

11.3.1 Energy Management

As depicted in Figure 11.4, energy management issues in a power system, particularly a microgrid, connect the source, load, storage system, and utility grid. Energy management is crucial in a variety of ways. To begin with, it

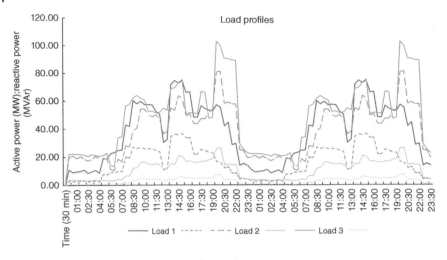

Figure 11.4 Load Profiles at nodes of modified IEEE-9 bus system.

can increase the rate of renewable energy utilization and monitor household appliance consumption. It can also develop a storage scheduling strategy and react to current electricity costs. It's worth noting that most energy management issues can be converted into sequential decision-making problems and solved well with DRL. Residential appliances require effective energy management solutions, which DRL can provide. The use of DRL in imagining an online optimization for the scheduling of power consumption in residential loads and aggregations of buildings is proposed in Mocanu et al. [49]. This energy management method can be used to deliver real-time feedback to consumers in order to encourage them to utilize electricity more wisely. In the paper [50], the authors proposed an effective DRL-based technique for minimizing overall power cost in domestic energy management problems without knowing real-time household load or electricity pricing. A home multicarrier energy system (MCES) with a PV array, battery, heat pump, and gas boiler is developed in Mbuwir et al. [51], and DRL is utilized to build a control strategy for optimal battery scheduling using real-world data.

Electric vehicle energy management issues have become increasingly prevalent in recent years, drawing the attention of DRL experts. An energy management method based on deep Q-learning (DQL) is suggested in Wu et al. [52] for electric vehicles to maximize fuel economy over a particular time horizon while keeping battery state of charge (SoC) steady. DQL outperforms Q-learning in terms of training time and convergence rate. Similarly, without any system model information, Wan et al. [53] provides a model-free DRL approach for determining an optimal strategy for real-time electric vehicle (EV) charging/discharging scheduling. In [54–56], a DRL-based data-driven control approach with a real-time learning architecture for a hybrid electric vehicle without any prediction or predetermined rules is established. When compared to typical control systems, it saves a significant amount of energy. The authors in [57] proposed a continuous control method for plug-in hybrid electric vehicles (PHEVs) based on deep deterministic policy gradient (DDPG), with the algorithm performing near to the global optimal for dynamic programming.

One of the most important concerns in a microgrid is energy management, and DRL has had a lot of success with this. DRL is being used by some researchers to solve the problem of efficiently operating a hybrid storage system in a microgrid with photovoltaics (PV), batteries, and hydrogen [58]. Using state-action value functions to design optimal battery scheduling, the article [59] presents a batch reinforcement learning application in microgrid energy management. The authors of [60] contribute to the use of a multi-agent reinforcement learning system to regulate freestanding microgrids in order to ensure electricity supply and dependability. In Tan et al. [61], a two-layer optimization for real-time optimal energy management (OEM) of a grid-connected microgrid is proposed. The top layer is a model-free Q-learning for decision-making and knowledge acquisition, while the bottom layer is a traditional convex optimization (interior point approach). In Mbuwir et al. [62], a data-driven RL technique for battery energy management in a home microgrid is used to maximize self-consumption of PV production by arranging battery operation scheduling. An energy management problem of an Energy Internet is presented as an optimal control problem and solved using a DRL methodology with greater performance than the optimal flow method in a work by Hua et al. [63]. To obtain online energy control strategies in large energy harvesting networks, [64] proposes a mean-field multi-agent DRL system. It can achieve performance comparable to futuristic centralized policies without requiring the state knowledge of other nodes.

As previously stated, DRL has a number of advantages over traditional methods for solving the problem: (i) DRL can perform online energy management optimization as well as real-time control and feedback. (ii) DRL can improve energy utilization efficiency, lower operating expenses, and increase profits. (iii) DRL's capacity to learn diverse levels of abstraction from data allows it to execute complex tasks with less prior information. However, energy management continues to face the following challenges: (i) Wind and solar power generation has a lot of ups and downs, as well as a lot of unknowns. The model is sophisticated with large data dimensions when incorporating energy storage systems and curtailable loads. (ii) Because the generations, capacity, efficiency, and prices of different energy storage devices vary, coordinated control is challenging. (iii) The charging and discharging status of electric vehicles and home appliances is unpredictably inconsistent, and the data is incomplete. (iv) Energy management integrates power generation, transmission, substation, distribution, and load into a single system.

Due to the aforementioned challenges, DRL should concentrate on the following: (i) Using historical data and physical models to convert challenges in real-world scenarios into sequence decision problems. (ii) Developing appropriate reward functions based on the aims and constraints of real-world problems. (iii) Using data-driven methodologies while considering traditional models and methods.

11.3.2 Power Systems' Demand Response (DR)

Demand response is a common concern in smart grids that maintains a price or incentive-based balance between consumer electricity demand and utility supply. Demand response must include consumer feedback and consumption in the control loop to increase grid stability and shift peak demand. To handle such difficulties, DRL is an effective optimum control strategy using data-driven support models [65, 66].

The major goal for power customers is to reduce expenses; however, utility companies' principal purpose is to maximize profits. A DRL solution is proposed in Lu and Hong [67] to handle a real-time incentive-based demand response problem by supporting the service provider in acquiring electricity from multiple users to balance power fluctuations and maintain grid dependability. This is similar to Zhang's [68], which proposes a DRL strategy for making sequential optimal decisions in heating, ventilation, and air conditioning (HVAC) systems that are subject to demand response. In Hao [69], multi-agent RL is used to plan an autonomous and optimal HVAC electricity consumption scheduling in order to reduce the societal cost of a game-theoretic methodology. In the paper [70], the authors have developed an optimal pricing system for a demand response program based on RL, and the balance between exploration and exploitation in learning processes leads to improved load serving entity (LSE) performance. In the paper [71], the authors present an effective model for handling home load scheduling by RL that considers consumer satisfaction, stochastic renewable energy, and cost. In addition, the model can be made to be more general. Using batch RL and a Bayesian neural network, Hou et al. [72] provides a unique demand response strategy to lower the long-term charging/discharging cost of plug-in electric vehicles.

It is feasible for DRL to create a game model that considers demand response between power companies and customers. The article [73] proposes and RL solves a two-stage game model between power companies and customers. The optimal power consumption of clients is established in the first stage, and the pricing of power companies is determined in the second stage. The authors in the paper [74] proposed a dynamic pricing demand response model that considers the profit of the service provider as well as the costs of the clients. RL determines the retail pricing based on electricity demand and wholesale electricity costs. In Lu et al. [75], a multi-agent RL is utilized for a decentralized control approach to find the best bidding strategy between power firms and customers when demand response is considered. Similarly, the authors in the paper [76] present an applied data-driven RL methodology for agile demand response in an unbundled power market using sophisticated bidding rules. To optimize the total payoff of smart grid agents, [77] proposes a virtual leader-follower Stackelberg game model based on deep transfer Q-learning (DTQ).

DRL offers advantages, as shown by the preceding references: (i) DRL can make decisions based on insufficient information, and these decisions can be made online. (ii) Using game theory, DRL can maximize system advantages while lowering transaction costs. (iii) DRL has a better transfer capability and can be used in a variety of situations.

Simultaneously, the challenges of demand response can be seen in the following areas:

(a) Incentive measures come in a variety of shapes and sizes, including economic, technological, environmental, and political incentives (different users react to incentives differently).
(b) Demand response is frequently accompanied by changes in multiple elements such as load and electricity price, and the effects of these factors vary.
(c) The electrical equipment engaged in the response has varied control methods and constraint circumstances, making the model more difficult.
(d) The demand response process is frequently accompanied by a game between consumers, service providers, and electricity corporations, resulting in varied optimization targets.

DRL techniques must consider the following factors to solve the aforementioned challenges:

(1) Extracting customer behavior traits and forecasting their behavior using deep neural network (DNN) and other methods as the foundation for optimal control.
(2) Choosing the proper state space, considering factors such as price, load, storage system SoC, and so on.
(3) To compensate for the lack of models, make full use of previous data and customer input.

11.3.3 Electricity Market

A wholesale electricity market and a retail electricity market are two types of hierarchical electrical markets. As demonstrated in Figure 11.3, it connects service providers, power companies, and customers through information and power. When rival power companies give their electricity to retailers who then sell it to the service provider, the wholesale electricity market is formed. Meanwhile, customers choose their providers from a pool of competing electrical retailers in the retail electricity market. Trading between these elements is a difficult game problem, and DRL can be used to find the best tactics when there isn't enough information.

An efficient bidding approach offers more benefit and lower cost for service providers, electricity companies, and customers. In the paper [78], the authors proposed an event-driven electricity market for energy trading in local distribution networks, with DRL determining prosumers' trading tactics to optimize their benefit. The authors in the article [79] investigate a distribution network's indirect customer-to-customer power market, and RL is used to determine energy trading strategies. Adaptive RL with partial knowledge is used to suggest a limited energy trading game among end-consumers in Wang et al. [80], and the bidding strategy eventually converges to Nash equilibrium. In Cao et al. [81], a DRL-based algorithm for energy storage systems (ESSs) to arbitrage in real-time electricity markets under price uncertainty is presented, with electricity price information gathered using an exponential moving average (EMA) filter and an recurrent neural network (RNN). The article [82] investigates an electrical market model with dynamic pricing and energy consumption in a microgrid, and RL is used to lower system costs for the service provider.

Equilibrium and social welfare are the overall objectives of game theory. In Wang et al. [83], the authors developed a multi-leaders and multi-followers Stackelberg game concept for electricity market, in which RL is used to attain equilibrium while maintaining anonymity. The authors in [84, 85] offer a microgrid energy trading game model that considers renewable energy production and demand, battery capacity, and trading history, with the Nash equilibrium computed using the DRL approach. In addition, a continuous real-time power market in a microgrid is established in Boukas and Ernst [86], and to attain optimal trading cost, a DRL approach with discrete high-level actions of action spaces is used. In the article [87], the authors proposed an hour-ahead energy market model for continuous renewable power penetration, and fuzzy Q-learning is used to create an IEEE 30-bus test system. The authors in [88] have developed a hierarchical electricity market with LSE bidding, and pricing is suggested, using DNN learning dynamical bid and price response functions and a DDPG algorithm generating state transition samples.

To summarize, DRL has the following advantages over traditional methods: (i) The majority of DRL algorithms are model-free and suitable for scenarios that cannot be modeled. (ii) Function approximators like neural networks can extract more data features that the models don't consider. (iii) In electricity markets, DRL can achieve Nash equilibrium between the supply and demand sides.

The power market, on the other hand, has the following key challenges: (i) The hierarchical electrical market has various entities with varied objectives, making reward functions challenging. (ii) In addition to the energy flow, there is insufficient information between the entities, necessitating the use of data-driven solutions. (iii) Unlike traditional discrete choice problems, energy trading is a continuous decision problem that necessitates real-time decision-making.

In order to address these issues, DRL's research objectives should include the following:

(1) Constructing multiple market entities as diverse game entities using game theory models.
(2) Using multi-agent RL with agents that represent various game entities.
(3) Due to the complexity of the game process, research should begin with small-scale scenarios and gradually scale them up.
(4) Improving the capacity to combine and extract data such as pricing and energy.

11.3.4 Operations and Controls

A classic power system challenge is the operational control problem. Operational control gets more complex and difficult as renewable energy sources grow more prominent. DRL can learn control techniques and optimization decisions online in large-scale settings with little knowledge [89].

Power generation control is a key aspect of operational control. The authors in [90, 91] propose a smart generation control technique for networked multi-area grids. When faced with complex operating scenarios that typical centralized automatic generation control cannot resolve, DRL can provide an optimal method (automatic generation control [AGC]). For the first time, Yin and Yu [92] and Yin et al. [93] describe a new architectural DQL method and utilize it to develop a smart generation controller for multi-agent systems with high-level robustness using the proposed algorithm. Similarly, [94] proposes a preemptive technique for smart generation control (SGC) in the face of huge continuous disturbances. Deep forest reinforcement learning (DFRL) outperforms standard AGC in terms of effectiveness. The first layer is to get the generating command, and the second layer is to use prior knowledge for optimal control using consensus transfer Q-learning (CTQ). In the article [95], the authors have proposed a unique optimal yaw control approach based on RL, and artificial neural network (ANN) is employed to avoid the big matrix quantification problem. This method considers the mechanical limitations in the yawing system as well as the mechanical stresses in order to achieve the quantity of instant wind turbine power and orientation change.

Further to reduce frequency deviation, faster response times and more flexibility are obtained. A load near-optimal control issue under sparse observations is proposed in Ruelens et al. [96]. DRL solves the problem using the function approximators CNN and Long Short-Term Memory (LSTM), with LSTM outperforming CNN in performance. The authors in [97] extract hidden state features of the residential load using CNN as a function approximator. Fitted Q-iteration is used to address the high-dimensional load control problem and cut power costs.

Appliances and system control, in addition to generation and load control, are complicated decision-making tasks. In the article [98], the authors propose an RL control strategy for using natural ventilation in an HVAC system that outperforms traditional rule-based heuristic control. Similarly, the authors in [99] design an optimal control model for an HVAC system for energy efficiency and thermal comfort. For DRL training in high-dimension data and continuous space, the A3C algorithm is used. During the transitory process, deep-Q networks (DQN) and double DQN are implemented in a restricted information situation, according to a smart grid emergency control approach presented in Liu et al. [100]. In Huang et al. [101], the authors have described a DRL-based emergency control strategy in a similar way. With non-linear generalization and high-dimensional feature extraction capabilities, grid security and resiliency are improved online. The authors in [102] develop the operating and maintenance concerns of a power grid. The information about the state and components of the grid is used by RL, which replaces the tabular representation of the value function with ANN. The authors in [103] proposed a real-time two-timescale voltage control technique. MDPs are used to mimic active power generation and load consumption, and DRL optimizes them to reduce bus voltage variations. The authors in [104] offer a novel Grid Mind architecture for efficiently mitigating voltage concerns. DRL can learn autonomous grid operational control strategies by interacting with offline simulations.

Therefore, DRL offers the following advantages in terms of operational control: (i) DRL can achieve continuous control in continuous state and action spaces. (ii) With inadequate data, DRL can make the control system more automated. (iii) DRL can handle some unexpected circumstances that most standard approaches are unable of handling. However, there are still certain operational control issues in the following areas:

(i)   Device control must be integrated with the physical structure and operational conditions of the device.
(ii) Operational control must address both steady state and transient stability, which necessitates the use of different time scales.
(iii) The synchronous operation status of the unit must be considered in the power producing system.

The following guidelines should be considered in order to solve these issues:

(1) To avoid system failures, use DRL with traditional control methods and strategies.
(2) Using a hierarchical strategy in which one layer employs control tactics and the other employs optimization strategies.
(3) Grid data and device model attributes are converged for more intelligent and flexible control.

11.4   Mathematical Formulation of Objective Function

The goal of this research is to manage a highly penetrated solar photovoltaic power network with low generation costs in an interconnected power system during an overgeneration scenario. The IP-OPF technique was used to find the best solution to the specified problem.

The objective function and associated security constraints have been shown in the following Eqs. (11.13)–(11.20):

$$\min f(P_{g_i}) = \sum_{u=1}^{U} \sum_{i=1}^{G} \{C_i(P_{g_i})\} \tag{11.13}$$

s.t.

$$\sum_{u=1}^{U} \left\{ \sum_{i=1}^{G} P_{g_i} \right\} - \sum_{i=1}^{D} P_{d_i} - P_{\text{loss}} = 0 \tag{11.14}$$

$$\underline{P_{g_i}} \leq P_{g_i} \leq \overline{P_{g_i}}; \; g_i \in \text{Coventional Gen} \tag{11.15}$$

$$\underline{Q_{g_i}} \leq Q_{g_i} \leq \overline{Q_{g_i}}; \; g_i \in \text{Coventional Gen} \tag{11.16}$$

$$P_{g_i} = P_{g_i}{}^{\text{MPP}}; \; g_i \in \text{PV units} \tag{11.17}$$

For this study, the insignificant value of PV units' reactive power production (less than 20% of active power output) has not been considered.

$$\underline{V_i} \leq V_i \leq \overline{V_i} \tag{11.18}$$

$$S_{ik} \leq \overline{S_{ik}} \tag{11.19}$$

subject to the state of overgeneration: $\sum_{u=1}^{U} \left\{ \sum_{i=1}^{G}(P_{g_i}) - \sum_{i=1}^{D} P_{d_i} - P_{\text{loss}} \right\} > 0$, Status of M2M tie-lines: on

$$S_{ik} \leq \overline{S_{ik}}; i \subset g^{\text{PV}}, k \in d \text{ and } S_{T_l} \leq \overline{S_{T_l}} \tag{11.20}$$

The direct interconnection of identified PV systems to identified demand nodes is depicted in Eq. (11.20).

The locational marginal pricing for active power was evaluated in this study; however, the LMPs for reactive power were ignored due to minimal values. The LMP values were derived from the Lagrange active power balancing equations, and the Lagrange multiplier at each bus is the Lagrange multiplier.

11.4.1 Locational Marginal Prices (LMPs) Representation

The imperative power flow optimality criterion is that the partial derivative of the Lagrange function with respect to the state variables (angles and voltages) must equal 0. The LMPs equation for the bus i_s is obtained by applying the constraint.

$$\lambda_{P_i} = \frac{\partial C_s(P_s)}{\partial P_s} - \frac{\partial P_{\text{loss}}}{\partial P_i} \frac{\partial C_s(P_s)}{\partial P_s} - \frac{\partial \mathcal{H}^T}{\partial P_i} \mu \tag{11.21}$$

Three components are included in Eq. (11.21): energy, loss, and congestion.

11.4.2 Relative Strength Index (RSI)

RSI uses 14 period historical data to generate an indicator for better trading in the wholesale market. This indicator sets two boundaries between 0 and 100, and below a certain value it implies buy signal and vice versa [38].

$$\text{RSI} = 100 - \left(\frac{100}{\left(1 + \frac{\text{average price increase(14 periods)}}{\text{average price decrease(14 periods)}}\right)} \right) \tag{11.22}$$

11.4.2.1 Autoregressive Integrated Moving Average (ARIMA)

The ARIMA stands for: AR—lags of the static series, I—a series which requires to be differenced to be formulated static, and MA—lags of estimated errors [37]

ARIMA forecasting equation is as following:

$$(1 - \emptyset_1 L)(1 - L)y_t = c + (1 + \theta_1 L)\varepsilon_t \tag{11.23}$$

where y_t is differenced time series, $(1 - \emptyset_1 L)(1 - L)$ are autoregressive terms (lagged values of y), c is constant, and $(1 + \theta_1 L)\varepsilon_t$ is lagged errors (MA terms)

11.5 Interior-point Technique & KKT Condition

The current optimization problem is non-linear. As a result, the non-linear problem was solved using a conventional non-linear optimization method. Karmarkar's IP approach is a sophisticated optimization algorithm for handling non-linear optimization problems. Conventional optimization strategies required more iterations and took longer to discover the best solution. Nonetheless, in IP-based optimization, this is a recursive method that cuts through the interior of the ideal solution. As a result, the IP approach produces a faster result after a substantial number of iterations. As a result, the IP method is better suited for large LPs [105]. The following is the main concept of the IP and its approach:

Let's say, the problem is outlined as

$$\text{min} \quad e(y)$$

$$\text{s.t.} \quad f(y) = 0$$

$$gl \leq g(y) \leq gu \tag{11.24}$$

where $e(y)$ denotes an objective function. $gl \leq g(y) \leq gu$ is a set of nonlinear inequality constraints, and $f(y)$ is a set of nonlinear equality requirements (power balance). The IP technique is used to construct the Lagrange function for Eq. (11.24) redrafted as OPF solutions.

$$L_g = e(y) - x^{\mathrm{T}}f(y) - z^{\mathrm{T}}[g(y) - l - gl] - w^{\mathrm{T}}[g(y) + u - gu] - \gamma \sum_{i=1}^{r} \ln l_i - \gamma \sum_{i=1}^{r} \ln u_i \tag{11.25}$$

where x, z, and w are Lagrange multipliers for equality and inequality constraints, respectively; l_i and u_i are slack variables; γ is boundary restrictions. After confirming the KKT condition for Eq. (11.25), the Jacobean matrices J_e, J_f, and J_g, as well as the Hessian matrices H_e, H_f, and H_g, are produced from $e(y)$, $f(y)$, and $g(y)$, respectively. A decomposed linear equation is then obtained for using a reduced Newton form Eq. (11.24).

11.5.1 Explanation of Karush–Kuhn–Tucker Conditions

Consider a general problem of optimization for a first- or second-order situation,

minimize e(y)

subject to

$$f_i(y) \leq 0, i = 1, 2, 3, \ldots, m$$

$$g_i(y) = 0, i = 1, 2, 3, \ldots, n \tag{11.26}$$

If y^* is a local minimum for a given case and the constraint is stable, exclusive vectors $\mu_i^* = 1, 2, 3, \ldots, m)$ and $\lambda_i^* = 1, 2, 3, \ldots, n)$ appear, which are known as KKT multipliers [106], are present, resulting in

$$\nabla e(y^*) + \nabla f_i(y^*)^{\mathrm{T}}\mu^* + \nabla g_i(y^*)^{\mathrm{T}}\lambda^* = 0 \tag{11.27}$$

$$\mu^* \geq 0 \tag{11.28}$$

$$f_i(y^*) \leq 0 \tag{11.29}$$

$$g_i(y^*) \geq 0 \tag{11.30}$$

$$(\mu^*)^T f_i(y^*) = 0 \tag{11.31}$$

The conditions (11.27)–(11.31) are known as the KKT conditions. All these conditions must be met to obtain the best pseudo-optimal solutions. For convex optimization problems with affinity constraints, the KKT requirements are sufficient for global optimality.

11.5.2 Algorithm for Finding a Solution

The following are the major stages for the OPF-based IP solution.

Stage (1) Initialization: Provide initial values to the established objective function.

Stage (2) Formulate the objective function's Jacobean and Hessian matrices, as well as equality, and inequality constraints: J_e, J_f, J_g, H_e, H_f, and, H_g.

Stage (3) Using a predictor-corrector technique, solve this linear system by formulating the linear equation for (11.24). Stop if the convergence conditions are met; otherwise, proceed to Stage 2.

11.6 Test Results and Discussion

11.6.1 Illustrative Example

The test has been performed in an interconnected network. Modified IEEE-9 bus system and modified IEEE-5 have been taken into the account. The modified IEEE-9 bus system shares a highly penetrated PV system in the network. As discussed earlier, direct interconnection of identified PV units to identified load buses provides highly economical market for the inter-connected electricity markets. Therefore, in this case study this method of interconnection has been adopted. Further, the load profile of both networks has been chosen same for this study.

Now the variation of the PV generation output in half-hourly period has been estimated. And the PV systems have been penetrated at the load buses of modified IEEE-9 bus system. The MPP of PV units is the 120% of minimum day time load of respective load buses (Figure 11.5).

Further, the LMPs have been forecasted using ARIMA and validated with RSI. Since the RSI for a specific time has been calculated using the historical values of LMPs. The results show that decreasing RSI values indicates the falling prices of energy (LMPs) whether the increasing RSI values show the rising prices.

Figure 11.6 is clearly showing the LMP and RSI relations at bus no. 7 of modified IEEE-9 bus system. As the RSI values are decreasing, the LMP values are falling too and for the increasing value of RSI, the LMP values are rising. Hence, it could be predicted that below the value 20, the RSI indicates that there are opportunities to purchase the electricity and above 45–50, the RSI indicates the rising value of LMPs.

Now, using historical LMP values, it has forecasted the future values of LMP and RSI using the ARIMA. The forecasted value of LMP also has been supported by forecasted value of RSI which shows that the RSI indicator would be useful for the future prediction of LMP values. The forecasted LMP values have been shown in Figure 11.7 (Table 11.1).

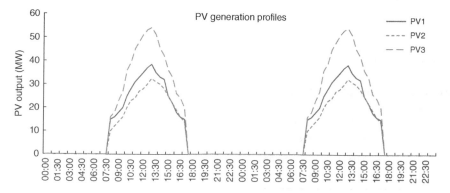

Figure 11.5 PV Generation Profiles in modified IEEE-9 bus System.

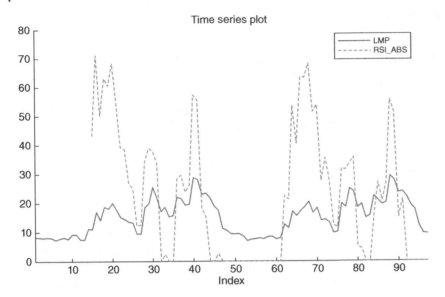

Figure 11.6 LMP and RSI Time series plot.

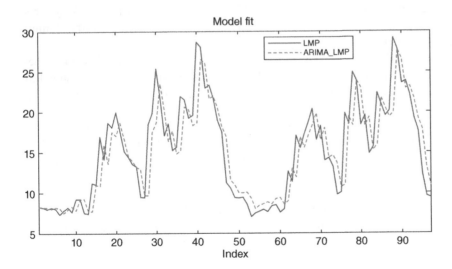

Figure 11.7 Forecasted values of LMPs with respective time series.

Table 11.1 Estimation results of ARIMA (Gaussian model) for LMP.

Parameter	Value	Standard error	*t* Statistic	*P*-value
Constant	0.016206	0.015459	1.0483	0.29451
AR {1}	0.86553	0.10883	7.9529	1.8219e-15
MA {1}	−1	0.062181	−16.0821	3.4078e-58
Variance	9.7561	1.2111	8.0559	7.8881e-16

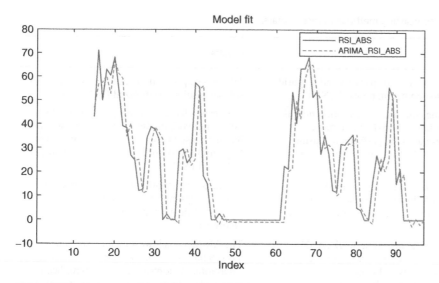

Figure 11.8 Forecasted RSI values using ARIMA.

Table 11.2 Estimation results of ARIMA (Gaussian model) for RSI.

Parameter	Value	Standard error	*t* Statistic	*P*-value
Constant	−1.647	2.6526	−0.6209	0.53467
AR {1}	−0.83126	0.1036	−8.0236	1.0269e-15
MA {1}	0.77313	0.14177	5.4533	4.9451e-08
Variance	171.6393	20.448	8.3939	4.701e-17

Similarly, the RSI has been calculated using the time series historical data obtained from the 14 period LMP values. Henceforth, using ARIMA has forecasted the future RSI values for the highly PV penetrated interconnected network. Figure 11.8 shows the forecasted RSI values (Table 11.2).

11.7 Comparative Analysis with Other Methods

Table 11.3 has described some similar works that have been published in recent years to explain the electricity price forecasting methods with examples. There have been more articles that have explained more methods of the forecasting the LMPs (electricity prices), and these articles are much similar to each other in prediction. But these works do not provide any visual signal to seize the opportunities in the wholesale market. Therefore, this work has used the ARIMA method that provides lowest mean error and significant standard deviation values. Further, this work has constituted the forecasted values in a relative stochastic indicator which has been classified in two signals (buy/exit signal) shown by mountain curve graph or line graph for more clarity. Since the visual signals are superlative to the numbers. Thus, this work can be distinguished with the earlier indicator published in similar works.

Table 11.3 Description of electricity forecasting methods in recent years.

Reference	Methodology	Results
[107]	Combination of wavelet transform (WT), and a hybrid forecast method based on neural network (NN)	The two-step selection based on NN provides sub-set results of optimized LMP values
[108]	Forecasted LMP using combined convolution neural network (CNN) and Long Short-Term Memory (LSTM). Further compared the performance using mean absolute error (MAE) and root-mean square (RMSE)	The results have been computed using deep neural network model EPNet for the next hour based on last 24 hours data. Lower values of MAE and RMSE indicate high performance of CNN and LSTM
[109]	Composed the WT and Takagi–Sugeno (TS) fuzzy rule-based algorithm followed by the stochastic search algorithm to optimize the clustering objective function	The improved hybrid fuzzy neural network (TS fuzzy system) demonstrates the capability of forecasting with the help of VEHBMO algorithm
[110]	Applied ARIMA, artificial neural network (ANN), and Trigonometric seasonal Box-cox Transformation with ARMA residual trends and seasonal components (TBATS). Further simple averaged the forecasted values of all three to improve the accuracy	The ARIMA has lowest mean error, and TBATS has lower mean error than the ANN. Therefore, the forecasted values have been averaged for better accuracy (lower mean error and standard deviation)

11.8 Conclusion

In the era of growing energy demand and high share of intermittent generation units, it is hard to find a right time to plug-in to electricity to get low-cost electricity. DRL has made rapid progress in solving sequential decision-making challenges in theoretical, methodological, and experimental disciplines over the last few years. DL, in particular, acquires the object's traits, categories, or features from the environment, while RL makes control strategy decisions based on the data. As a result, DRL is capable of solving problems in high-dimensional states and action spaces. However, there are various digital platforms to provide real-time value of electricity. The consumers are unable to find suitable data from the big data factories. Therefore, this study has introduced the RSI technique in the electricity market so that the consumers and retailers could get alerts during fall of electricity prices in the wholesale market. The test performed in this work shown that the low value of RSI indicates the low prices of electricity at a certain time and vice versa. Further this work also estimated the future values of LMPs using ARIMA and validated the RSI method over the estimated result. Hence, the RSI method would be a useful indicator in the electricity market.

The principle, development, algorithms, and characteristics of DRL, as well as its applications in the power system, such as energy management, demand response, electricity markets, and operational control, are discussed in this chapter. There are still a lot of things to talk about. Conclusively, there are many potentials and problems for DRL and its applications in the power system. These will elicit additional interest and investigation, and there will undoubtedly be more unexpected occurrences in the future.

11.9 Assignment

1. Define reinforcement learning with real world examples.
2. Define the deep reinforcement learning and explain how it is different from deep learning. Validate the difference with examples.
3. Explain reinforcement learning algorithms.

4. What is the algorithm of deep reinforcement learning?
5. What is the role of deep reinforcement learning in the energy management? Please do mention some examples.
6. Explain the electricity market and what is its future with deep reinforcement learning?
7. Compare the ARIMA with other forecasting tools.
8. Deep reinforcement learning's are playing a vital role to make decisions in modern power systems. Validate the statement with examples.

Acknowledgment

The authors extend their appreciation to the Deputyship for Research and Innovation, Ministry of Education in Saudi Arabia, for funding this research work through the project number (IFPRC-048-135-2020), and King Abdulaziz University, DSR, Jeddah, Saudi Arabia. The authors also acknowledge the support provided by King Abdullah City for Atomic and Renewable Energy (K.A.CARE) under K.A.CARE-King Abdulaziz University Collaboration Program.

References

1 Fu, R., Feldman, D., and Margolis, R. (2018). U.S. solar photovoltaic system cost benchmark: Q1 2018, *Nrel*.
2 Obi, M. and Bass, R. (2016). Trends and challenges of grid-connected photovoltaic systems – A review. *Renewable and Sustainable Energy Reviews* 58: 1082–1094. https://doi.org/10.1016/j.rser.2015.12.289.
3 Haegel, N.M., Atwater, H. Jr., Barnes, T. et al. (2019). Terawatt-scale photovoltaics: transform global energy. *Science* 364 (6443): 836–838. https://doi.org/10.1126/science.aaw1845.
4 Nwaigwe, K.N., Mutabilwa, P., and Dintwa, E. (2019). An overview of solar power (PV systems) integration into electricity grids. *Materials Science for Energy Technologies* https://doi.org/10.1016/j.mset.2019.07.002.
5 Hoke, A., Nelson, A., Chakraborty, S. et al. (2015). *Inverter Ground Fault Overvoltage Testing Inverter Ground Fault Overvoltage Testing*. NREL, Technical Report no.: NREL/TP-5D00-64173.
6 R. Mcallister, D. Manning, L. Bird, M. Coddington, and C. Volpi, *New Approaches to Distributed PV Interconnection: Implementation Considerations for Addressing Emerging Issues*, 2018.
7 Bryan Palmintier, Elaine Hale, Timothy M. Hansen, Wesley Jones, David Biagioni, Kyri Baker, Hongyu Wu, Julieta Giraldez, Harry Sorensen, Monte Lunacek, Noel Merket, Jennie Jorgenson, and Bri-Mathias Hodge (2016). Final Technical Report: Integrated Distribution-Transmission Analysis for Very High Penetration Solar PV Final Technical Report: Integrated Distribution-Transmission Analysis for Very High Penetration Solar PV, NREL, Technical Report no.: NREL/TP-5D00-65550.
8 Sun, H., Guo, Q., Qi, J. et al. (2019). Review of challenges and research opportunities for voltage control in smart grids. *IEEE Transactions on Power Systems* 34 (4): 2790–2801. https://doi.org/10.1109/TPWRS.2019.2897948.
9 Kovacs, A. (2018). On the computational complexity of tariff optimization for demand response management. *IEEE Transactions on Power Systems* 33 (3): 3204–3206. https://doi.org/10.1109/TPWRS.2018.2802198.
10 Cheng, L. and Yu, T. (2019). A new generation of AI: a review and perspective on machine learning technologies applied to smart energy and electric power systems. *International Journal of Energy Research* 43 (6): 1928–1973. https://doi.org/10.1002/er.4333.
11 Huo, Y., Prasad, G., Lampe, L., and Leung, V.C.M. (2020). Advanced smart grid monitoring: intelligent cable diagnostics using neural networks. In: *2020 IEEE International Symposium on Power Line Communications and its Applications (ISPLC), Malaga, Spain*, 11–13 May 2020, INSPEC Accession Number: 19688271. IEEE.

12 Han, M., May, R., Zhang, X. et al. (2019). A review of reinforcement learning methodologies for controlling occupant comfort in buildings. *Sustainable Cities and Society* 51 (101748). https://doi.org/10.1016/j.scs.2019.101748.

13 François-Lavet, V., Henderson, P., Islam, R. et al. (2018). An introduction to deep reinforcement learning. *Foundations and Trends in Machine Learning* 11 (3–4): 219–354. https://doi.org/10.1561/2200000071.

14 Hua, J., Zeng, L., Li, G., and Ju, Z. (2021). Learning for a robot: deep reinforcement learning, imitation learning, transfer learning. *Sensors (Switzerland)* 21 (4): 1–21. https://doi.org/10.3390/s21041278.

15 Nguyen, H. and La, H. (2019). Review of deep reinforcement learning for robot manipulation. In: *2019 Third IEEE International Conference on Robotic Computing (IRC), 25–27 February 2019, Naples, Italy*, 590–595. IEEE. https://doi.org/10.1109/IRC.2019.00120.

16 Benhamou, E., Saltiel, D., Ohana, J.J., and Atif, J. (2020). Detecting and adapting to crisis pattern with context based deep reinforcement learning. In: *Proceedings – International Conference on Pattern Recognition*, 10050–10057. Cornell University/ Quantitative Finance/ Portfolio Management, https://arxiv.org/abs/2009.07200v2 https://doi.org/10.1109/ICPR48806.2021.9412958.

17 Li, Z., Liu, X.-Y., Zheng, J. et al. (2021). FinRL-podracer: high performance and scalable deep reinforcement learning for quantitative finance. In: *ICAIF '21: Proceedings of the Second ACM International Conference on AI in Finance*, Article No.: 48, 1–9. Association for Computing Machinery https://doi.org/10.1145/3490354.3494413.

18 Soleymani, F. and Paquet, E. (2020). Financial portfolio optimization with online deep reinforcement learning and restricted stacked autoencoder—DeepBreath. *Expert Systems with Applications* 156 (113456). https://doi.org/10.1016/j.eswa.2020.113456.

19 Siddhant, A. and Lipton, Z.C. (2018). Deep Bayesian active learning for natural language processing: results of a large-scale empirical study. http://arxiv.org/abs/1808.05697 (last revised 24 September 2018).

20 Misra, D., Langford, J., and Artzi, Y. (2017). Mapping instructions and visual observations to actions with reinforcement learning. In: *EMNLP 2017 – Conference on Empirical Methods in Natural Language Processing, Proceedings, Copenhagen, Denmark*, 1004–1015. Association for Computational Linguistics https://doi.org/10.18653/v1/d17-1106.

21 Shi, W., Li, N., Chu, C.C., and Gadh, R. (2017). Real-time energy management in microgrids. *IEEE Transactions on Smart Grid* 8 (1): 228–238. https://doi.org/10.1109/TSG.2015.2462294.

22 Zachar, M. and Daoutidis, P. (2017). Microgrid/macrogrid energy exchange: a novel market structure and stochastic scheduling. *IEEE Transactions on Smart Grid* 8 (1): 178–189. https://doi.org/10.1109/TSG.2016.2600487.

23 Ordoudis, C., Pinson, P., and Morales, J.M. (2019). An integrated market for electricity and natural gas systems with stochastic power producers. *European Journal of Operational Research* 272 (2): 642–654. https://doi.org/10.1016/j.ejor.2018.06.036.

24 Duchaud, J.L., Notton, G., Darras, C., and Voyant, C. (2018). Power ramp-rate control algorithm with optimal state of charge reference via dynamic programming. *Energy* 149: 709–717. https://doi.org/10.1016/j.energy.2018.02.064.

25 Zéphyr, L. and Anderson, C.L. (2018). Stochastic dynamic programming approach to managing power system uncertainty with distributed storage. *Computational Management Science* 15 (1): 87–110. https://doi.org/10.1007/s10287-017-0297-2.

26 Nguyen, H.T., Le, L.B., and Wang, Z. (2018). A bidding strategy for virtual power plants with the intraday demand response exchange market using the stochastic programming. *IEEE Transactions on Industry Applications* 54 (4): 3044–3055. https://doi.org/10.1109/TIA.2018.2828379.

27 Naidu, I.E.S. and Sudha, K.R. (2018). Dynamic stability margin evaluation of multi-machine power systems using genetic algorithm. In: *International Proceedings on Advances in Soft Computing, Intelligent Systems and Applications*, vol. 628 (ed. M.S. Reddy, K. Viswanath and K.M. Shiva Prasad), 1–16. Singapore: Springer.

28 Megantoro, P., Wijaya, F.D., and Firmansyah, E. (2017). Analyze and optimization of genetic algorithm implemented on maximum power point tracking technique for PV system. In: *2017 International Seminar on Application for Technology of Information and Communication (iSemantic), Semarang, Indonesia*, 79–84. https://doi.org/10.1109/ISEMANTIC.2017.8251847.

29 Raviprabakaran, V. and Subramanian, R.C. (2018). Enhanced ant colony optimization to solve the optimal power flow with ecological emission. *International Journal of System Assurance Engineering and Management* 9 (1): 58–65. https://doi.org/10.1007/s13198-016-0471-x.

30 Srikakulapu, R. and Vinatha, U. (2018). Optimized design of collector topology for offshore wind farm based on ant colony optimization with multiple travelling salesman problem. *Journal of Modern Power Systems and Clean Energy* 6 (6): 1181–1192. https://doi.org/10.1007/s40565-018-0386-4.

31 Gu, H., Yan, R., and Saha, T.K. (2018). Minimum synchronous inertia requirement of renewable power systems. *IEEE Transactions on Power Systems* 33 (2): 1533–1543. https://doi.org/10.1109/TPWRS.2017.2720621.

32 Li, H., Yang, D., Su, W. et al. (2019). An overall distribution particle swarm optimization MPPT algorithm for photovoltaic system under partial shading. *IEEE Transactions on Industrial Electronics* 66 (1): 265–275. https://doi.org/10.1109/TIE.2018.2829668.

33 Rai, A. and Nunn, O. (2020). On the impact of increasing penetration of variable renewables on electricity spot price extremes in Australia. *Economic Analysis and Policy* 67: 67–86. https://doi.org/10.1016/j.eap.2020.06.001.

34 PJM Interconnection, 155 FERC ¶ 61,282 United States of America Federal Energy Regulatory Commission, 2016.

35 PJM Interconnection (2011). Amended and restated operating agreement of PJM Interconnection, L.L.C, no. 24.

36 Yadav, D., Mekhilef, S., Singh, B., and Rawa, M. (2021). Carbon trading analysis and impacts on economy in market-to-market coordination with higher PV penetration. In: *IEEE Transactions on Industry Applications*, vol. 57, no. 6, 5582–5592. https://doi.org/10.1109/TIA.2021.3105495.

37 Box, G.E.P., Jenkins, G.M., Reinsel, G.C., and Ljung, G.M. (2017). *Time Series Analysis: Forecasting and Control*. Wiley ISBN: 978-1-118-67502-1.

38 Bhargavi, S.G. and Anith, R. (2017). Relative strength index for developing effective trading strategies in constructing optimal portfolio. *International Journal of Applied Engineering Research* 12: 8926–8936.

39 Li, Y. (2018). Deep reinforcement learning. http://arxiv.org/abs/1810.06339.

40 Bertsekas, D.P. (2022). Class notes for ASU course CSE 691; Spring 2022 topics in reinforcement learning. http://www.mit.edu/~dimitrib/RL_CLASS NOTES_2022.pdf (accessed 12 March 2022).

41 Fan, J., Wang, Z., Xie, Y., and Yang, Z. (2019). A theoretical analysis of deep Q-learning. http://arxiv.org/abs/1901.00137 (last revised 24 February 2020).

42 Sutton, R.S. and Barto, A.G. (1998). *Reinforcement Learning: An Introduction*. MIT Press.

43 Busoniu, L., Babuska, R., De Schutter, B., and Ernst, D. (2010). *Reinforcement Learning and Dynamic Programming using Function Approximators*. CRC Press, ISBN 9781439821084.

44 Agostinelli, F., Hocquet, G., Singh, S., and Baldi, P. (2018). From reinforcement learning to deep reinforcement learning: an overview. In: *Braverman Readings in Machine Learning. Key Ideas from Inception to Current State: International Conference Commemorating the 40th Anniversary of Emmanuil Braverman's Decease* (ed. L. Rozonoer, B. Mirkin and I. Muchnik), 298–328. Boston, MA, USA: Cham: Springer International Publishing https://doi.org/10.1007/978-3-319-99492-5_13.

45 Szepesvári, C. (2010). *Algorithms for Reinforcement Learning*. Springer ISBN: 978-3-031-01551-9.

46 Lange, S., Gabel, T., and Riedmiller, M. (2012). Batch reinforcement learning. In: *Reinforcement Learning: State-of-the-Art* (ed. M. Wiering and M. van Otterlo), 45–73. Berlin, Heidelberg: Springer Berlin Heidelberg https://doi.org/10.1007/978-3-642-27645-3_2.

47 Maxim Lapan (2020) Deep Reinforcement Learning Hands-On; Packt, ISBN 9781838826994.

48 Nachum, O., Norouzi, M., Xu, K., and Brain, G. (2017). Bridging the gap between value and policy based reinforcement learning. 31st Conference on Neural Information Processing Systems (NIPS 2017), Long Beach, CA, USA., https://arxiv.org/abs/1702.08892 https://github.com/tensorflow/models/tree/.

49 Mocanu, E., Mocanu, D.C., Nguyen, P.H. et al. (2019). On-line building energy optimization using deep reinforcement learning. *IEEE Transactions on Smart Grid* 10 (4): 3698–3708. https://doi.org/10.1109/TSG.2018.2834219.

50 Wan, Z., Li, H., and He, H. (2018). Residential energy management with deep reinforcement learning. In: *2018 International Joint Conference on Neural Networks (IJCNN), Milano, Italy*, 1–7. IEEE. https://doi.org/10.1109/IJCNN.2018.8489210.

51 Mbuwir, B.V., Kaffash, M., and Deconinck, G. (2018). Battery scheduling in a residential multi-carrier energy system using reinforcement learning. In: *2018 IEEE International Conference on Communications, Control, and Computing Technologies for Smart Grids (SmartGridComm), Aalborg, Denmark*, 1–6. IEEE. https://doi.org/10.1109/SmartGridComm.2018.8587412.

52 Wu, J., He, H., Peng, J. et al. (2018). Continuous reinforcement learning of energy management with deep Q network for a power split hybrid electric bus. *Applied Energy* 222: 799–811. https://doi.org/10.1016/j.apenergy.2018.03.104.

53 Wan, Z., Li, H., He, H., and Prokhorov, D. (2019). Model-free real-time EV charging scheduling based on deep reinforcement learning. *IEEE Transactions on Smart Grid* 10 (5): 5246–5257. https://doi.org/10.1109/TSG.2018.2879572.

54 Hu, Y., Li, W., Xu, K. et al. (2018). Energy management strategy for a hybrid electric vehicle based on deep reinforcement learning. *Applied Sciences* 8 (2). https://doi.org/10.3390/app8020187.

55 Qi, X., Luo, Y., Wu, G. et al. (2019). Deep reinforcement learning enabled self-learning control for energy efficient driving. *Transportation Research Part C: Emerging Technologies* 99: 67–81. https://doi.org/10.1016/j.trc.2018.12.018.

56 Qi, X., Luo, Y., Wu, G. et al. (2017). Deep reinforcement learning-based vehicle energy efficiency autonomous learning system. In: *2017 IEEE Intelligent Vehicles Symposium (IV), Los Angeles, CA, USA*, 1228–1233. IEEE. https://doi.org/10.1109/IVS.2017.7995880.

57 Wu, Y., Tan, H., Peng, J. et al. (2019). Deep reinforcement learning of energy management with continuous control strategy and traffic information for a series-parallel plug-in hybrid electric bus. *Applied Energy* 247: 454–466. https://doi.org/10.1016/j.apenergy.2019.04.021.

58 V. François-Lavet, D. Taralla, D. Ernst, and R. Fonteneau, *Deep Reinforcement Learning Solutions for Energy Microgrids Management*.

59 Mbuwir, B.V., Ruelens, F., Spiessens, F., and Deconinck, G. (2017). Battery energy management in a microgrid using batch reinforcement learning. *Energies* 10 (11). https://doi.org/10.3390/en10111846.

60 Kofinas, P., Dounis, A.I., and Vouros, G.A. (2018). Fuzzy Q-learning for multi-agent decentralized energy management in microgrids. *Applied Energy* 219: 53–67. https://doi.org/10.1016/j.apenergy.2018.03.017.

61 Tan, Z., Zhang, X., Xie, B. et al. (2018). Fast learning optimizer for real-time optimal energy management of a grid-connected microgrid. *IET Generation, Transmission & Distribution* 12: 2977–2987. https://doi.org/10.1049/iet-gtd.2017.1983, Print ISSN 1751-8687, Online ISSN 1751-8695.

62 Mbuwir, B.V., Spiessens, F., and Deconinck, G. (2018). Self-learning agent for battery energy management in a residential microgrid. In: *2018 IEEE PES Innovative Smart Grid Technologies Conference Europe (ISGT-Europe), Sarajevo, Bosnia and Herzegovina*, 1–6. IEEE. https://doi.org/10.1109/ISGTEurope.2018.8571568.

63 Hua, H., Qin, Y., Hao, C., and Cao, J. (2019). Optimal energy management strategies for energy Internet via deep reinforcement learning approach. *Applied Energy* 239: 598–609. https://doi.org/10.1016/j.apenergy.2019.01.145.

64 Sharma, M.K., Zappone, A., Debbah, M., and Assaad, M. (2019). Multi-agent deep reinforcement learning based power control for large energy harvesting networks. In: *2019 International Symposium on Modeling and Optimization in Mobile, Ad Hoc, and Wireless Networks (WiOPT)*, 1–7. IEEE. https://doi.org/10.23919/WiOPT47501.2019.9144098.

65 Siano, P. (2014). Demand response and smart grids—A survey. *Renewable and Sustainable Energy Reviews* 30: 461–478. https://doi.org/10.1016/j.rser.2013.10.022.

66 Vázquez-Canteli, J.R. and Nagy, Z. (2019). Reinforcement learning for demand response: a review of algorithms and modeling techniques. *Applied Energy* 235: 1072–1089. https://doi.org/10.1016/j.apenergy.2018.11.002.

67 Lu, R. and Hong, S.H. (2019). Incentive-based demand response for smart grid with reinforcement learning and deep neural network. *Applied Energy* 236: 937–949. https://doi.org/10.1016/j.apenergy.2018.12.061.

68 X. Zhang, S. Rahman, M. Pipattanasomporn, R. Broadwater, G. Yu, and C.-T. Lu, *A Data-driven Approach for Coordinating Air Conditioning Units in Buildings during Demand Response Events*, https://vtechworks.lib.vt.edu/bitstream/handle/10919/87517/Zhang_X_D_2019.pdf?sequence=1&isAllowed=y, 2018.

69 Hao, J. (2019). Multi-agent reinforcement learning embedded game for the optimization of building energy control and power system planning. http://arxiv.org/abs/1901.07333.

70 Lu, R., Hong, S.H., Zhang, X. et al. (2017). A perspective on reinforcement learning in price-based demand response for smart grid. In: *2017 International Conference on Computational Science and Computational Intelligence (CSCI), Las Vegas, NV, USA*, 1822–1823. IEEE. https://doi.org/10.1109/CSCI.2017.327.

71 Wan, Y., Qin, J., Yu, X. et al. (2022). Price-based residential demand response management in smart grids: a reinforcement learning-based approach. *IEEE/CAA Journal of Automatica Sinica* 9 (1): 123–134. https://doi.org/10.1109/JAS.2021.1004287.

72 Hou, L., Ma, S., Yan, J. et al. (2020). Reinforcement mechanism design for electric vehicle demand response in microgrid charging stations. In: *2020 International Joint Conference on Neural Networks (IJCNN), Glasgow, UK*, 1–8. IEEE. https://doi.org/10.1109/IJCNN48605.2020.9207081.

73 Apostolopoulos, P.A., Tsiropoulou, E.E., and Papavassiliou, S. (2021). Demand response management in smart grid networks: a two-stage game-theoretic learning-based approach. *Mobile Networks and Applications* 26 (2): 548–561. https://doi.org/10.1007/s11036-018-1124-x.

74 Lu, R., Hong, S.H., and Zhang, X. (2018). A Dynamic pricing demand response algorithm for smart grid: reinforcement learning approach. *Applied Energy* 220: 220–230. https://doi.org/10.1016/j.apenergy.2018.03.072.

75 Najafi, S., Talari, S., Gazafroudi, A.S. et al. (2018). *Decentralized Control of DR using a Multi-Agent Method*, 233–249. Springer, Sustainable Independent Networks https://doi.org/10.1007/978-3-319-74412-4_13.

76 Babar, M., Nguyen, P.H., Ćuk, V. et al. (2018). The evaluation of agile demand response: an applied methodology. *IEEE Transactions on Smart Grid* 9 (6): 6118–6127. https://doi.org/10.1109/TSG.2017.2703643.

77 Zhang, X., Bao, T., Yu, T. et al. (2017). Deep transfer Q-learning with virtual leader-follower for supply-demand Stackelberg game of smart grid. *Energy* 133: 348–365. https://doi.org/10.1016/j.energy.2017.05.114.

78 Chen, T. and Su, W. (2018). Local energy trading behavior modeling with deep reinforcement learning. *IEEE Access* 6: 62806–62814. https://doi.org/10.1109/ACCESS.2018.2876652.

79 Chen, T. and Su, W. (2019). Indirect customer-to-customer energy trading with reinforcement learning. *IEEE Transactions on Smart Grid* 10 (4): 4338–4348. https://doi.org/10.1109/TSG.2018.2857449.

80 Wang, H., Huang, T., Liao, X. et al. (2017). Reinforcement learning for constrained energy trading games with incomplete information. *IEEE Transactions on Cybernetics* 47 (10): 3404–3416. https://doi.org/10.1109/TCYB.2016.2539300.

81 Cao, J., Harrold, D., Fan, Z. et al. (2020). Deep reinforcement learning-based energy storage arbitrage with accurate lithium-ion battery degradation model. *IEEE Transactions on Smart Grid* 11 (5): 4513–4521. https://doi.org/10.1109/TSG.2020.2986333.

82 Kim, B.-G., Zhang, Y., van der Schaar, M., and Lee, J.-W. (2016). Dynamic pricing and energy consumption scheduling with reinforcement learning. *IEEE Transactions on Smart Grid* 7 (5): 2187–2198. https://doi.org/10.1109/TSG.2015.2495145.

83 Wang, H., Huang, T., Liao, X. et al. (2016). Reinforcement learning in energy trading game among smart microgrids. *IEEE Transactions on Industrial Electronics* 63 (8): 5109–5119. https://doi.org/10.1109/TIE.2016.2554079.

84 Hau, C., Radhakrishnan, K.K., Siu, J., and Panda, S.K. (2020). Reinforcement learning based energy management algorithm for energy trading and contingency reserve application in a microgrid. In: *2020 IEEE PES Innovative Smart Grid Technologies Europe (ISGT-Europe), The Hague, Netherlands*, 1005–1009. IEEE. https://doi.org/10.1109/ISGT-Europe47291.2020.9248752.

85 Xiao, X., Dai, C., Li, Y. et al. (2017). Energy trading game for microgrids using reinforcement learning. In: *Proceedings- GameNets 2017: Game theory for Networks*, 131–140. Springer, https://doi.org/10.1007/978-3-319-67540-4_12.

86 Boukas, I. and Ernst, D. (2018). *Real-Time Bidding Strategies from Micro-Grids Using Reinforcement Learning, Ljubljana, Slovenia*, 440. AIM http://dx.doi.org/10.34890/163.

87 Salehizadeh, M.R. and Soltaniyan, S. (2016). Application of fuzzy Q-learning for electricity market modeling by considering renewable power penetration. *Renewable and Sustainable Energy Reviews* 56: 1172–1181. https://doi.org/10.1016/j.rser.2015.12.020.

88 Xu, H., Sun, H., Nikovski, D. et al. (2019). Deep reinforcement learning for joint bidding and pricing of load serving entity. *IEEE Transactions on Smart Grid* 10 (6): 6366–6375. https://doi.org/10.1109/TSG.2019.2903756.

89 Glavic, M., Fonteneau, R., and Ernst, D. (2017). Reinforcement learning for electric power system decision and control: past considerations and perspectives. *IFAC-PapersOnLine* 50 (1): 6918–6927. https://doi.org/10.1016/j.ifacol.2017.08.1217.

90 Xi, L., Chen, J., Huang, Y. et al. (2018). Smart generation control based on multi-agent reinforcement learning with the idea of the time tunnel. *Energy* 153: 977–987. https://doi.org/10.1016/j.energy.2018.04.042.

91 Xi, L., Chen, J., Huang, Y. et al. (2018). Smart generation control based on deep reinforcement learning with the ability of action self-optimization. *SCIENTIA SINICA Informationis*.

92 Yin, L. and Yu, T. (2018). Design of strong robust smart generation controller based on deep Q learning. In: *Dianli Zidonghua Shebei/Electric Power Automation Equipment*, vol. 38, 12–19. Dianli Zidonghua Shebei/Electric Power Automation Equipment https://doi.org/10.16081/j.issn.1006-6047.2018.05.002.

93 Yin, L., Yu, T., and Zhou, L. (2018). Design of a novel smart generation controller based on deep Q learning for large-scale interconnected power system. *Journal of Energy Engineering* 144 (3): 04018033. https://doi.org/10.1061/(asce)ey.1943-7897.0000519.

94 Yin, L., Zhao, L., Yu, T., and Zhang, X. (2018). Deep forest reinforcement learning for preventive strategy considering automatic generation control in large-scale interconnected power systems. *Applied Sciences* 8 (11): 2185. https://doi.org/10.3390/app8112185.

95 Saenz-Aguirre, A., Zulueta, E., Fernandez-Gamiz, U. et al. (2019). Artificial neural network based reinforcement learning for wind turbine yaw control. *Energies* 12 (3). https://doi.org/10.3390/en12030436.

96 Ruelens, F., Claessens, B.J., Vrancx, P. et al. (2019). Direct load control of thermostatically controlled loads based on sparse observations using deep reinforcement learning. *CSEE Journal of Power and Energy Systems* 5 (4): 423–432. https://doi.org/10.17775/CSEEJPES.2019.00590.

97 Claessens, B.J., Vrancx, P., and Ruelens, F. (2018). Convolutional neural networks for automatic state-time feature extraction in reinforcement learning applied to residential load control. *IEEE Transactions on Smart Grid* 9 (4): 3259–3269. https://doi.org/10.1109/TSG.2016.2629450.

98 Chen, Y., Norford, L.K., Samuelson, H.W., and Malkawi, A. (2018). Optimal control of HVAC and window systems for natural ventilation through reinforcement learning. *Energy and Buildings* 169: 195–205. https://doi.org/10.1016/j.enbuild.2018.03.051.

99 Zhang, Z., Chong, A., Pan, Y. et al. (2018). A deep reinforcement learning approach to using whole building energy model for HVAC optimal control. In: *2018 Building Performance Modeling Conference and SimBuild, Chicago, USA*. ASHRAE and IBPSA-USA.

100 Liu, W., Zhang, D., Wang, X. et al. (2018). A decision making strategy for generating unit tripping under emergency circumstances based on deep reinforcement learning. *Zhongguo Dianji Gongcheng Xuebao/Proceedings of the Chinese Society of Electrical Engineering* 38: 109–119. https://doi.org/10.13334/j.0258-8013.pcsee.171747.

101 Huang, Q., Huang, R., Hao, W. et al. (2020). Adaptive power system emergency control using deep reinforcement learning. *IEEE Transactions on Smart Grid* 11 (2): 1171–1182. https://doi.org/10.1109/TSG.2019.2933191.

102 Rocchetta, R., Bellani, L., Compare, M. et al. (2019). A reinforcement learning framework for optimal operation and maintenance of power grids. *Applied Energy* 241: 291–301. https://doi.org/10.1016/j.apenergy.2019.03.027.

103 Yang, Q., Wang, G., Sadeghi, A. et al. (2020). Two-timescale voltage control in distribution grids using deep reinforcement learning. *IEEE Transactions on Smart Grid* 11 (3): 2313–2323. https://doi.org/10.1109/TSG.2019.2951769.

104 Diao, R., Wang, Z., Shi, D. et al. (2019). Autonomous voltage control for grid operation using deep reinforcement learning. In: *2019 IEEE Power Energy Society General Meeting (PESGM), Atlanata, GA, USA*, 1–5. IEEE. https://doi.org/10.1109/PESGM40551.2019.8973924.

105 Yadav, D., Chauhan, A.S., and Singh, B. (2018). Contingency analysis and security constraint based optimal power flow in power network. In: *3rd International Conference on Innovative Applications of Computational Intelligence on Power, Energy and Controls with their Impact on Humanity, CIPECH 2018, Ghaziabad, India*, vol. 201206, 210–214. IEEE. https://doi.org/10.1109/CIPECH.2018.8724298.

106 Wu, Y.C., Debs, A.S., and Marsten, R.E. (1994). A direct nonlinear predictor-corrector primal-dual interior point algorithm for optimal power flows. *IEEE Transactions on Power Systems* 9 (2): 876–883. https://doi.org/10.1109/59.317660.

107 Ebrahimian, H., Barmayoon, S., Mohammadi, M., and Ghadimi, N. (2018). The price prediction for the energy market based on a new method. *Economic Research-Ekonomska Istraživanja* 31 (1): 313–337. https://doi.org/10.1080/1331677X.2018.1429291.

108 Kuo, P.-H. and Huang, C.-J. (2018). An electricity price forecasting model by hybrid structured deep neural networks. *Sustainability* 10 (4). https://doi.org/10.3390/su10041280.

109 Gao, W., Sarlak, V., Parsaei, M.R., and Ferdosi, M. (2018). Combination of fuzzy based on a meta-heuristic algorithm to predict electricity price in an electricity markets. *Chemical Engineering Research and Design* 131: 333–345. https://doi.org/10.1016/j.cherd.2017.09.021.

110 Karabiber, O.A. and Xydis, G. (2019). Electricity price forecasting in the Danish day-ahead market using the TBATS, ANN and ARIMA methods. *Energies* 12 (5). https://doi.org/10.3390/en12050928.

12

Power Quality Conditioners in Smart Power System

Zahira Rahiman[1], Lakshmi Dhandapani[2], Ravi Chengalvarayan Natarajan[3], Pramila Vallikannan[1], Sivaraman Palanisamy[4], and Sharmeela Chenniappan[5]

[1] Department of Electrical and Electronics Engineering, B.S. Abdur Rahman Crescent Institute of Science & Technology, Chennai, India
[2] Department of Electrical and Electronics Engineering, Academy of Maritime Education and Training (AMET), Chennai, India
[3] Department of Electrical and Electronics Engineering, Vidya Jyothi Institute of Technology, Hyderabad, Telangana, India
[4] World Resources Institute (WRI) India, Bengaluru, India
[5] Department of Electrical and Electronics Engineering, Anna University, Chennai, India

12.1 Introduction

The IEC standard 61000-4-30 defines the power quality as characteristics of the electricity at a given point on an electrical system, evaluated against a set of reference technical parameters. In simple words, the quality of electrical power supply refers to comparing or evaluating the electric parameters at a particular site with a set of reference values during the normal operating conditions. The standard IEEE 1159-2019 specifies the various power quality characteristics and their typical magnitude and duration. Based on the characteristics, power quality problems are classified as transients (impulsive and oscillatory), short-duration voltage variation (sag, swell, and interruption), long-duration voltage variation (under-voltage, overvoltage, and sustained interruption), unbalance, voltage fluctuations, waveform distortion (harmonics, inter-harmonics, DC injection, notching, and noise), and power frequency variations [1].

The exceedance of any of the above-mentioned power quality characteristics may affect the performance of the end-users' equipment like tripping, malfunctions, overloading, overheating, failures, etc.

The advantages of smart power systems are effective energy management, ICT, smart metering, wide-area monitoring and control, renewable energy integration including distributed generation, and energy storage systems. The introduction of power electronics-based switching devices is required for sophisticated operation and control. The use of these power electronics-based devices is negatively impacting the power system by introducing the power quality problems like harmonics, DC injection, flicker, rapid voltage variation, etc. [2]. Due to the fast depletion of conventional resources, renewable energy-based generations (like solar PV and wind turbine generators) are widely used for power and energy requirements. These renewable energy systems also use power electronics-based components and are distributed across the power systems.

Voltage sag is one of the most common power quality problems that occur due to faults, starting of large motors, energization of higher rating transformers, arcing faults, etc. in the power systems. It has a negative impact on the power systems like malfunctions or tripping of voltage-sensitive devices (programmable logic controllers, variable frequency drives, process controllers, etc.) [3]. Hence, the quality of power needs to be improved for satisfactory and trouble-free operation. The power quality conditioners play a major role in power quality enhancement [4]. These power quality conditioners will monitor the voltage, current, and frequency of the supply system and if any characteristics deviate from the reference values, it will compensate for those deviations. Hence the quality of

Artificial Intelligence-based Smart Power Systems, First Edition.
Edited by Sanjeevikumar Padmanaban, Sivaraman Palanisamy, Sharmeela Chenniappan, and Jens Bo Holm-Nielsen.
© 2023 The Institute of Electrical and Electronics Engineers, Inc. Published 2023 by John Wiley & Sons, Inc.

power delivered to the end-user equipment is within the limits. This chapter discusses the need for power quality enhancement for smart power systems.

Thus the objectives of this chapter are summarized as the study of the need for a smart grid and the power quality in a smart grid and to simulate the smart system with nonlinear loads, which produces voltage and current distortion. To reduce THD levels in the distribution system, an active shunt filter were designed using several techniques such as SPWM controller, SMC controller, Fuzzy based PI controller, and GWO-based PI controller. Finally, a comparison of the different control techniques and suggestions for suitable control techniques for the system is made [5].

The definitions of the various power quality characteristics are as follows:

12.1.1 Voltage Sag

Voltage sag is a reduction in RMS voltage from 0.9 to 0.1 pu time duration of less than 60 seconds. The reason for causes of voltage sag is system faults, switching ON the large motor, and switching OFF the large capacitor banks. Due to voltage sag, all equipments receive reduced nominal voltage which can affect the customer equipments.

12.1.2 Voltage Swell

Voltage swell is an increase of RMS voltage from 1.1 to 1.8 pu for a time duration of less than 60 seconds. The causes of voltage swell are removal/switching OFF of large loads and capacitor bank switch ON.

12.1.3 Interruption

Interruption is a decrease in RMS voltage magnitude less than 0.1 pu for less than 60 seconds and is called a momentary interruption. The decrease in RMS voltage magnitude less than 0.1 pu for greater than 1 minute is called sustained interruption.

12.1.4 Under Voltage

Under-voltage is a reduction in RMS voltage from 0.9 to 0.8 pu greater than 1 minute. The causes of under-voltage are a permanent fault, overloading, failure in reactive power compensation, and incorrect tap position of the transformer. Due to under-voltage, constant power loads draw higher current than the rated current from mains supply voltage resulting in increased power losses and a reduction in the equipment performance.

12.1.5 Overvoltage

Overvoltage is an increase in RMS voltage from 1.1 to 1.2 pu greater than 1 minute. The causes of overvoltage are lightning, fixed reactive power compensation, and incorrect tap position of the transformer. The overvoltage leads to premature insulation failure of pieces of equipment.

12.1.6 Voltage Fluctuations

Voltage fluctuation is defined as cycle-to-cycle voltage variation due to nonlinear characteristics of connected load (arching load) such as arc furnace and arc welding. Voltage fluctuations cause continuous voltage variation which affects other loads connected in the same bus. The voltage fluctuation affects the intensity of illumination given by the lamp.

12.1.7 Transients

Transients are sudden changes or variations in voltage or current or both due to internal or external causes. Based on changes, transients are classified into two types. They are impulsive transients and oscillatory transients.

12.1.8 Impulsive Transients

Impulsive transients are transients whose magnitude changes are unipolar in nature either changes positive or negative. The main causes of impulsive transients are lightning. However, transmission lines are exposed to open sky and subjected to lightning.

12.1.9 Oscillatory Transients

Oscillatory transients are transients whose changes are bi-polar in nature both positive and negative. The main causes of oscillatory transients are switching operations and capacitor switching.

12.1.10 Harmonics

Harmonics is a distortion of a voltage or current or both from a sinusoidal waveform. Multiple frequencies are added with fundamental frequency resulting in harmonics. The harmonics are produced due to nonlinear loads.

12.2 Power Quality Conditioners

The power quality conditioners are used to enhance one or more power quality characteristics whenever they deviate from the reference value [6]. They are effective and dynamically enhance the power quality. Some of the power quality conditioners used for power quality enhancements are as follows:

I. STATCOM
II. SVC
III. Harmonic filters
IV. UPS systems
V. Dynamic voltage restorer (DVR)

12.2.1 STATCOM

STATCOM is a power quality conditioner device connected parallel to the system. It controls the reactive power absorbed from or injected into the power system through the voltage source converter (VSC) to provide voltage regulation at its connection point. VSC is the power electric device process voltage and current waveform thereby it generates or absorbs required reactive power. The STATCOM provides reactive power when the system voltage is low (capacitive) and absorbs the same when the system voltage is high (inductive). It is a fast-acting device that controls the reactive power balance and the voltage level at the point of connection.

12.2.2 SVC

Static var compensator (SVC) is a fast-acting shunt-connected FACTS device that consists of a passive reactor and capacitor and power electronic switches. It gets the feedback of voltage level in the connection point and based on that, it generates or absorbs reactive power to maintain bus voltage magnitude. It adjusts bus voltage to adjust for changes in reactive power loading over time. It is mainly used in transmission lines for regulating the voltage, harmonics, and power factor and to maintain the stability of the system.

12.2.3 Harmonic Filters

A harmonic filter is the power quality conditioner to filter out the unwanted waveforms. Odd harmonics in the power system create an undesirable effect and need to eliminate. Harmonic limitations are developed based on power quality standards. If the limit is exceeded, the consumers' equipment could be harmed. Various filters are required to compensate for waveform distortions in order to keep the equipment from malfunctioning. Different types of filters used are passive, active, and hybrid filters.

12.2.3.1 Active Filter

Harmonic currents are injected through the filters to compensate for existing harmonic components. An active filter is meant to overcome the disadvantages of a passive filter. These active filters are classified as series, shunt, and hybrid filters. Among them shunt type has been explained and used for the simulation.

12.2.4 UPS Systems

Uninterrupted power supply (UPS) is a power quality conditioner used to maintain the voltage level within the tolerance limit. It consists of a rectifier, an energy storage device, and an inverter. UPS regulates the voltage by generating additional reactive power or absorbing reactive power based on the utility supply voltage. The control of reactive power and voltage levels is carried out by the power electronics devices in the UPS. Apart from the power quality conditioning it also gives out the emergency power supply when the utility is failed.

12.2.5 Dynamic Voltage Restorer (DVR)

It is a series-connected power quality conditioner. DVR is a solid-state device that injects reactive power and voltage component into the point of connection [7]. This voltage component is added with the utility voltage vector and thereby it controls the voltage sag or swell during the steady-state and transient conditions.

These power quality conditioners are used for enhancing one or more power quality characteristics based on requirements or application. The power quality conditioner used for enhancing the various power quality characteristics is listed in Table 12.1.

12.2.6 Enhancement of Voltage Sag

The voltage sag is defined as a reduction in RMS voltage from 0.9 to 0.1 pu for a time duration of less than 60 seconds. The causes of voltage sags are electrical faults, switching ON large motors, energization of the higher capacity transformer, etc.

Example 12.1: The industrial power system receives the power supply at 11 kV from DISCOM and it has 11 kV/433 V step-down transformer as shown in Figure 12.1.

Table 12.1 Power quality conditioner for power quality enhancement.

Device for power quality improvement	Minimize voltage sag/swell	Reactive power support	Harmonic reduction
STATCOM	Yes	Yes	No
SVC	Yes	Yes	No
Harmonic filter	No	No	Yes
UPS system	Yes	No	No
DVR	Yes	Yes	No

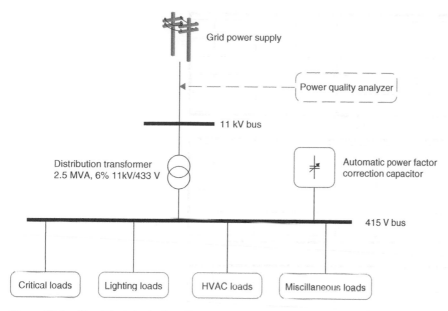

Figure 12.1   Simplified single line diagram.

The Dranetz power quality analyzer is connected at 11 kV grid supply from 13:39 hours to the next day at 07:20 hours. During the monitoring duration, voltage sag was experienced multiple times due to fault/transformer energization. Figure 12.2 shows the reduction in voltage due to fault/transformer energization resulting in the voltage sag at 11 kV grid supply in phase-to-phase voltage in the RMS trend.

Figure 12.3 shows the reduction in voltage at 23:02:59 hours due to fault/transformer energization resulting in the voltage sag at 11 kV grid supply in voltage in phase-to-phase instantaneous voltage waveform and Figure 12.4 shows the reduction in voltage in phase-to-phase RMS voltage.

Due to voltage sag, equipment connected to the faulted bus has been reduced below the nominal voltage which may affect the customer equipments, i.e., malfunction of equipment, tripping or failures of devices, etc. In order to enhance the voltage during the fault or transformer energization, etc., power quality conditioners like STATCOM or DVR shall be used. The installation of STATCOM at 11 kV bus is shown in Figure 12.5.

These power quality conditioners will inject the reactive power into the system and hence the voltage shall be improved.

12.2.7   Interruption Mitigation

Interruption is a decrease in RMS voltage magnitude less than 0.1 pu for less than 60 seconds and is called a momentary interruption. The decrease in RMS voltage magnitude less than 0.1 pu for greater than 1 minute is called sustained interruption.

Example 12.2: The industrial power system receives the power supply at 11 kV from DISCOM and it has 11 kV/433 V step-down transformer as shown in Figure 12.1.

Any faults in the transmission or distribution line lead to power supply interruption to the connected loads. But this industrial plant has some critical business loads which require a continuous power supply without any supply interruption. In order to provide an uninterrupted power supply to the critical business loads, the UPS system is one of the best options. This UPS system shall be sized based on the connected critical business loads

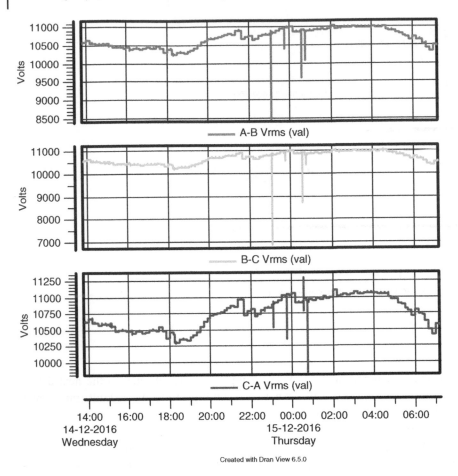

Figure 12.2 Voltage sag in R & Y phase.

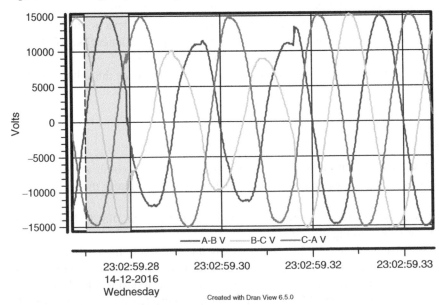

Figure 12.3 Instantaneous voltage waveform at 11 kV.

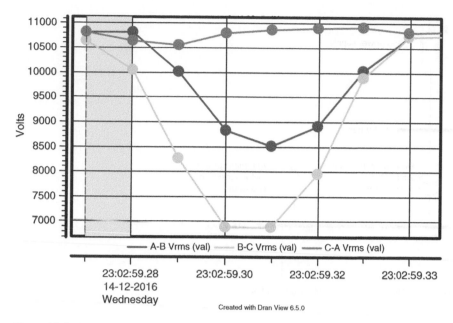

Figure 12.4 Voltage in RMS trend at 11 kV.

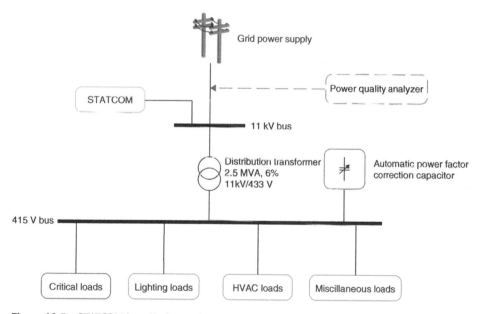

Figure 12.5 STATCOM installation at 11 kV bus.

of the industrial plant. Figure 12.6 shows the single line diagram (SLD) of the industrial plant with a UPS system powering the critical loads.

During the normal operation, power to the critical business loads flows through the rectifier and inverter. At the same time, batteries are being charged as shown in Figure 12.7.

In the event of interruption, the energy stored in the batteries is used to power the critical business loads as shown in Figure 12.8.

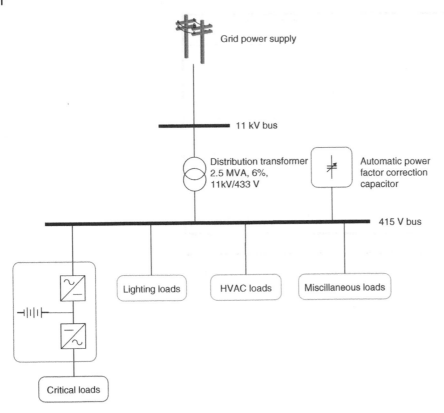

Figure 12.6 UPS system powering the critical loads.

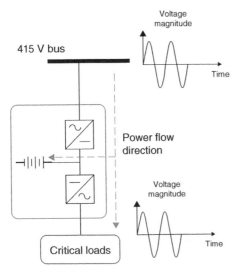

Figure 12.7 Power flow direction during normal operation.

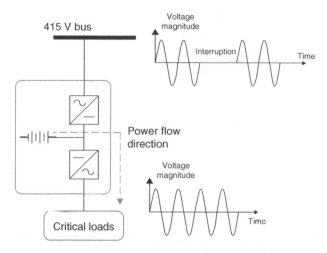

Figure 12.8 Power flow direction during interruption.

12.2.8 Mitigation of Harmonics

Harmonics is a distortion of a voltage or current or both from a sinusoidal waveform. Multiple frequencies are added with fundamental frequency resulting in harmonics. The harmonics are produced due to nonlinear loads [8].

Example 12.3: The industrial system receives the power from the electricity board at 22 kV and uses the transformer of 22 kV/433 V to step down the voltage for low voltage application. A Dranetz power quality analyzer is connected at 22 kV grid supply voltage between 11:47 hours to 18:50 hours at 3Φ delta monitoring mode used. The harmonics are measured at a 22 kV grid power supply due to nonlinear loads connected on the low voltage side. The phase-to-phase voltage in RMS in all three phases is 21342/21162/21172 V. The current in RMS in all three phases is 49.2/49.3/49.8 A.

The phase-to-phase voltage waveform in instantaneous voltage is shown in Figure 12.9.

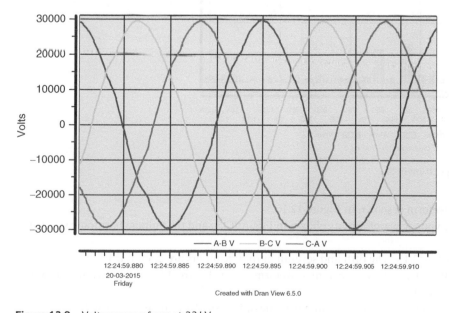

Figure 12.9 Voltage waveform at 22 kV.

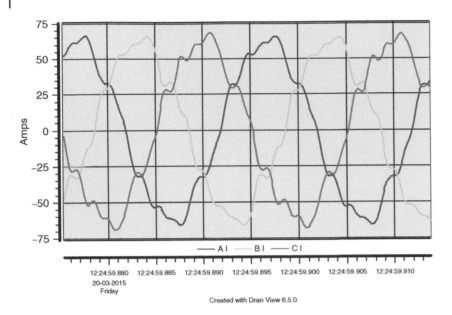

Figure 12.10 Current waveform at 22 kV.

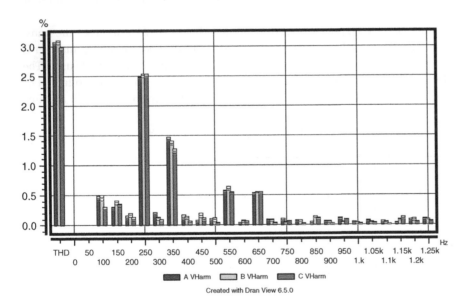

Figure 12.11 Voltage harmonic spectrum in %.

The instantaneous current waveform is shown in Figure 12.10.

The voltage harmonics for the voltage waveform are shown in Figure 12.9, and the voltage harmonic spectrum in % are shown in Figure 12.11.

The voltage THD is approximately 3.1%. The fifth and seventh harmonic orders are 2.5% and 1.5%, respectively.

The current harmonics for the instantaneous current waveform are in Figure 12.10, the current harmonic spectrum in % is shown in Figure 12.12.

The current THD is around 11%. The fifth and seventh harmonic orders are 7.3% and 6%, respectively.

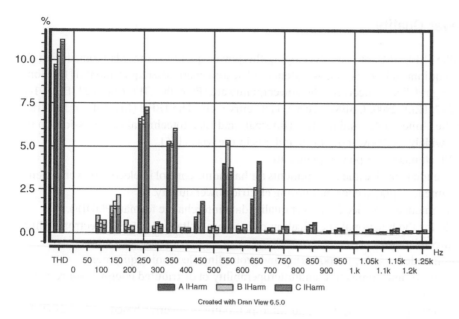

Figure 12.12   Current harmonic spectrum in %.

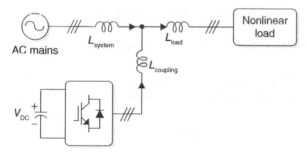

Figure 12.13   Model of shunt active filter.

Identification of harmonics limit was made based on the standard. Whenever the harmonic limit exceeds the recommended value, it violates the grid code and leads to the penalty imposed by DISCOMs. Also, higher harmonic levels have an impact on plant normal operations like overloading, overheating, electromagnetic interference, equipment failure or damage, etc. To protect the equipment damage due to higher harmonic levels, harmonic filters are used for harmonic mitigation. There are different types of harmonic filters available in the market. They are passive, active, and hybrid filters.

Shunt active power filter (SAPF) is used for the mitigation of harmonic current propagation into the system. Also, SAPF is used for reactive power compensation and to balance the unbalanced currents [9]. Injection of compensating current opposite in phase to nullify the harmonics produced by the nonlinear load is done by SAPF [10, 11]. It is also used to improve voltage profile and stabilization [12, 13]. The basic model of SAPF is shown in Figure 12.13.

12.3 Standards of Power Quality

The standards organizations like IEEE, IEC, etc. are developing the power quality standards (considering the power quality characteristics at equipment level as well as system level, testing methods/requirements) [14]. Some of the standards or regulations or guidelines concerning the power quality are IEC 61000, IEC 61400-21, EN 50160, IEEE 1159, IEEE 519, IEEE 1547, IEEE 2800, CBEMA curve, ITIC curve, etc. The IEEE (Institute of Electrical and Electronics Engineers) in the United States and the IEC (International Electrotechnical Commission) in the European Union have the most widely used power quality standards in the world [15].

Some of the majorly used IEEE standards for power quality are:

IEEE 519 – 2014 IEEE recommended practice and requirements for harmonic control in electric power systems.

IEEE 1159 – 2019 IEEE recommended practice for monitoring electric power quality.

IEEE 1159.3 – 2019 IEEE recommended practice for power quality data interchange format (PQDIF).

IEEE 1623 – 2020 IEEE guide for the functional specification of medium voltage (1–35 kV) electronic shunt devices for dynamic voltage compensation.

IEEE 1100 – 2005 IEEE recommended practice for powering and grounding electronic equipment.

IEEE 1547 – 2018 IEEE standard for interconnection and interoperability of distributed energy resources with associated electric power systems interfaces.

IEEE 2800 – 2022 IEEE standard for interconnection and interoperability of inverter-based resources (IBRs) interconnecting with associated transmission electric power systems.

Some of the majorly used IEC standards for power quality are:

IEC TR 61000-3-7:2008, Electromagnetic compatibility (EMC)—Part 3-7: Limits—Assessment of emission limits for the connection of fluctuating installations to MV, HV, and EHV power systems.

IEC 61000-4-3, Electromagnetic compatibility (EMC)—Part 4-3: Testing and measurement techniques—Radiated, radio-frequency, electromagnetic field immunity test.

IEC 61000-4-5, Electromagnetic compatibility (EMC)—Part 4-5: Testing and measurement techniques—Surge immunity test.

IEC 61000-4-7, Electromagnetic compatibility (EMC)—Part 4-7: Testing and measurement techniques—General guide on harmonics and interharmonics measurements and instrumentation, for power supply systems and equipment connected thereto.

IEC 61000-4-15, Electromagnetic compatibility (EMC)—Part 4-15: Testing and measurement techniques—Flickermeter—Functional and design specifications.

IEC 61000-4-30, Electromagnetic compatibility (EMC)—Part 4-30: Testing and measurement techniques—Power quality measurement methods.

IEC 61000-6-2, Electromagnetic compatibility (EMC)—Part 6-2: Generic standards—Immunity for industrial environments.

12.4 Solution for Power Quality Issues

The major issues of the power quality are over-voltage, under-voltage, voltage swells, voltage sag, and harmonic distortion. The power quality conditioning device has to mitigate the power quality issues and provide a quality power supply to the consumers. Voltage-related issues such as over/under voltage and voltage sag/swell are solved by reactive support devices such as STATCOM, SVC, UPS, and DVR [16]. The sensitive loads are sensitive to harmonics and these devices are not suitable to remove the harmonics. The harmonic filters on other hand remove the harmonics and cannot support the reactive power [17]. The vulnerable odd harmonics may remove using the harmonics filters [18].

12.5 Sustainable Energy Solutions

A new paradigm for the energy supply chain is evolving, leading to the creation of smart grids, to improve the design of power infrastructure and the services provided by distributed and renewable energy sources [19]. This increases the local availability of renewable energy resources, improves efficiency, and educates the public. Government has to encourage R&D for alternative renewable energy resources, all energy resources should be allowed to compete in a free market with no government intervention, and to promote change, the government should use constructive subsidies rather than destructive subsidies, which will result in resource conservation and reduced overconsumption [20].

12.6 Need for Smart Grid

A smart grid entails technology applications that will allow easier integration and higher penetration of renewable energy [21]. It will be essential for accelerating the development and widespread usage of plug-in hybrid electric vehicles (PHEVs) and their potential use as storage for the grid.

A smart grid serves several purposes and the movement from traditional electric grids to smart grids is driven by multiple factors, including the deregulation of the energy market, evolutions in metering, changes in the level of electricity production, decentralization (distributed energy), the advent of the involved "prosumer," changing regulations, the rise of microgeneration and (isolated) microgrids, renewable energy mandates with more energy sources and new points where and purposes for which electricity is needed (e.g., electrical vehicle charging points).

12.7 What Is a Smart Grid?

A smart grid is an electricity network enabling a two-way flow of electricity and data with digital communications technology enabling to detect, react, and pro-act to changes in usage and multiple issues.

Smart grids have self-healing capabilities and enable electricity customers to become active participants. They increase local availability of renewable energy resources.

An electrical network that efficiently integrates all the stakeholders – from generators to consumers, economic efficiency (time-based pricing), and sustainable network (demand response).

12.8 Smart Grid: The "Energy Internet"

Figure 12.14 shows the smart grid environment which consists of generation, transmission, communication network, security, and distribution network.

- Improve electrical power generation and distribution system.
- Integration of electric infrastructure and ICT infrastructure.
- More efficient and better management of power infrastructure.
- Increase the use of renewable energy sources.
- Alternate energy sources – Wind, solar generation, power storage.
- Integration of distributed energy sources into power infrastructure.
- Wind and solar generation by nature are variable.
- Matching or supply and demand to reduce the traditional bulk generation.
- Better management of energy usage.
- Use of smart meters and demand response systems to reduce and balance energy usage.
- Enable use of plug-in electric vehicles – more friendly to the environment, also as energy storage.

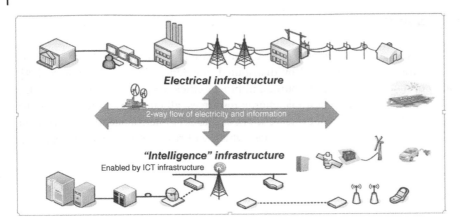

Figure 12.14 Smart grid environment.

12.9 Standardization

A standard for exchanging location-based data that addresses the geographic data needs of many smart grid applications. This family of standards defines information security for power system control operations.

The two-way flow of electricity and data that is the essential characteristic of a smart grid enables to feed information and data to the various stakeholders in the electricity market which can be analyzed to optimize the grid, foresee potential issues, react faster when challenges arise, and build new capacities – and services – as the power landscape is changing [22].

The electricity market, the consumption of electricity, regulations, demands of various stakeholders, and the very production of electricity are all changing. So, smart grid initiatives exist across the globe, albeit sometimes with different approaches and goals. Figure 12.15 represents the networks connected to form a smart grid.

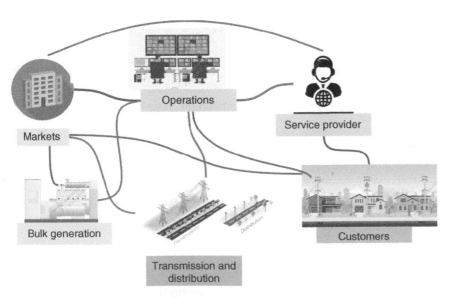

Figure 12.15 Network of smart grid.

12.10   Smart Grid Network

Transformed Power System Network – Utilities are poised to move from the traditional power system to a highly flexible, secured, and green power system by using integrated two-way communications and advanced control technology. Entire networks of the smart grid are shown in Figure 12.16.

12.10.1   Distributed Energy Resources (DERs)

- Distributed generations using renewable technologies.
- Photovoltaic (India aims at > 100 GW by 2022).
- Solar thermal.
- Small wind systems (up to 50 kW).
- Large wind systems (ranging up to 1–2 MW).
- Biomass, etc.

12.10.2   Optimization Techniques in Power Quality Management

Since most DGs employ power electronics, the power injection by the DG also injects the current harmonics in a smart grid. To prevent penalties from DISCOMS, equipment failure, etc., these harmonics must be reduced to the minimum level. To reduce harmonics, harmonic filters are employed. The performance of the harmonic filter depends on the proportional (P), and integrator (I) gain constants of the control unit. The optimum value of these constants is determined using both conventional and intelligent algorithms. The conventional algorithm adjusts the control variables using the gradient technique until the negative slope is achieved. The optimal location for the local minima will be reached using conventional method, but they cannot achieve the global minima. In order to overcome this, intelligent algorithms are used to determine the control variable values for performance improvement. Intelligent algorithms that are population-based and employ multiple agents in the search space will overcome the trap of local minima and arrive at the global minima. This intelligent algorithm determines the complete or optimum control variable values to minimize harmonics and provide consumers with high-quality power.

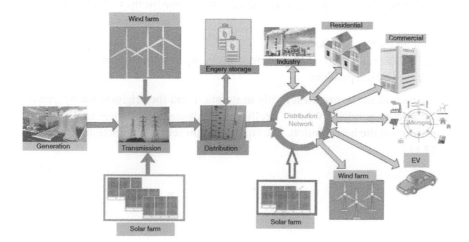

Figure 12.16   Entire smart grid network.

12.10.3 Conventional Algorithm

Conventional algorithms utilize search techniques based on mathematical procedures to provide results based on a starting point. The most prominent traditional optimization procedures include linear programming (LP), nonlinear programming (NLP), quadratic programming (QP), Newton-based solution, and interior point optimization (IP). The LP optimization uses the simplex technique. Optimization procedures can be solved using LP, NLP, QP, Newton-based solutions, and IP techniques. NLP approaches are used because power quality problems are nonlinear. QP is a kind of NLP that employs sensitivity-based approaches. In conventional optimization, the search ends with the nearest local minima point and cannot overcome it. They cannot solve nonlinear and discontinuous objective functions or involve rigorous techniques [22].

On the other hand, intelligent algorithms find the global minima and are easy to implement to get optimal global values of the control variables. QP is a kind of NLP that makes use of sensitivity-based techniques. In traditional optimization, the search is restricted to the nearest local minimum point and cannot be exceeded [22]. On the other hand, intelligent algorithms discover global minima and are simple to implement to obtain optimal global values for the control variables.

12.10.4 Intelligent Algorithm

According to recent studies and literature, intelligent algorithms can be utilised to solve power quality problems. Numerous algorithms are available, including the firefly algorithm (FA), ant colony algorithm (ACA), bee algorithm (BA), differential evolution (DE), harmony search (HS), simulated annealing (SA), genetic algorithm (GA), simulated annealing (SA), ant colony algorithm (SA), flower pollination (FP) algorithm, spider monkey optimization (SMO), and several others. The FA algorithm and SMO are examined in this chapter as harmonic filter tuning solutions.

12.10.4.1 Firefly Algorithm (FA)

The firefly algorithm is based on the intelligence of the insect firefly [23]. Firefly has the swarm intelligence and moves toward the brighter spot for the search of food or to find a mating partner. It is a population-based swarm-based intelligent algorithm to find the global optimal solution. The goal function is used to formulate a flashing light for optimization. The brightest firefly is the most ideal answer for the challenge at hand. Firefly is a collection of control variables for the problem under consideration in this case. The brightness of the firefly is calculated by calculating the optimal objective function [24]. This algorithm can be used to solve problems that need maximum or minimization. FA has idealization when compared to a genuine firefly, with the following assumptions:

- Firefly is unisex and attracted to another firefly regardless of sex.
- Firefly flies toward the brightest fly; if there is no brighter fly, it moves randomly in solution space.
- The brightness of the firefly is influenced by the problem.

Figure 12.17 depicts the firefly moving toward the brightest firefly in iteration and the brightest firefly moves toward the best optimal solution iteration by iteration. When a firefly cannot discover a brighter firefly, it traverses the search space at random in search of the best solution. The equation represents this function, which is a monotonically declining function (12.1)

$$\beta = \beta_0 \exp(-\gamma r^2) \tag{12.1}$$

where

β is attractiveness of a firefly
β_0 is initial attractiveness
γ is absorption coefficient
r is distance between fireflies

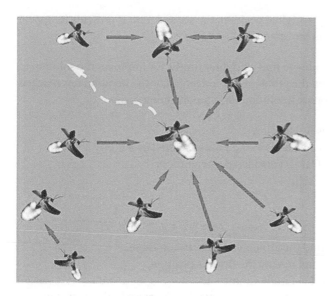

Figure 12.17   Firefly movement.

Figure 12.18   Distance between two fireflies.

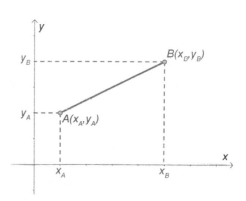

Distance   The distance between firefly A and B is computed using the equation for Cartesian distance (2).
  Figure 12.18 shows the position of firefly A and B and their two-dimensional position (x,y).

$$r_{AB} = \|A - B\| = \sqrt{\sum_{k=1}^{D}(A - B)^2} \tag{12.2}$$

In the preceding equation, k is the dimension, and the distance between A and B firefly may be computed using the equation for 2-dimensional solution space (3).

$$r_{AB} = \sqrt{(x_A - x_B)^2 + (y_A - y_B)^2} \tag{12.3}$$

r_{AB} is the distance between fireflies. (x_A,y_A) is the position of firefly A and (x_B,y_B) is the position of firefly B in the 2-dimensional space.

Movement   The attraction of "A" firefly toward "B" brighter firefly is depending on their distance from each other, as shown below:

$$x_i^k = x_i^k + \beta_{0*} \exp(-\gamma r^2) * \left(x_j^k - x_i^k\right) + \alpha^* \varepsilon_i^k \tag{12.4}$$

If the first term on the left side represents the ith firefly's original position, the second term represents its appeal to the jth firefly, and the third term represents its random movement.

Stopping Criterion Fireflies move about aimlessly, attempting to find a brighter firefly. FA improves the solution to issues iteratively, and iterations must be terminated when the problem has converged or the iteration has reached its maximum value. It is critical to cease iterating to propose a solution for temporal complexity.

Algorithm Firefly algorithm for solving power quality is listed below:

Step 1: Harmonic filter controller tuning parameter of constants P and I is considered as a firefly.
Step 2: The objective function is the minimization of THD is considered a brightness.
Step 3: Create the basic firefly population $\{x_1, x_2, x_3 \dots x_{20}\}$.
Step 4: Set the number of iterations to one ($itr = 1$).
Step 5: Calculate each firefly's brightness.
Step 6: Change the Firefly Count to 1.
Step 7: For each firefly, choose the brightest one and find the attractiveness between them.
Step 8: Move each firefly closer to the brightest firefly and update the firefly's position.
Step 9: Increase the Firefly Count by one, Firefly Count = Firefly Count + 1.
Step 10: Check the firefly count; if Firefly count $\leq$ Num. of fireflies, go to step 7, otherwise to the next step.
Step 11: Increase the Iteration count $itr = itr + 1$
Step 12: Check itr, if $itr \leq$ maximum number of iteration then go to step 5, else go to next step.
Step 13: The best answer is the brightest firefly, which may be printed as a consequence.

12.10.4.2 Spider Monkey Optimization (SMO)

Many studies on the lives, intellect, and behavior of monkeys were undertaken at Central Washington University in Washington and the National Autonomous University of Mexico. Monkeys have considerable intellect, which they employ to detect changes in the environment to choose the best food source. The intelligence of monkeys is founded on their social and collective interactions. They are gregarious creatures who live in packs. They use a variety of sounds and body language to communicate with one another. The majority of monkey species live in groups, which they retain during their travels and in their quest for food [25].

SMO was created based on spider monkey food search behavior. For sustenance, this spider monkey employs two basic processes: fission and fusion. Fission is a process that divides a group into subgroups and is guided by a local leader. If there is insufficient food for its subgroup, the subgroup summons the other subgroup. Each subgroup will locate its food source. Even when the subgroup has access to food, the subgroup is divided into smaller subgroups. The fission process continues until the minimal number of monkeys is reached, which is three. The food search procedure continues with this tiny subgroup for a set amount of time to cover a specific area of food source space. If the food supply is still unavailable after these steps, the group's primary leader will fusion or merge all of the subgroups into a single group [26]. This solitary group will now relocate to another location where food is available. This fission and fusion process continues until a food supply is discovered.

SMO processes are modeled into six phases to find the global optimal solution [27]. The phases involved in the SMO are listed below.

Local Leader Phase (LLP) The local leader in the monkey has minimum THD in the sub-group and is separated from the main group. Monkeys in the group will update their position using the Eq. (12.5)

$$M_{ij}^{new} = M_{ij} + U(0,1) \times (MLL_j - M_{ij}) + U(0,1) \times (M_{rj} - M_{ij}) \qquad (12.5)$$

where

M_i – ith spider monkey in the subgroup,
"j" – control variable in the M,

MLL – a local leader in the subgroup,

M_r – another spider monkey in the subgroup,

$U(0,1)$ – uniformly distributed random number.

Global Leader Phase (GLP)   All the monkeys in the group will observe the direction given by the global leader apart from the local leader. The position updation is also based on the position and directions given by the global monkey and it is given by the Eq. (12.6). The new position of the monkey is updated from its old position and the position of the global monkey and two other members in the same group.

$$M_{ij}^{new} = M_{ij} + U(0,1) \times (MGL_j - M_{ij}) + U(-1,1) \times (M_{rj} - M_{ij}) \tag{12.6}$$

Learning Phase   SMO has two learning phases one for local leaders and another one for global leaders phase. They are named as Local Leader Learning (LLL) phase and the Global Leader Learning (GLL) phase. In this phase the number of leader updates is counted if the count is more than the maximum limit, then the action is to be taken in the decision phase. The other two important phases are as follows.

Local Leader Decision (LLD) Phase   To come out of the local minima/maxima point, the local_limit_count is compared with the maximum value of the count. If the local_limit_count value is equal to the maximum count then it is evident that the search is locked in the local minima/maxima point. To unlock this local leader decision phase uses the random update of subgroup members based on the position of global and local leaders. Equation (12.7) and (12.8) give the team members position update based on local and global leaders, minimum and maximum limit

$$M_{ij}^{new} = M_{ij} + U(0,1) \times (MGL_j - M_{ij}) + U(0,1) \times (M_{ij} - MLL_j) \tag{12.7}$$

$$M_{ij}^{new} = M_{ij}^{min} + U(0,1) \times \left(M_{ij}^{max} - M_{ij}^{min} \right) \tag{12.8}$$

where M_i is the ith spider monkey considered to update, "j" is the decision variable in the spider monkey, $U(0,1)$ is the uniformly distributed random number, MGL is the global leader, MLL is the local leader of the subgroup, M_{ij}^{min} is the minimum limit of the spider monkey, and M_{ij}^{max} is the maximum limit of the spider monkey.

To update the spider monkey in the LLDP either Eq. (12.3) or Eq. (12.4) is used based on the random number and the perturbation constant (PC). The LLDP pseudo-code is given below for the update of spider monkey in the subgroup.

Global Leader Decision (GLD) Phase   In this phase, the global limit count count is verified to determine whether it has reached its maximum value. The global leader then splits the for the first time into two subgroups and assigns the global limit count to zero. The phases are repeated, and the count is compared to the previous time. If the global limit count reaches its maximum value, the global leader increases the number of subgroups from two to three. This technique is repeated until the maximum number of subgroups is reached. When the subgroup's division reaches its maximum and global limit count reaches its maximum value, the global leader merges all subgroups into a single group. The following is a graphical illustration of the grouping and sub-grouping.

SMO in a Single Group   All the monkeys are part of a single group, headed by the wise lady monkey. The global leader has the experience and decision-making ability to search for food sources. The control variables representing the monkey with the minimum objective function or THD are considered the global monkey in the optimization algorithm. In Figure 12.19, X_i is the monkey with the minimum objective function and is considered as the global monkey.

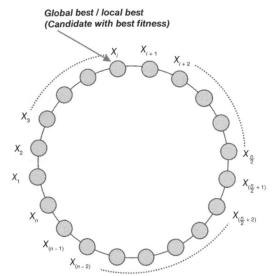

Figure 12.19 SMO in a single group.

SMO Is Divided into Two Groups When the global_limit_count reaches its maximum value, a single group is divided into two subgroups, as shown in Figure 12.20. Each sub-group has one local leader, X_i and $X_{(n-3)}$. Among the local leader, the minimum is selected as the global leader. In this case, X_i is the minimum among two and selected as the global leader.

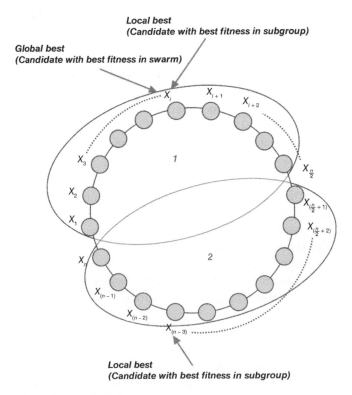

Figure 12.20 SM in two group.

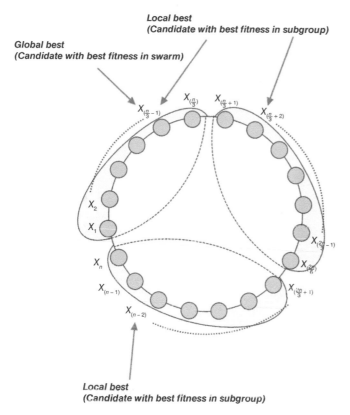

Figure 12.21 SMO in three group.

SMO Is Divided into Three Groups Figure 12.21 shows the three sub-groups in the SMO population. When the global_limit_count reaches its maximum value after the two subgroups the population is again subdivided into three groups as shown in Figure 12.20. Each sub-group has one local leader and the minimum objective function leader becomes the global leader, in this case, $X_{(n/3-1)}$.

SM Is Divided into Minimum Size Group The group is divided up to the minimum size of three members in the group. Further, the group cannot divide. If the global_limit_count reaches its maximum value after this situation then all the groups will be merged into one single group. Figure 12.22 shows the minimum size groups which cannot be divided further. Each group has one local leader and one among them who has minimum objective function is considered the global leader.

Spider Monkey Optimization (SMO) Algorithm The power quality problem in the SMO algorithm is optimized by following the steps below.

Step 1: Convert the power quality problem into SMO by considering the PI constants of the controller as a monkey.

Step 2: Create an initial population of spider monkeys in the solution space.

Step 3: Determine the fitness of all spider monkeys.

Step 4: Set the iteration count = 1.

Step 5: Choose the global leader with the best fitness out of all of them.

Step 6: Choose a subgroup's local leader which has high fitness in the subgroup.

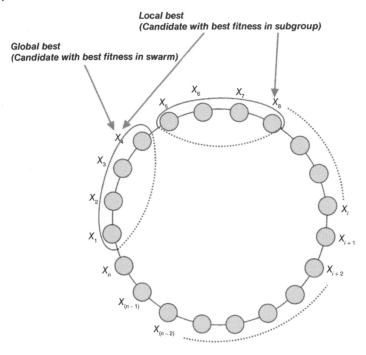

Figure 12.22 SM with min size group.

Step 7: Create new positions for all members of the group by using self-experience, local leader experience, and group members' experience.

Step 8: Apply the greedy selection process between the existing position and the newly generated position, based on fitness, and select the better one.

Step 9: Re-direct all members of that particular group for foraging if any local group leader does not update her position after a certain number of times (local leader limit).

Step 10: If the global leader does not update her position a set number of times (global leader limit), the group is divided into smaller groups.

Step 11: Repeat Steps 4–11 until the stopping criterion has been met.

Step 12: If the iteration count is fewer than the maximum number of iterations then go to Step 8, else go to the next step.

Step 13: The global leader which consists of the best control variables is the solution to the considered problem.

12.11   Simulation Results and Discussion

The simulation is performed using MATLAB Simulink software, Non-linear RL load is used to simulate the system. The conventional PI controller is used to tune the controller. The same is compared with the firefly algorithm and spider monkey optimization.

Figure 12.23 shows the total harmonic distortion level system with PI controller in which the THD level is around 30%.

Figure 12.24 shows the total harmonic distortion level system with firefly algorithm tuned PI controller in which the THD level is around 1.35%.

Figure 12.25 shows the total harmonic distortion level system with SMO algorithm tuned PI controller in which the THD level is around 1.05%. Table 12.2 gives THD for the system without filter, with PI controller without

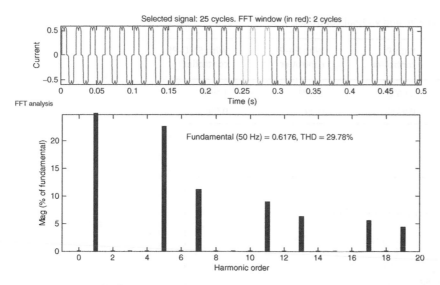

Figure 12.23 THD of the system with PI controller.

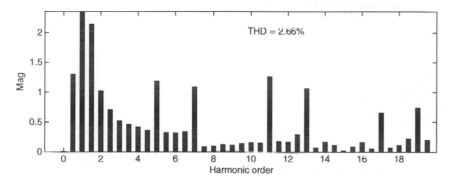

Figure 12.24 THD of the system with PI tuned by firefly algorithm.

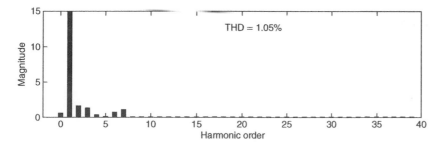

Figure 12.25 THD of the system with PI tuned SMO.

optimization, firefly optimized controller, and SMO optimized controller is given in the table. THD in SMO optimized PI controller gives a better result.

Figure 12.26 shows the THD level of the system with and without a controller, from which it is understood that the proposed PI tuned SMO gives better performance. Figure 12.27 shows the line diagram of THD reduction by harmonics filter without filter and with a filter which is tuned by firefly algorithm and SMO optimization.

Table 12.2 Comparison of THD levels with and without controller.

Parameters	Without FILTER	With PI controller	With FA-PI controller	With SMO-PI controller
Overall THD	29.78%	2.66%	1.35%	1.05%
Power factor	0.6591	0.9318	0.99	0.99

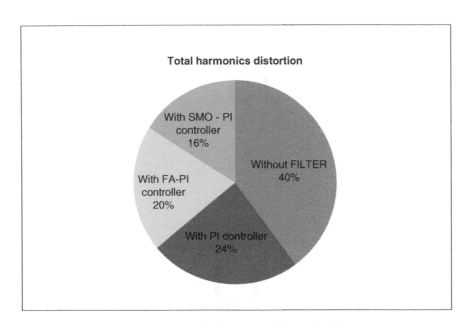

Figure 12.26 Comparison of THD reduction by intelligent algorithms.

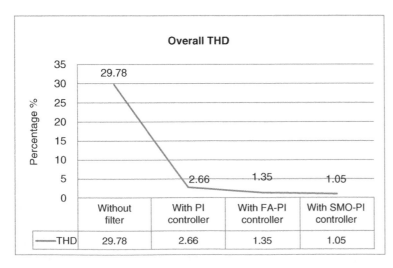

Figure 12.27 PI-controller chart of various controllers.

12.12 Conclusion

Power quality issues and the conditioners to mitigate the issues are analyzed and presented. The example issues provide as an illustration of how STATCOM can be utilised to reduce voltage sag. In addition to reducing waveform distortion and harmonics, UPS also addresses the problem of transient power outages. IEEE standards for power quality are listed for reference to calibrate the simulation result given in the chapter. The requirement of a smart grid for the modern age is discussed. Standardization and networking of smart grids are explained to understand the function of smart grids. The inclusion of distributed generators in the smart grid invites new power quality issues; the major one is the harmonics. This harmonic is removed using the shunt active power filter. The function of the filter is based on the proportional and integrator gain constant. To determine the value of these gain constants, either conventional or intelligent optimization techniques may be used. The intelligent optimization techniques have advantages in finding the optimal global solution and are used in this chapter. Two main intelligent algorithms are developed to optimize the shunt active power filter: the firefly algorithm and spider monkey optimization. The procedural implementation and simulation results are presented and compared. With MATLAB/Simulink, the wind energy system connected to a nonlinear load has been simulated to implement an intelligent algorithm. The THD value has been measured without compensation as 29.78%, which is too high compared with the standard. The active filter in the system will reduce the THD using different control algorithms such as PI, GWO-PI, firefly algorithm tuned PI, and SMO-PI the THD has been reduced to 2.66%, 1.64%, 1.35%, and 1.05%, respectively. From the results, it is concluded that SMO tuned PI controller in the filter reduce the THD and consumers receive a quality power supply.

References

1 Teke, A., Saribulut, L., and Tumay, M. (2011). A novel reference signal generation method for power-quality improvement of unified power-quality conditioner. *IEEE Transactions on Power Delivery* 26 (4): 2205–2214.

2 dos Santos, R.J.M., da Cunha, J.C., and Mezaroba, M. (2014). A simplified control technique for a dual unified power quality conditioner. *IEEE Transactions on Industrial Electronics* 61 (11): 5851–5860.

3 Zahira, R., Lakshmi, D., and Ravi, C.N. (2021). *Power Quality Issues in Microgrid and its Solutions. Microgrid Technologies* 255–286.

4 Huang, W., Lu, M., and Zhang, L. (2011). Survey on microgrid control strategies. *Energy Procedia* 12: 206–212.

5 Ghosh, A. and Ledwich, G. (2002). *Power Quality Enhancement Using Custom Power Devices.* London: Kluwer.

6 Zahira, R., Lakshmi, D., Ezhilarasi, G. et al. (2022). Stand-alone microgrid concept for rural electrification: a review. *Residential Microgrids and Rural Electrifications* 109–130.

7 Aredes, M. and Watanabe, E.H. (1995). New control algorithms for series and shunt three-phase four-wire active power filters. *IEEE Transactions on Power Delivery* 10 (3): 1649–1656.

8 Domijan, A. and Embriz-Santander, E. (1992). A novel electric power laboratory for power quality and energy studies: training aspects. *IEEE Transactions on Power Systems* 7 (4): 1571–1578.

9 Zahira, R. and Fathima, A.P. (2012). A technical survey on control strategies of active filter for harmonic suppression. *Procedia Engineering* 30: 686–693.

10 Daehler, P. and Affolter, R.H. (2000). Requirements and solutions for dynamic voltage restorer, a case study. In: *2000 IEEE Power Engineering Society Winter Meeting. Conference Proceedings (Cat. No.00CH37077)*, vol. 4, 2881–2885.

11 Ch, K., Phani, S.V., and Kumar, R.D. (2016). Power quality improvement of grid integrated type I wind turbine generation system operating as DSTATCOM by dq control method. *Journal of Electrical Systems* 12 (2): 278–290.

12 Zahira, R., Peer Fathima, A., Lakshmi, D., and Amirtharaj, S. (2021). Modeling and simulation analysis of shunt active filter for harmonic mitigation in islanded microgrid. In: *Advances in Smart Grid Technology*, 189–206. Singapore: Springer.

13 Al-Zamil, A. and Torrey, D. (2001). A passive series, active shunt filter for high power applications. *IEEE Transactions on Power Electronics* 16 (1): 101–109.

14 Zahira, R. (2018). Design and performance analysis of a shunt active filter for power quality improvement. https://shodhganga.inflibnet.ac.in:8443/jspui/handle/10603/251849.

15 De la Rosa, F. (2006). *Harmonics and Power Systems*. CRC press.

16 Afonso, J.L., Aredes, M., Watanabe, E., and Martins, J.S. (2000). *Shunt Active Filter for Power Quality Improvement: Electricity for a Sustainable Urban Development*, 683–691.

17 Sivaraman, P. and Sharmeela, C. (2021). Power quality and its characteristics. In: *Power Quality in Modern Power Systems* (ed. P. Sanjeevikumar, C. Sharmeela and J.B. Holm-Nielsen), 1–60. Academic Press.

18 Thentral, T.M.T., Palanisamy, R., Usha, S. et al. (2022). Analysis of power quality issues of different types of household applications. *Energy Reports* 8: 5370–5386.

19 Akagi, H. (1996). New trends in active filters for improving power quality. In: *International Conference on Power Electronics Drives and Energy Systems for Industrial Growth*, 417–425. IEEE.

20 Verdelho, P. and Marques, G.D. (1997). An active power filter and unbalanced current compensator. *IEEE Transactions on Industrial Electronics* 44 (3): 321–328.

21 Billinton, R. and Sachdeva, S.S. (1972). Optimal real and reactive power operation in a hydro thermal system. *IEEE Transactions on Power Apparatus and Systems* PAS 91: 1405–1411.

22 Li, Q., Li, D., Zhao, K. et al. (2022). State of health estimation of lithium-ion battery based on improved ant lion optimization and support vector regression. *Journal of Energy Storage* 50: 104215.

23 Yang, X.S. (2010). Firefly algorithm, Levy flights and global optimization. In: *Research and Development in Intelligent Systems*, vol. XXVI, 209–218. London: Springer.

24 Zhao, J., Chen, D., Xiao, R. et al. (2022). Multi-strategy ensemble firefly algorithm with equilibrium of convergence and diversity. *Applied Soft Computing* 108938.

25 Al-Azza, A.A., Al-Jodah, A.A., and Harackiewicz, F.J. (2015). Spider monkey optimization: a novel technique for antenna optimization. *IEEE Antennas and Wireless Propagation Letters* 15: 1016–1019.

26 Sharma, A., Sharma, N., and Sharma, K. (2022). Enhancing the social learning ability of spider monkey optimization algorithm. In: *International Conference on Intelligent Vision and Computing*, 413–435. Cham: Springer.

27 Rao, J.M. and Narayan, B.H. (2022). Novel coronavirus (COVID-19) prediction using deep learning model with improved meta-heuristic optimization approach. In: *2022 4th International Conference on Smart Systems and Inventive Technology (ICSSIT)*, 935–943. IEEE.

13

The Role of Internet of Things in Smart Homes

Sanjeevikumar Padmanaban[1], Mostafa Azimi Nasab[2,3], Mohammad Ebrahim Shiri[4],
Hamid Haj Seyyed Javadi[5], Morteza Azimi Nasab[2], Mohammad Zand[2], and Tina Samavat[2]

[1] Department of Electrical Engineering, Information Technology, and Cybernetics, University of South-Eastern Norway, Porsgrunn, Norwny
[2] CTIF Global Capsule, Department of Business Development and Technology, Aarhus University, Herning, Denmark
[3] Department of Electrical and Computer Engineering, Boroujerd Branch, Islamic Azad University, Boroujerd, Iran
[4] Mathematics and Computer Science Department, Amirkabir University of Technology, Tehran, Iran
[5] Department of Mathematics and Computer Science, Shahed University, Tehran, Iran

13.1 Introduction

Internet technology is a vast network of objects such as humans, machines, and the Internet that interact intelligently with one another. Using Internet technology from objects, different parts of a home can be intelligently controlled and managed. One of the most important goals of smart homes is to establish more control and smart security in a home or even a large building. In fact, by using Internet technology, objects, and web-based applications or mobile-based applications, we can make the devices and devices in a smart home in the form of objects under more supervision and control and have better security in the smart home [1]. Advances in the field of smart homes are related to the developments of society, which are directly related to the progress of society. In addition, to create added value, the focus should be placed on environmental smart homes, as well as creating smart environments to support the elderly and disabled has a great potential that includes different stakeholders for a full life. There's a lot of smart home technology emerging. The Internet of Emerging Objects, which effectively integrates physical-shadow space to create intelligent environments, will have countless applications shortly. Due to the expansion of the smart home concept, we are witnessing the intelligence productions presented to the international market. For this purpose, smartphones play a significant role in managing various devices at home while users do not have physical access. Equipment such as rice cookers, vacuum cleaners, and security systems are great examples of productions that can be used in smart homes and Internet of Things (IoT). Smartphones mainly link them to their owners through a local network or Internet. Considering such applications, they are connected by the Internet and can be turned on or off from remote locations with the help of an installed application on a smartphone. In addition, it is a great technology enabler for building smart homes that provide many benefits to our community [2]. Therefore, research on the IoT and smart homes is very important and has a special place and a lot of research should be done in this field. This chapter presents the evaluation of performance of Internet-based distribution systems. This study determines the performance of IoT-based distributed systems for things like smart energy management and smart home security [3].

Artificial Intelligence-based Smart Power Systems, First Edition.
Edited by Sanjeevikumar Padmanaban, Sivaraman Palanisamy, Sharmeela Chenniappan, and Jens Bo Holm-Nielsen.
© 2023 The Institute of Electrical and Electronics Engineers, Inc. Published 2023 by John Wiley & Sons, Inc.

These are our goals from this review:

1. Intelligent energy management using distributed systems based on Internet objects technology.
2. Intelligent management of smart home equipment using distributed systems based on Internet object technology.
3. Maintaining security using distributed systems based on Internet object technology [4–6].

13.2 Internet of Things Technology

The technology of the IoT and the Internet all mean the formation of vast networks that make up the objects around us. Based on this model, objects and systems are intelligent, automatically receiving useful information from around them and sending them to one another or central control systems. This technology is one of the main adverbs of fifth generation communication and data macro technologies [7]. Based on this technology, very useful services can be provided to improve the quality of life. According to reliable news provided by reputable companies such as Sisco and Erickson, by 2025, about 50 billion smart objects will be included in this model, and will support this technology. The services provided by this technology are used in all important industries such as health, oil, and gas, transportation, etc. Up to $3,000 billion per year is guaranteed by the financial turnover of services provided by the technology at the international level. Considering the considerable market share of this technology, the development and application of this technology in the country are very important from an economic perspective. Considering that this technology interacts with people's physical life, it is very important from a social and security perspective. On the other hand, products based on this technology such as covers, smart meters, connected cars, smart homes based on M2M device communication platform, etc. have entered the market with the epidemic of Internet objects. In the not too distant future, any device will be able to connect to the Internet. When this technology is equipped with a variety of sensors and activators, a real virtual system will be formed that will have a variety of capabilities such as smart grid coverage, smart homes, smart transport networks, and smart cities. Each object, through its embedded computer system, has a unique identity and the possibility of interaction on the Internet. According to experts, the IoT will have 50 billion objects by 2020 [8].

One of the significant consequences of Internet implementation of objects in the country is the ability of automatic activity of electronic and digital devices. According to predictions made by 2025, more than 25 billion different devices will be available in the world through Internet-based services or other information networks. The transmission will be connected and the price of machines and electronic systems will be connected through the Internet without the need for humans and even in restrictions beyond the current limits. Although achieving this goal can be accompanied by concerns, it should be noted that research institutes have predicted that governments, industrial factories, and service institutions are more concerned than those seeking to meet the needs. As a new revolution, the IoT allows objects to recognize themselves. This technology has an important place to make changes in industries and lifestyles. In a world where objects are based on the Internet, lack of access to data is meaningless and such a platform will eventually lead to an increase in the efficiency of labor forces [9]. In the modern business world, access to more and better information means better control of the future. To have in control information leads to the production of products and services that facilitate human life; but facilitating human life today is associated with the special complexities of the modern and contemporary world. Also, the IoT is also based on the same goal of receiving, registering, and sending information for the use, analysis, and improvement of products and services, and finally the use of users [10] has been proposed and developed. As long as, based on a program for exercise and physical activity in Have you considered analyzing the changes that have taken place in your body or what you intend to do about living things? Research differently, you can use the technology of the IoT to get information to the database instantly Send the data to which it is attached. Information about the amount of food in the refrigerator or the amount of gasoline in your car tank is another IoT application. With regard to IoT, there is no need for human–human interaction or human–computer interaction, data is sent. Another example is the rise of the IoT phenomena, the result is the spread of the Internet and the development of wireless technologies. The IoT is a new concept in the world of information and communication technology, through which humans, animals, and objects can send information data through communication networks. One of the concerns raised about the IoT is security and privacy. The IoT has had its problems and limitations before, but the right and principled approach in this field can play a very important role in eliminating the problems and limitations [11]. Vinit Cerf, who has been dubbed the "father of the Internet," is among those who have expressed concern about the development of the IoT and believes that the IoT is a combination of equipment and software, and this feature can cause concern and fear. According to Winter, with the increase in the number of software-based equipment, people's ability to write good computer code increases. However, this information should be prevented from being made available to individuals who intend to use it for improper purposes. Hackers can access information about people's presence at home using the IoT. Modulators have seriously pursued the security of this technology. Although good efforts have been made in this area, more work needs to be done to ensure security and privacy as two important issues in the field of the IoT. Promoting health, comfort, and well-being in life and organizing information in more useful ways for people and organizations are positive aspects of the development of the IoT. In contrast, the development of this technology will be accompanied by challenges to privacy. One of the largest parts of the IoT is smart home objects. According to the studies, about 40% of the 6.2 billion networked objects in smart cities are placed in smart homes. In smart homes, everything from locking and opening the entrance, ordering food needed at home, and many other everyday tasks is done automatically [12]. In the future world, houses will be a suitable space for the growth and development of the IoT, and accordingly, many household tools and appliances will be connected to the Internet, and access to information about their latest situation will be provided in time and space. Wearing technologies and planting technologies will be developed in this direction. People can also be another platform for establishing Internet objects. These are just a part of the developments that are provided in the context of Internet objects and as a result of these developments, people's relationships will also undergo significant changes. Human progress in modern times has brought with it crises that are the result of neglect of nature, but the same progress can be used purposefully and in a planned way to address some of the crises that have arisen. Let's repair [13, 14]. Consequently, concerns like sustainable architecture were brought forward, to achieve which different methods have been proposed, each of these methods in a way tries to excel sustainable architecture. Using today's technologies, we can carry out effective activities in the discussion of sustainable architecture, most of these activities can be seen in smart buildings. Helping the environment and conserving its resources while reducing its consumption costs creating an atmosphere of comfort for humans can be considered the main and common goal of sustainable architecture and smart buildings [15].

13.2.1 Smart House

Smart homes are now making their way into the industry as part of the IoT phenomenon. The terms smart home and connected homes are more commonly used interchangeably, meaning the concept of home communication tools to create value for the people living in the home (Figure 13.1). Ericsson defines smart homes as a home in which home appliances are improved by an Internet connection [16]. According to Frost and Sullivan, the penetration rate for security measures, including home surveillance, is currently rising among Western European users, followed by home entertainment.

A proposal on energy savings among homeowners in the report "Creating Growth from Connected Homes" suggests that smart lights and smart thermometers in the German market will be more attractive in the coming

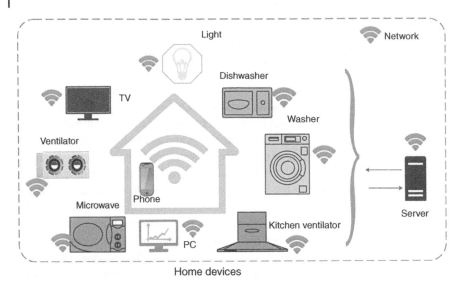

Home devices

Figure 13.1 Smart home structure.

years. The report also stated that there will be no full smart home market until 2020–2025, but 30% of Western European homes may have some kind of smart system in their homes by 2019 [17].

A major challenge in maintaining the efficiency of smart home technology is internal interaction, which means that tools from different manufacturers must work well together. Internal interaction is also essential from the user's point of view to increase ease of use and ensure fast and hassle-free implementation. With technical advances and the reduction of hardware costs and the presence of sensors, the acceptance of devices by end consumers is an important issue that should be considered and facilitated by prominent brands. As smartwatches, wristbands, and connected consumer electronics make their way, the concept of a fully connected home needs to be properly defined. This requires that the complexity of consumer needs be reduced while at the same time the functions, interaction, and convenience that ultimately lead to value creation are increased. In general, building decoration is a set of tools that are designed and implemented to increase efficiency and create a suitable environment for people who live in homes, called home intelligence. The purpose of implementing smart home projects can be to turn a building into a building by reducing energy consumption or to turn a home into a luxury home with smart management [18].

13.3   Different Parts of Smart Home

There are different categories for different parts of smart homes. One of these categories in the smart home is divided into three sections: devices, communications, and services (Figure 13.2). Devices: smart home devices are hardware devices that typically include sensors, actuators, busses, and smart objects. Sensor: measures the physical properties of an environment or physical existence. Sensors can be divided into wearable and non-wearable. Activator: it is the responsibility of the activator to perform actions such as turning the lights on, off, or dimming, closing the windows, activating the alarm, etc. [19].

Bus: usually acts as an access point to the home, allowing the homeowner or other entity to remotely monitor, control, and manage home applications or sensors. It also serves as an aggregation point for sending measured

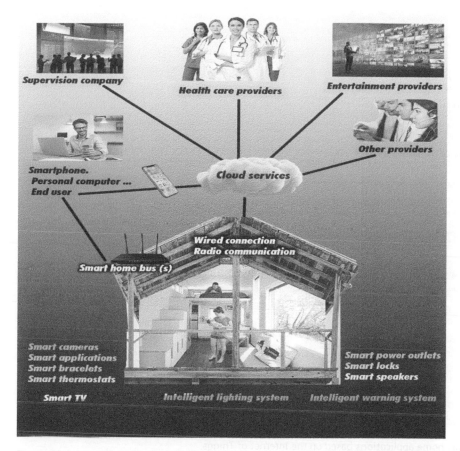

Figure 13.2 A sample of smart home architecture. Source: Rawpixel.com/Adobe Stock.

data to an external network such as a software company. Smart objects: consists of sensors or actuators that are connected to the home network. Examples of smart objects include smart applications such as smart locks [20]. Connections to a typical smart home use a variety of communication protocols, ranging from wired to radio. The general communication of sensors used in home automation is through Z-Wave, Zigbee, KNX, and DASH7 protocols or network communication protocols such as Wi-Fi, Bluetooth 3.1, IEEE 802.15.4, Low PAN, or cellular network technology. Reconnaissance systems with radio waves and near-field communication technology are used to monitor and track residents in the healthcare sector and are commonly used in smart door locks. Services: applications hosted on the cloud or in the home environment, which are responsible for automation, device management, decision making, etc. A special category of services is controllers that allow management to connect to devices. Such software runs on smartphones or tablets of home residents to communicate locally or remotely with the device [21].

Fixing security vulnerabilities requires remote planning. This is not possible for all devices such as operating systems, stack protocols, or software or hardware because it is possible that they do not support dynamic patches. In addition, some devices, such as smart meters, expect online systems such as those that do not require replacement components or are stored directly. Figure 13.3 shows the Internet application areas of objects for managing and controlling smart homes or buildings [22].

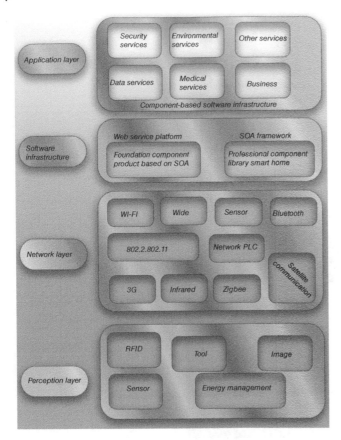

Figure 13.3 Architecture of smart home applications based on the Internet of Things.

13.4 Proposed Architecture

For the proposed network architecture, according to the existing literature, a three-layer architecture, which includes the physical layer, the application layer, and the control layer, is considered. The physical layer includes a variety of physical devices such as sensors, coders, switches, and gateways [23]. These tools are responsible for environmental protection and data transmission. However, these tools alone cannot perform these tasks. Therefore, decision-making protection is left to the network control department. The communication of these two layers is done using the standard interface of software-based networks (SOA). The control layer is located between the physical layer and the application layer. The physical tool manages the network using the network interface and provides services such as data processing and transfer to the application layer. This connection is made using the standard interface below. In the application layer, programmers create programs such as the environment search program using a standard relationship provided by the control layer. Programmers can change the properties of the network, and in addition to various applications can use the physical equipment of the network. A network controller architecture is a bilayer in which the sublayer or local controller is directly connected to the switches and is responsible for detecting physical devices and connections. In this layer, the

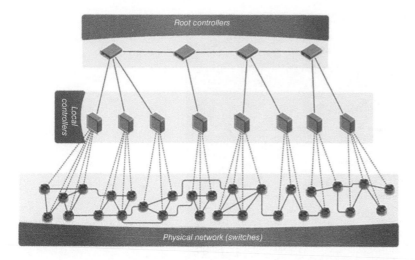

Figure 13.4 Proposed architecture.

view of the network below it is protected and this view is shared with the root controller. The top layer also has root controllers. Distribution in this layer is a solution for geographical distribution. Figure 13.4 shows the overall architecture of the controller [24, 25].

Local controllers are constantly updating the status of the physical network under their control. They share the available information with their associated root controllers, and finally, the root controllers create an overview of the network across all root controllers by sharing views between each other. This preserves the centrality of the control section of the network and the unchanged nature of the software-centric control of these controllers is distributed throughout the network. In current network architecture, normal communication begins when an entity on the network requests the network to send a packet of data or message of any other type to another entity on the network, the respondent. Before this, the applicant must know the respondent's ID and use it in the transaction. This model can be used for most protocols, including IoT protocols. For such communication, the network establishes a path from the requester to the respondent. Such a path may be virtual or logical. Therefore, software-based is used and allows entities to rely on different protocols to communicate with each other; therefore, software-based mechanisms can be used to create a path between two endpoints. This is called sending routing, in which all the necessary sending rules are placed in the sending tool on the network. To do this, the various components in the network controller go hand in hand.

13.5 Controller Components

On the IoT to collect, analyze, and present data using the proposed multi-component controller must be present in the controller. These components determine the goals of architectural design. The condition of logical centrality of the control section is observed in the software-based, a physically distributed control section is provided to observe the scalability of the controllers. Also, requests that can be managed on local controllers should not be referred to as root controllers. Figure 13.5 shows the components and relationships of the proposed architectural layers.

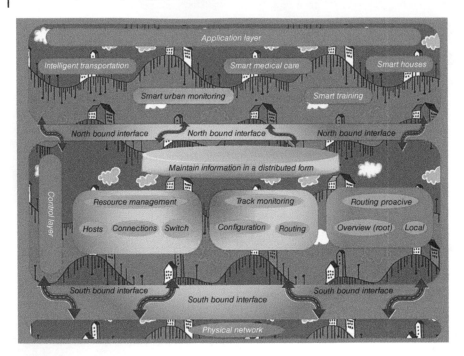

Figure 13.5 Components of architectural layers.

13.6 Proposed Architectural Layers

The proposed architectural layers are as follows:

13.6.1 Infrastructure Layer

In this layer are the infrastructures which are composed of the following parts::

13.6.1.1 Information Technology
Information technology changes the evolution of cities. The presence of the Internet has made it possible for planners to consider not only physical issues when planning for a city, but also information technology for greater efficiency in government, the economy, and the mobility and dynamism of the city. It consists of communication networks, operating systems, business systems, hardware, integrators, and data storage systems.

13.6.1.2 Information and Communication Technology
It includes person-to-person, person-to-machine, and machine-to-machine communications and plays a key role in controlling and monitoring the urbanization process. It can be said to be synonymous with information technology, but it focuses more on seamless communications between systems, telecommunications, and the use of computers as software, processors, middleware, and storage environments.

13.6.1.3 Electronics
A city's electronic infrastructure includes high-performance computing resources, advanced storage systems, network infrastructures, electronic access tools, high-speed networks, and supports services. Sharing information resources, collecting data in real-time for fast processing, and computing large data sets all depend on the good implementation of electronic infrastructure.

13.6.2 Collecting Data

The second layer, which is related to collection, consists of the IoT, a network of wired sensors, and built-in systems that communicate with each other. The city is divided into several cells, and in each of these cells, there is a local access point. Various devices send their information to each other and access points through wired networks and technologies such as 3G, 4G, Wi-Fi, etc. These access points send data to central access points for storage. To send data over networks, we need a data model. The data model organizes data components and standardizes and is used to communicate between different data components. A data model shows the structure of data used in information systems and for storing information. Because each device or network in which information flows has a specific protocol for sending information over the communication network, the protocols are adapted to coordinate between different protocols on the network. In this layer, Internet geographic information services and mobile geographic information systems can also be used [26].

13.6.3 Data Management and Processing

In the third layer, first, the information coming from the second layer is stored as new data. Then, the information is processed for processing. Intelligent communication is closely related to fog processing because fog processing can be a subset of cloud processing. With the difference, that data can be processed locally on smart devices, without being sent to the cloud for processing. Fog processing is suitable for the IoT, in which smart devices with different demands are connected internally. Because objects produce a lot of data, transferring these data to the cloud to process and return these responses to objects requires high bandwidth, high-time networks, and the problem of data latency arises. In fog processing, most of the processing is done in routers instead of being transferred to the cloud. Then, data processing and data management among routers is performed using fog processing to classify and move them to the next section in the same layer [27]. There are four sub-sections as follows, which play a major role in the performance of this section:

13.6.3.1 Service Quality Management
This section is related to the efficiency of information networks, which ensures that the information is flowing correctly and completely in the network and has its standards. Some of the things that this section measures are: error rates, throughput, bandwidth, transmission delays, and network availability.

13.6.3.2 Resource Management
Due to the expansion of networks, information and processing resources available at the network level alone cannot plan and control related activities. As a result, a management department is needed to do this coordination. This section is also used to allocate resources on the network that are used for storage and processing [28].

13.6.3.3 Device Management
This section configures and evaluates the performance of all devices and objects that are in the network and generate or transmit information.

13.6.3.4 Security
This section establishes data security in the second and third layers because, after this section, the information is distributed to the cloud database and stored there. It then encrypts, authenticates, and so on. The data are then transferred to the cloud for storage, where it is stored in a database and on existing servers. In this section, some processing takes place.

13.7 Services

This layer consists of two parts called service environment and infrastructure environment. These two departments use the information they obtain from their previous layers to create services for urban development in the smart home. Infrastructure environment refers to transportation, natural resources, mobile services, water, waste, construction, etc. In other words, in this environment, there are opportunities for urban planning and development, and it makes the settings of a city based on the data it receives and the behaviors that it should show. Information and communication technology plays an important role in this sector. In contrast, the service environment includes health, security, economics, education, government, media, and so on. The services of this sector directly affect the citizens. Information and communication technology plays a more limited role in this area because the purpose, processing, and integration of information are not real-time [29].

13.8 Applications

This layer includes applications that are directly connected to the smart home. These include applications on smartphones, social networks, websites, and more. In this layer, services are in confrontation with individuals. In addition to all these layers that have been described, the management layer and the network integrator play a major role in all of this architecture. As we can see in Figure 13.6, this layer covers the three middle layers of architecture, i.e., the second to fourth layers [30].

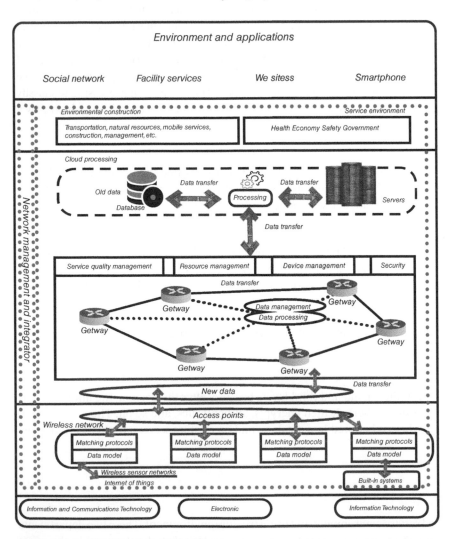

Figure 13.6 Environment and application.

13.9   Conclusion

The IoT brings a new era in information technology and can change our lives and jobs to a modern and intelligent stage. By creating the right infrastructure to use this technology, smart homes, smart residential areas, and more applications will emerge in the future. With thousands of objects capable and willing to cooperate, it will be impossible to adhere to the traditional view, especially about side effects such as the use of unnecessary networks. Therefore, the investment should be made toward a sharing/dissemination model in which the necessary institutions can receive or publish only the events that are acceptable to them. Due to the high growth of nanotechnology, network and communication, modeling and simulation, decision systems and resource planning of the organization, integration, etc., a new generation of equipment, innovative services, and commercial applications is expected to date it has not been predicted to be possible. The IoT is a pioneering technology with the use of various communication technologies and today the effect of this technology can be seen in different sectors such as agriculture, industry, smart homes, health, and other areas. Studies show the number of smart objects around us is increasing day by day, and this makes it possible to connect everything in the world to come. Using IoT technology, a huge communication network is created that has billions of nodes, and each node has several sensors to interact with its environment. One of the important applications of the IoT is its use in the field of smart cities and smart homes, so understanding the technology of the IoT and its implementation will help to provide a better understanding of future smart homes. Interdependence of light and ventilation system on the presence of the person and optimal planning of room temperature at different hours of the day and night are examples of this management of energy consumption under smart home control with the help of available factors such as telephone, cell phone, landline, remote control, Internet and e-mail can be done. The rapid development and learning of information and communication technology has led to the creation of various and diverse fields of its application in all aspects of human life. One of the new applications of information and communication technology is the development of building automation systems. The building automation system provides energy management solutions along with more comfort and security in the form of an integrated system. By applying for the latest technological advances in the residential and office complexes of hotels, universities and hospitals benefit from the widest range of building automation services and make the lives of residents and users of services easier and safer. By providing an architecture suitable for smart homes, the IoT makes it possible for homes to develop software based on cloud-based processing platforms to produce software that can be managed, monitored, controlled, and finally, an increase in interest rates is possible.

Exercise

- A method of a distributed system for smart grid using the IoT.
- Energy management for IoT via smart grid systems.

References

1 Zand, M., Nasab, M.A., Padmanaban, S., and Khoobani, M. (2022). Big data for SMART Sensor and intelligent Electronic devices–Building application. In: *Smart Buildings Digitalization*, 11–28. CRC Press.

2 Zand, M., Nasab, M.A., Hatami, A. et al. (2020). Using adaptive fuzzy logic for intelligent energy management in hybrid vehicles. In: *2020 28th Iranian Conference on Electrical Engineering (ICEE)*, 1–7. IEEE.

3 Ahmadi-Nezamabad, H. et al. (2019). Multi-objective optimization based robust scheduling of electric vehicles aggregator. *Sustainable Cities and Society* 47: 101494.

4 Zand, M., Nasab, M.A., Sanjeevikumar, P. et al. (2020). Energy management strategy for solid-state transformer-based solar charging station for electric vehicles in smart grids. *IET Renewable Power Generation* 14 (18): 3843–3852.

5 Ghasemi, M., Akbari, E., Zand, M. et al. (2019). An efficient modified HPSO-TVAC-based dynamic economic dispatch of generating units. *Electric Power Components and Systems* 47 (19–20): 1826–1840.

6 Nasri, S., Nowdeh, S.A., Davoudkhani, I.F. et al. (2021). Maximum Power point tracking of Photovoltaic Renewable Energy System using a New method based on turbulent flow of water-based optimization (TFWO) under Partial shading conditions. In: *Fundamentals and Innovations in Solar Energy*, 285–310. Singapore: Springer.

7 Rohani, A., Joorabian, M., Abasi, M., and Zand, M. (2019). Three-phase amplitude adaptive notch filter control design of DSTATCOM under unbalanced/distorted utility voltage conditions. *Journal of Intelligent & Fuzzy Systems* 37 (1): 847–865.

8 Zand, M., Nasab, M.A., Neghabi, O. et al. (2019). Fault locating transmission lines with thyristor-controlled series capacitors by fuzzy logic method. In: *2020 14th International Conference on Protection and Automation of Power Systems (IPAPS)*, 62–70. IEEE.

9 Zand, Z., Hayati, M., and Karimi, G. (2020). Short-channel effects improvement of carbon nanotube field effect transistors. In: *2020 28th Iranian Conference on Electrical Engineering (ICEE), Tabriz, Iran*, 1–6. https://doi.org/10.1109/ICEE50131.2020.9260850.

10 Tightiz, L. et al. (2020). An intelligent system based on optimized ANFIS and association rules for power transformer fault diagnosis. *ISA Transactions* 103: 63–74, ISSN 0019-0578, https://doi.org/10.1016/j.isatra.2020.03.022.

11 Zand, M., Neghabi, O., Nasab, M.A. et al. (2020). A hybrid scheme for fault locating in transmission lines compensated by the TCSC. In: *2020 15th International Conference on Protection and Automation of Power Systems (IPAPS)*, 130–135. IEEE.

12 Zand, M., Azimi Nasab, M., Khoobani, M. et al. (2021). Robust speed control for induction motor drives using STSM control. In: *2021 12th (PEDSTC)*, 1–6. https://doi.org/10.1109/PEDSTC52094.2021.9405912.

13 Sanjeevikumar, P., Zand, M., Nasab, M.A. et al. (2021). Spider community optimization algorithm to determine UPFC optimal size and location for improve dynamic stability. In: *2021 IEEE 12th Energy Conversion Congress & Exposition - Asia (ECCE-Asia)*, 2318–2323. https://doi.org/10.1109/ECCE-Asia49820.2021.9479149.

14 Azimi Nasab, M., Zand, M., Eskandari, M. et al. (2021). Optimal planning of electrical appliance of residential units in a smart home network using cloud services. *Smart Cities* 4: 1173–1195. https://doi.org/10.3390/smartcities4030063.

15 Nasab, M.A., Zand, M., Padmanaban, S. et al. (2022). An efficient, robust optimization model for the unit commitment considering renewable uncertainty and pumped-storage hydropower. *Computers and Electrical Engineering* 100: 107846.

16 Azimi Nasab, M., Zand, M., Padmanaban, S., and Khan, B. (2021). Simultaneous long-term planning of flexible electric vehicle photovoltaic charging stations in terms of load response and technical and economic indicators. *World Electric Vehicle Journal* 12 (4): 190.

17 Sundaram, C., Shanmuga, M.S., and Ajay-D-Vimal Raj, P. (2017). Tabu search-enhanced artificial bee colony algorithm to solve profit-based unit commitment problem with emission limitations in deregulated electricity market. *International Journal of Metaheuristics* 6 (1–2): 107–132.

18 Khoa, T.A., Nhu, L.M.B., Son, H.H. et al. (2020). Designing efficient smart home management with IoT smart lighting: a case study. *Wireless Communications and Mobile Computing*, 2020.

19 AlJanah, S., Zhang, N., and Wah Tay, S. (2021). A survey on smart home authentication: toward secure, multi-level and interaction-based identification. *IEEE Access* 9: 130914–130927.

20 Dashtaki, M.A., Nafisi, H., Khorsandi, A. et al. (2021, Energies). Dual two-level voltage source inverter virtual inertia emulation: a comparative study. 14 (4): 1160.

21 Burhan, M., Rehman, R.A., Khan, B., and Kim, B.S. (2018). IoT elements, layered architectures and security issues: a comprehensive survey. *Sensors 2018* 18 (9): 2796.

22 Samuel, S.S.I. (2016). A review of connectivity challenges in IoT-smart home. In: *2016 3rd MEC International conference on big data and smart city (ICBDSC)*. IEEE.

23 Sivaraman, V., Gharakheili, H.H., Vishwanath, A. et al. (2015). Network-level security and privacy control for smart-home IoT devices. In: *2015 IEEE 11th International conference on wireless and mobile computing, networking and communications (WiMob)*, 163–167. IEEE.

24 O'Connor, T.J., Mohamed, R., Miettinen, M. et al. (2019). HomeSnitch: behavior transparency and control for smart home IoT devices. In: *Proceedings of the 12th Conference on Security and Privacy in Wireless and Mobile Networks*, 128–138.

25 Asadi, A.H.K., Jahangiri, A., Zand, M. et al. (2022). Optimal design of high density HTS-SMES step-shaped cross-sectional solenoid to mechanical stress reduction. In: *2022 International Conference on Protection and Automation of Power Systems (IPAPS)*, vol. 16, 1–6. IEEE.

26 Gupta, B. (2021). Analysis of IoT concept applications: smart home perspective. Future access enablers for ubiquitous and intelligent infrastructures. In: *5th EAI International Conference, FABULOUS 2021, Virtual Event, May 6–7, 2021, Proceedings*, vol. 382. Springer Nature.

27 Dashtaki, M.A., Nafisi, H., Pouresmaeil, E., and Khorsandi, A. (2020). Virtual Inertia Implementation in dual two-level voltage source inverters. In: *2020 11th Power Electronics, Drive Systems, and Technologies Conference (PEDSTC)*, 1–6. IEEE.

28 (a) Kim, J., Choi, S.C., Yun, J., and Lee, J.W. (2018). Towards the oneM2M standards for building IoT ecosystem: analysis, implementation and lessons. *Peer-to-Peer Networking and Applications*. (b) Qian, Y., Wu, D., Bao, W., and Lorenz, P. (2018). The Internet of Things for Smart Cities. *Technologies and Applications. IEEE Network* 33 (2): 4–5.

29 Saqlain, M., Piao, M., Shim, Y., and Lee, J.Y. (2019). Framework of an IoT-based industrial data management for smart manufacturing. *Journal of Sensor and Actuator Networks* 8 (2): 25.

30 He, J., Kunze, K., Lofi, C. et al. (2014). Towards mobile sensor-aware crowdsourcing: architecture, opportunities and challenges. In: *International Conference on Database Systems for Advanced Applications*, 403–412. Berlin, Heidelberg: Springer.

14

Electric Vehicles and IoT in Smart Cities

Sanjeevikumar Padmanaban[1], Tina Samavat[2], Mostafa Azimi Nasab[2], Morteza Azimi Nasab[2], Mohammad Zand[2], and Fatemeh Nikokar[2]

[1]*Department of Electrical Engineering, Information Technology, and Cybernetics, University of South-Eastern Norway, Porsgrunn, Norway*
[2]*CTIF Global Capsule, Department of Business Development and Technology, Aarhus University, Herning, Denmark*

14.1 Introduction

The smart city is emerging due to the increasing development of information technology in cities and meeting the new needs of citizens for information and hardware facilities in their urban life. What drives a city to be smart is not just the use of that city's electronic tools and communication system; Rather, the use of this tool is to improve the "quality of life" of the citizens of a city [1].

In many existing sources and research about smart cities, the first smart city is described, and different definitions are mentioned, and then a contractual definition is provided to make the content the same. Some of these definitions are:

The smart city is the city of "knowledge," "digital," "cyber," [2] and "economy" [3]; or an open concept for all kinds of interpretations and mental perceptions and related to goals not set by smart city planners. Today, our perception of the smart city may be cities developed in both performance and structure, sometimes using information and communication technology (ICT) as infrastructure. It is a city that makes good use of forward-looking strategies in the economy, people, government, mobility, transportation, environment, and life. Also, it is built on a delegate combination of talent, sustainable action, and independent and knowledgeable components. It has monitored and integrated its vital infrastructure, including roads, bridges, tunnels, railways, subways, airports, communications, water, electricity, and even its main buildings. This city has better efficiency in its resources and for it plans maintenance and prevention activities, monitors security aspects, and at the same time provides full service to citizens.

Building a smart city is emerging as a strategic goal to reduce the problems created by urban population growth and increasing urbanization, while research is underway to develop and implement it.

One of the most important needs for creating an intelligent structure in large cities is a suitable platform and infrastructure for communications and telecommunications [4].

The Internet of Things (IoT) is one of the new technologies in the present era that its application in cities can be considered a major innovation for sustainable urban growth and development. In short, the IoT is a modern technology in which any creature (human, animal, or object) can send data through communication networks, whether the Internet or intranet [5].

The term IoT refers to a set of technologies and researches that aim to bring the Internet into the world of physical objects [6].

IoT is typically associated with technologies such as radio frequency identification (RFID) systems, short-range wireless communications, and sensor networks [7].

Artificial Intelligence-based Smart Power Systems, First Edition.
Edited by Sanjeevikumar Padmanaban, Sivaraman Palanisamy, Sharmeela Chenniappan, and Jens Bo Holm-Nielsen.

This environment aims to overcome key issues such as data extraction from various sources, integration of different technologies, and the ability to reconfigure the system, which has received much attention today. The IoT allows objects to be remotely controlled across existing network infrastructures and provides an opportunity to integrate directly from the physical world into computer-based systems while improving human intervention, efficiency, accuracy, and economic benefits [8].

Today, information technology has led to the production of many powerful computers that make it possible to collect, transmit, combine, and store large volumes of data at a low cost. Increasing the size of databases leads organizations to extract information from stored data. Nevertheless, the data in the database alone cannot be used as valuable treasures to extract confidential information [8]. In many cases, humans are also unable to detect and extract information contained in large volumes of data, so they need the help of database learning algorithms. Nowadays, among the learning algorithms used for data processing, deep learning approaches have attracted the attention of researchers more than other methods [9]. Deep neural networks are one of the newest methods of machine learning that have made it possible to solve complex problems [10]. In recent years, artificial neural networks (including recursive networks) have won numerous competitions in pattern recognition and machine learning.

High fuel consumption by motor vehicles such as cars motorcycles, alike in the world due to the reduction of fossil fuel sources and high environmental pollution by these consumers, has led to increasing attention to equipment using other energy sources. According to studies, if the trend of energy consumption continues, the amount of carbon dioxide in the environment by 2050 will double its level in 2005, which is not acceptable from an environmental point of view and the existing roadmap. According to global plans, this amount should reach half of its level in 2005 by 2050. One of the most appropriate methods to achieve this goal, in addition to methods such as the use of distributed energy sources in the field of consumption and simultaneous production of electricity and heat, is the use of motor vehicles whose driving force is provided by electricity or main batteries. This issue is particularly important in developed countries, especially in the United States and Japan. In addition, countries such as China and India have made significant progress in this area.

In countries like Iran, due to the cheap energy carriers, fossil fuels have become one of the reasons for excessive consumption. Also, severe air pollution as the result of daily consumption of 60 million liters of gasoline, along with the low quality of cars and their presence in the country's transportation despite reaching the end of their expected life, which increases gasoline consumption in them, all emphasize the need to coordinate with global efforts to achieve electric vehicle technology.

Thus, the importance of using electric vehicles and their impact on mitigating environmental pollution and reducing fuel consumption is not irrefutable. However, an important issue is that the presence of electric vehicles in the transportation network will require a substructure and a place to charge the batteries of this equipment. For this purpose, electric vehicle charging stations are in different places, such as gas stations [11].

In such cases, connecting electric vehicles at the charging station to the national grid is considered a load for the distribution network. If the number of these vehicles in the network is limited, their presence will not affect the network's dynamic behavior; But by replacing electric cars with gasoline cars, a large number of these cars will enter the electric grid. In this case, if proper management is not done on how and at what time the vehicles are connected to the grid for charging, the electrical characteristics of the grid such as voltage, current, and active grid power level will be severely affected, and the grid will experience severe voltage drops [12].

On the other hand, electric vehicles are used to reduce the costs of fuel consumption and pollution. However, if the presence and operation of these vehicles are not properly managed, the cost of operating these vehicles will not only not be reduced but will also cause additional costs to the network operator. Therefore, a study on better understanding the structure and performance of electric vehicles, their presence on the behavior of the network, and its electrical characteristics seems to be essential. However, to achieve the above goals, it is necessary to restructure the existing traditional electricity network into an intelligent network in which telecommunication, computing, and control services and technologies are combined with power system infrastructure [13].

14.2 Smart City

"Smart city" is a city based on telecommunication technology that seeks to guarantee the revolution of lifestyles, respond to the needs of citizens through planning, novel design, development, and renovation of communities, preserve natural resources, increase ecological integration in the short and long term, as well as increasing the quality of life through the development of transport, employment, and housing.

Soon we will see the development of large networked and integrated cities. By 2025, more than 60% of the world's population is expected to live in urban areas. This urbanization process will have divergent effects and influence the future of personal life and dynamism. It has swallowed up its surroundings and formed huge cities with more than 10 million people. These conditions will lead to the evolution of smart cities with eight intellectual features: smart economy, smart buildings, smart mobility, smart energy, smart information, and communication technology, smart planning, smart citizen, and smart government [14].

The role of governments in establishing the IoT will be crucial. Up-to-date urban operations are underway, and the development of urban development strategies will lead to the IoT.

Therefore, by considering regulations, cities and their services represent an almost ideal platform for IoT research and become the solutions provided by IoT technology [15].

14.2.1 Internet of Things and Smart City

The new IoT technology can carry out smart city initiatives around the world. This feature provides the ability to remotely monitor, manage, and control devices by creating a platform for having practical information at the same time. The main features of a smart city are integrating information technology and the comprehensive use of information resources. Essential components of urban development for a smart city should include smart technology, smart industry, smart services, smart management, and smart living [16].

14.3 The Concept of Smart Electric Networks

According to the US Department of Energy, a smart grid is an extensive automated energy distribution network capable of transmitting electrical power and exchanging information in two ways. This network can monitor and respond to any changes in the network, from production sources to consumer preferences and even personal equipment. The smart grid structure emphasizes environmental protection by using variable renewable energy products such as wind and solar, load response, and controllable distributed generation sources and energy storage while maintaining reliable grid performance and the need to increase subscriber choice. Table 14.1 shows the factors associated with an intelligent network. Environmental protection is one of the most important issues that has received much attention in recent years.

Also, due to the factors such as the growing trend of network consumption load, the expansion of distributed generation sources with variable generators and consumers, considering the decay of the current production units, more attention will be needed to the stability of the power system. More efficient and economical operation of the network by creating the ability to instantly monitor consumption and more accurate modeling of production resources in the smart grid is another aspect in this area. Intelligent networks with the use of new monitoring and control technology will play an important role in the performance of system reliability and efficiency [17].

Table 14.1 Factors associated with an intelligent network.

	21st century smart grid				
Controlability	Demand and production	Self-repair network	Building automation	Distributed automation	Vehicles connected to the network
	Advanced load management	Ancillary services	Customer gateway	Rates vary with time	Renewable energy-based markets
	Load response			Advanced measurement infrastructure	
	Load interruption	Exit management	Energy management systems	Automated readings	Connect renewable resources
	Capacity	Power quality and reliability	Work efficiency	Optimal operation	Clean technology
	The basis of infrastructure				

(Visibility — marked along the right margin)

14.4 IoT Outlook

The IoT is a concept that has a wide presence in various environments, including wireless and wired connections, and can interact and collaborate to create new programs or services and achieve common goals. There are many R&D challenges for creating an intelligent world. Real, digital, and virtual worlds are integrating to create smart environments that generate energy, transportation, cities, and many more. The IoT aims to enable objects to connect to any other object or user anytime, anywhere. The IoT is a new revolution in which objects make themselves recognizable. They are also gaining intelligence by making and providing relevant decisions. Thanks to this fact, they can communicate and access information collected by other objects. Alternatively, they can be part of a larger and more complex service. This development coincides with the emergence of cloud computing capabilities and the transition from the traditional Internet to the new Internet protocols with almost unlimited addressing capacity.

14.4.1 IoT Three-layer Architecture

The IoT has three layers that have been widely used in a great deal of research based on the features of the IoT. As shown in Figure 14.1, this outline consists of three layers named Observation, Network, and Application Layer [18].

14.4.2 View Layer

The main purpose of this layer is to measure and collect information in intelligent network systems and the IoT, which is done using different devices. It includes a variety of IoT devices such as RFID tags, cameras, wireless sensor networks (WSNs), Global Positioning System (GPS), and Machine to Machine devices to collect information on smart grids. This layer is divided into two sub-layers, the observation control and the communication sub-layers. On the other hand, the communication sub-layer has a communication module that connects IoT devices to the network layer.

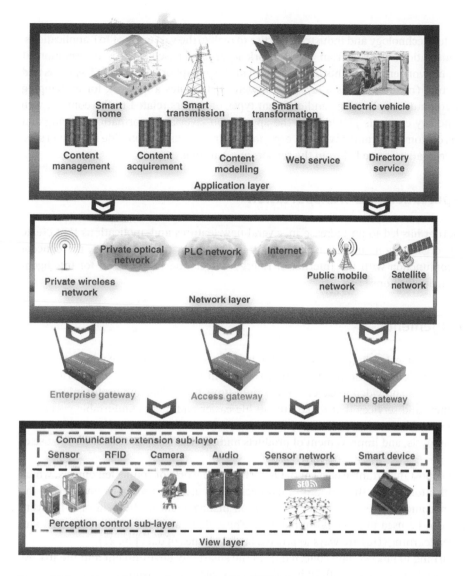

Figure 14.1   IoT architecture. Source: scharfsinn86/Adobe stock; scharfsinn86/Adobe stock; leszekglasner/Adobe stock.

14.4.3   Network Layer

The network layer is a convergent network formed by different telecommunication networks and the Internet. The network layer has expanded widely due to its advanced technologies. This layer maps the information collected by IoT devices in the observation layer to telecommunication protocols; And then transmits the mapped data through the telecommunication network to the relevant application layer. The main network, the Internet, is responsible for routing and transmitting information and controlling access to the network based on other telecommunications networks. IoT management and information centers also belong to the network layer. The network layer can be public or rely on industry-specific communication networks.

14.4.4 Application Layer

The application layer integrates IoT technology and industry expertise to realize the vast array of IoT applications. This task is to process the information received from the network layer, and based on this information, it monitors IoT devices and smart network environments in real-time. This layer supports various IoT systems and smart grids. Software infrastructure (middleware) is a layer of software that creates a network for exchanging information between different computer applications and different types of servers related to the content, web services, and direct services. The key elements provided by the application layer are information sharing and security. The application layer is customizable and advanced, especially for networks that provide large data sets. These programs specify what information is needed from the sensors at what time.

14.5 Intelligent Transportation and Transportation

Connecting vehicles to the Internet has led to an increase in several new features and applications that create new benefits for individuals or organizations in easier and safer transportation. In this context, the concept of the Internet of Vehicles in conjunction with the Internet of Energy reflects the future trend for intelligent transportation and transportation applications [19].

14.6 Information Management

The modern world and communication technology have shown that in the business world, those who have access to more and better data and information will control the future; Having information in the New World is like an apple falling in Newton times. When provided to experts at the right time, up-to-date and useful information leads to the production of products and services that make human life easier and more comfortable every day. IoT technology has been developed with the same philosophy. Receive, record, and send up-to-date and instantaneous information for the use, analysis, and improvement of products and services and, of course, for the use of consumers.

Data management is one of the most important aspects of the IoT. Given the world of connected objects, which constantly exchange all kinds of information, the amount of data generated and the processes involved in the information flow process have become a major challenge. Information management in the field of IoT is very important to use methods that can be used in the field of big data. These days, artificial intelligence (AI) and machine learning (ML), and deep learning (DL) are ways to process large volumes of data [20, 21].

This chapter uses in-depth learning to process data collected from the smart city environment and the IoT. As shown in Figure 14.2, deep learning is a subset of artificial intelligence and machine learning.

14.6.1 Artificial Intelligence

Artificial intelligence means machine intelligence or the ability to perceive and learn a machine. The machine here means any smart device with a processor that can take input data from the environment and process it or even make decisions. The purpose of artificial intelligence is to simulate and understand human behavior; That is, the ultimate goal is for the machine or intelligent device to comprehend or be able to do what man does in the same way or even better than man.

Artificial intelligence refers to systems that can respond similarly to intelligent human behaviors, including understanding complex situations, simulating and responding to human thought processes and reasoning techniques, learning, and the ability to acquire knowledge and reasoning to solve problems.

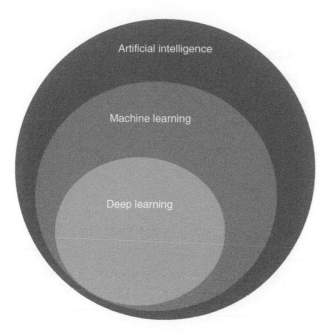

Figure 14.2   The structure of artificial intelligence.

14.6.2   Machine Learning

One of the most important artificial intelligence tools is machine learning, which is the core of artificial intelligence, i.e., the learning part. Machine learning involves a variety of methods and algorithms. There are three common steps in all of these methods. In the first step, our information from the subject is entered into the algorithm as data. In the second step, our algorithm learns and is trained to achieve the desired application or goal using this data. After completing the training phase, the algorithm prepared for the desired application should be used, which is the purpose of prediction in this chapter.

14.6.3   Artificial Neural Network

Before entering the main topic, i.e. deep learning, it is necessary to be familiar with the topic of neural networks or artificial neural networks. As shown in Figure 14.3, artificial neural networks are generally derived from the structure of the human neural network. In the human neural network, we have a series of operational units called cells or neurons, in which information or data enters and leaves the cell in the form of pulses or electrical signals so that these electrical pulses travel through the branch. Dendrites enter the cell nucleus from different sources or cells and are collected there with all the inputs (ini), and so-called processing is performed on it, and a new pulse is generated, which this pulse through a branch called axon from. The nucleus of the cell leaves and enters the next cell or tissue, so in this process, a pulse passes through the cell, and a series of changes are applied to it. These changes and processing are learned throughout human life, and the so-called neural network structure is taught.

For example, suppose a three-dimensional input (Figure 14.4). If we multiply these input dimensions by a factor and then pass their sum through a nonlinear function, a new output is generated. This process is similar to the same process. Is that there is inside the nerve cell, that is, our input changes the passage of this cell, which here are changes in our weights, and the nonlinear function that leads to a new output into these three stages, namely input, weights, and the so-called nonlinear function. It is called a layer.

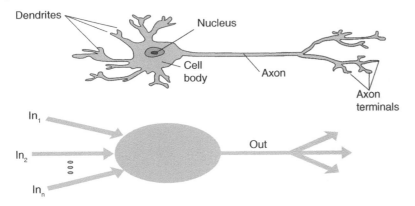

Figure 14.3 The structure of the human nerve cell.

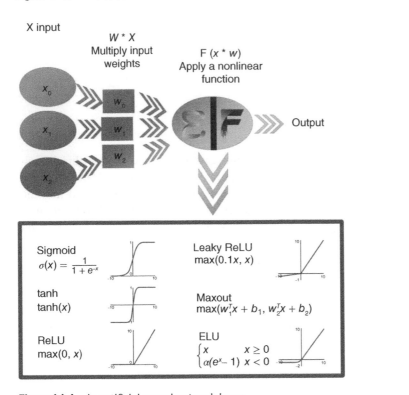

Figure 14.4 An artificial neural network layer.

In artificial networks, our job is to teach this layer. Training means finding weights that convert the input to our desired output, so these are the weights that are taught, and the nonlinear function is usually added to increase the network capability.

14.6.4 Deep Learning

Deep learning, in other words, deep machine learning, deep structure learning, or hierarchical learning, is a subfield of machine learning based on a set of algorithms that attempt to model high-level abstract concepts in data. The process is modeled using a deep graph with multiple processing layers consisting of several layers of linear and nonlinear transformations. Machine learning is commonly used for classification and prediction. For this reason, this model of learning is called deep learning because learning is done in several layers and so-called deep. It has different models, such as Auto Decoder, Generator autoencoder, Recurrent neural network, Convolutional neural network.

14.7 Electric Vehicles

Over the past hundred years, the release of acute greenhouse gas emissions from factories and internal combustion vehicles has caused unprecedented changes in the Earth's climate. One of these changes is global warming and endangering the planet's flora and fauna. Governments and environmentalists have made extensive efforts to reduce air pollution over the past decades [22].

Today, cars in countries due to the use of internal combustion engines and the use of fossil fuels, with low efficiency, are one of the factors increasing air pollution. Therefore, finding new solutions to change the source of energy production for cars is essential. Accordingly, governments have tried to increase the efficiency and effectiveness of automobiles by encouraging and supporting them to build more efficient vehicles. In this regard, companies such as Toyota, DaimlerChrysler, General Motors, etc., have started to build electric and hybrid vehicles. The main problem of these cars was the program of charging and controlling these cars at different distances. With the advancement of technology and the use of more advanced batteries, and the production of more efficient internal combustion engines, these cars were launched to solve some of the problems. With the advent of multi-directional electric drives, it became possible to charge and discharge these vehicles through the power grid, which led to the production of vehicles with grid connection technology. These cars were more capable and efficient than traditional cars, which, as a result, found many enthusiasts in the world.

Increasing the presence of distributed generation sources, including electric vehicles, will lead to changes in the formulation of planning. Also, by increasing the capacity of electric vehicles in the power network and providing them with the possibility of exchanging energy with the network, a suitable framework for economic exchanges in this field should be developed, considering network security. This is important because the significant presence of electric vehicles with the ability to connect to the network can cause fundamental changes in the network load curve, which will affect how the network is operated. Connecting an electric vehicle to a power grid creates a kind of charge with a new definition so that these vehicles can be considered as consumers of electrical power while charging and as a kind of "generator" of electricity for the power grid when charging [23].

Due to the flexibility of this technology, designing a suitable model according to the characteristics of the electric vehicle seems necessary for studies related to the power grid. On the other hand, according to the definition of authoritative references from smart grids, the concept of the vehicle to the grid is an integral part of these definitions.

14.7.1 Definition of Vehicle-to-Network System

An intelligent network provides the tools needed to control electric vehicles so that electric vehicles are discharged with a charging delay during peak demand or charged using a more sophisticated control strategy due to local demand or real-time energy prices. Therefore, when vehicles are connected to electricity, they provide scattered mobile storage for the network. Two-way power charging capacity and two-way communication between electric vehicles and the grid are called vehicle technology (V2G).

14.7.2 Electric Cars and the Electricity Market

With the emergence of electricity and energy markets in many countries, the reduction of energy consumption has shifted its position to the optimal use of primary resources. Accordingly, many companies have shifted their efforts to the optimal use of primary sources, and also, many companies have focused their efforts on the optimal production and consumption of energy. They have planned to supply their surplus energy to the electricity market. On the other hand, due to the small size of some manufacturers, such as automobiles, to produce energy, some companies tried to gather resources and coordinate between them to provide services to the electricity grid. These companies introduced themselves as integrators in the power system and provided various services to the electricity network and energy customers. Among these aggregators, we can name public parking lots, which can charge and discharge incoming vehicles with a special program, and the vehicle when leaving the parking lot, according to the contract with the owner of the vehicle in a capacity of Optimal charge spent. These parking lots have also been used in optimal planning to reduce costs and pollution with the presence of vehicle-to-network. With the increasing use of renewable energy sources to generate electricity and the unpredictability of these resources, electric vehicles have been proposed as a suitable solution to solve the problems caused by the accidental nature of sources such as wind and solar energy [24].

On the other hand, when these cars are collected in a place such as a public parking lot, it is possible for them to exchange much power with the electricity grid or to store it in themselves.

14.7.3 The Role of Electric Vehicles in the Network

Electric cars have three controllable load modes, manufacturer mode, and neutral mode. Since electric vehicles in the V2G concept can be considered mobile power plants, they will be one of the main sources of distributed storage in the future. Electric vehicles are charging in controllable load mode, receiving energy from the grid and storing it in the battery. Therefore, the status and charge speed can be controlled and considered a controllable load. In a neutral mode, the car is connected to the power grid but, for some reason, has no energy exchange with the grid. For example, the car owner wants the maximum battery capacity at a certain time. Logically, if the car's battery charging capacity reaches the desired level shortly before the scheduled time, the car will no longer be exchanged with the network until the specified time and will be neutral unless a special situation occurs in the network.

The energy stored in the car battery is returned to the grid in power generator mode. Power flow from the car to the grid may be due to technical and economic reasons. Sometimes a network operator is required to reduce or increase the load on the part of the network for security reasons. These operators can use these resources if they have enough electric vehicles in the concept of vehicle-to-network. On the other hand, electric vehicles equipped with vehicle-to-network technology can participate in the ancillary services market and maximize their profit by providing various services [25].

Different cars can connect to different charge levels. In other words, each type of car is charged with a certain voltage and current, which limits the power transferred between the network and electric vehicles. To show the effect of the type of connection of vehicles on the amount of delivered power (received) (set of variables that represent the charge and discharge rate), this variable is equivalent to the rate of the increase–decrease slope of power plants (set of electric vehicles), is defined.

With the expansion of the electric vehicles utilization and the presence of charging stations in smart grids in cities and their role in the communities' future progress, ensuring the applicability of the proposed plans is considered very important. In this regard, studies have been conducted on charging stations, among which we can mention the study conducted in Sydney. In this study, as mentioned in reference [26], several charging time planning techniques have been considered, and in the results, the degree of effectiveness has been discussed.

14.7.4 V2G Applications in Power System

In the case of connecting cars to the global network, V2G applications can be divided into six categories:

- Provide baseload
- Courier supply

- Auxiliary services of voltage control, frequency adjustment, etc.
- Backup power supply
- Ensuring network balance and reliability requirements

14.7.5 Provide Baseload Power

The baseload is determined on a 24-hour (full day) basis, usually provided by large nuclear or coal units that cost less per kilowatt-hour. Research has shown that electric vehicles cannot be supplied at a competitive price. Base loads are economical because the baseload price is very low, but they can increase the baseload and thus reduce the total cost so that electric vehicles can be charged at night when the price and demand for electricity are lower. This can cause the car batteries to be charged with cheap electricity.

14.7.6 Courier Supply

The peak load occurs at times of the day when consumption is high. To provide this load, units are needed that turn on and off quickly and provide peak load, but this is not cost-effective due to the high cost of these units [27]. Due to their rapid response, a large set of electric vehicles can be a significant source of support for the operator through which it can delay the lighting of large courier units, as shown in Figure 14.5.

14.7.7 Extra Service

The primary function of ancillary services is to support network reliability. Different conditions cause a mismatch between power generation and load consumption. One of the main causes of this inconsistency is the uncertainty related to the system load schedule, i.e. it is not possible to determine the exact time of disconnection and connection of subscribers to the power system, and as a result, there is an unbalanced supply–demand balance. Rotating reservations and reactive power are important.

14.7.8 Power Adjustment

Power regulation is responsible for keeping the frequency constant at 50 or 60 Hz, which is done in different countries by direct control telecommunication signals controlled by the network operator. If the load is more than

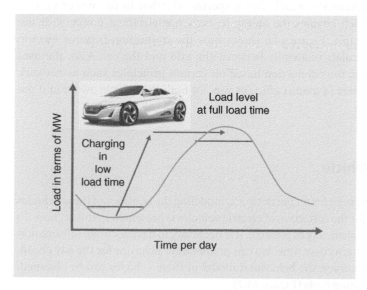

Figure 14.5 Fill the valleys and remove the peaks with electric vehicles.

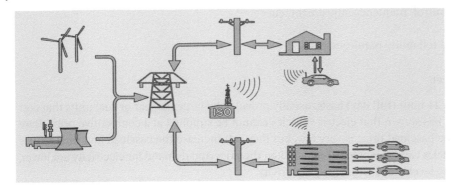

Figure 14.6 Schematic view of the connection between the electric vehicle and the power grid.

the output, the frequency and voltage decrease, and if the load is less than the output, the frequency and voltage increase, and in both cases, they need to be readjusted. Power adjustment contracts are set on an hourly basis. Ready capacity is provided to the operator for one hour [28].

14.7.9 Rotating Reservation

Rotating reservation is provided by additional output synchronized with the system. Rotating booking should be done quickly and available in less than 10 minutes when needed. The main difference between rotating and non-rotating backups is that rotary backup generators must be ready and able to help the network prevent frequency decay when there is a sudden drop from other sources.

On the other hand, because cars can quickly inject power into the network, they can easily enter the circuit in less than 10 minutes and be used to set up or down.

14.7.10 The Connection between the Electric Vehicle and the Power Grid

The most important feature of EV and PHEV vehicles, which draws special attention to the power grid, is the concept of grid connection technology, which enables the ability to exchange electrical power with the grid for these vehicles based on a specific principle. Figure 14.6 shows how the connection between electric vehicles and the power grid. Electricity can circulate bilaterally between the grid and the car. Also, the user and the supervisor of this network must manage the connection based on certain principles such as network status, connection duration, and battery parameters in a most efficient way for both the vehicle owner and the network [29].

14.8 Proposed Model of Electric Vehicle

To investigate the effect of the presence of electric vehicles on the network, modeling the behavior of these vehicles is necessary. In this regard, the proposed model for the collection of electric vehicles is presented in this section. To model the behavior of electric car collections as realistically as possible, it is necessary to have accurate information about them over time. Using their behavioral patterns over time, we can predict their behavior for the day ahead. As a result, with proper forecasting and high accuracy, the behavioral model of these vehicles can be obtained, and proper planning can be done to reduce operating costs (Figure 14.7).

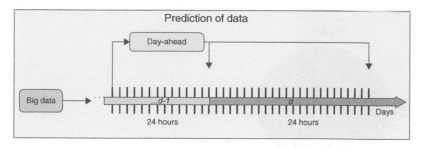

Figure 14.7   Forecast of the day ahead for electric vehicles.

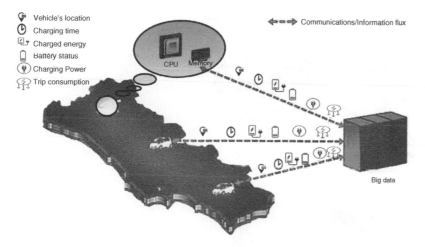

Figure 14.8   The concept of the presence of electric vehicles in the context of the Internet of Things.

The energy consumption of electric vehicles and their impact on the network depends on the number of electric vehicles, the initial charge level of electric vehicles, the start time of charging, the duration of connection to the network, and the location of the vehicle; So that all these variables are predicted using LSTM prediction method which is a sub-branch of deep learning. The previous sections gave an introduction to neural networks, machine learning, and deep learning, and this section will explain the LSTM prediction method (Figure 14.8).

The research presented in this chapter presents a cloud-based operation with scattered sensor devices in the field of IoT and big data. It is assumed that every electric vehicle has a suitable measuring tool for collecting and producing data. The data can be stored locally in the electric vehicle or accessed by cloud applications and stored remotely [29]. Figure 14.9 shows the structure described in the chapter that we can use this model to provide information such as vehicle location, battery charge level, charge time, charge energy, amount of power charged, and battery consumption during movement. Furthermore, use them as a solution to reduce costs. Moreover, it can predict the variables required for electric vehicles using the LSTM method [30].

14.9   Prediction Using LSTM Time Series

Recursive neural networks face sequence problems because their connection forms an efficient cycle; In other words, they can use their output as input for the next step. One of the major problems with simple recursive neural networks is that they cannot store long-term dependencies in the sequence. This is a fundamental problem because

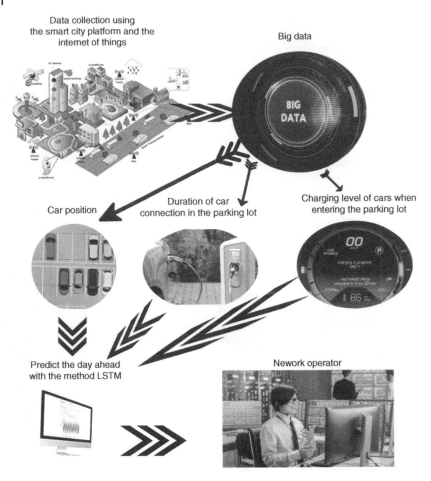

Figure 14.9 Information extraction steps and one day forecast for the network operator. Source: photon_photo/Adobe Stock; Intpro/Adobe Stock; scharfsinn86/Adobe stock; Gorodenkoff/Adobe stock.

it will be difficult to examine long-term dependent data. To solve this problem, a suitable machine learning model is needed to examine the history of a sequence of data and accurately predict the future elements of the sequence. For this reason, the LSTM method has been used for prediction.

14.9.1 LSTM Time Series

One of the most widely used and popular models in deep learning for time-consuming data processing is the LSTM model (Figure 14.10). The memory of this model is a special type of recursive neural network in which instead of the simple tanh function, more complex and combined functions are used, which allows us to adjust the short-term memory of the model in a situation where more parameters we have for this job. The purpose of LSTMs is to be able to process data that is long-term dependent on previous data. LSTM models typically perform better on timed issues than standard RNNs.

In the operation of standard RNN networks, as described in the previous section, a series of modes, the sum of the weights of the inputs and the previous modes in the tanh function, creates a new model. Nevertheless, in LSTM networks, this is different, and we have a more complex model. In addition to the tanh function, there are a series of gates that check how much of the new input or previous state is entered in each of these paths. Another major

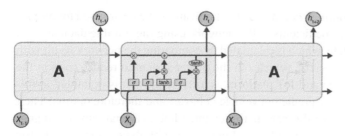

Figure 14.10 Standard LSTM model.

difference between the LSTM model and the simple RNN network is that we have two outputs, each a combination of previous modes and inputs instead of one output per step.

An LSTM module or cell has several essential components to model long-term and short-term data.

- **Cell state**: This cell represents internal memory, which stores short-term and long-term memory.
- **Hidden mode**: Output mode is calculated through current input information, previous hidden state, and current cell input, in addition, the cache mode can decide to store only short- or long-term memory or both types of memory in cellular mode (c_t) for further prediction.
- **Input gate**: This gate determines how much current input information is transferred to the cellular state.
- **Amnesia gate**: Decides how much information from the current input and the previous cell state to enter the current cell state.
- **Output gate**: Determines how much information is hidden from the current cellular state, so that, if necessary, LSTM can only choose long-term memories or short-term memories and long-term memories.

In an LSTM neural network, you not only give data to the network but also the status of the data in the earlier time enters the network, for example in video processing where you need an LSTM network.

What happens in the current framework is heavily dependent on the latest film. Over time, in this case, a neural network tries to learn how much of the past information and how much information to retain from the current situation, which makes it very powerful compared to a simple neural network. LSTM models are very powerful time series models, they can predict the desired number of stages in the future, so this method is used to predict this method because the behavior of electric vehicles depends on their behavioral patterns in the past and is the prediction for the next 24.

14.9.2 Predicting the Behavior of Electric Vehicles Using the LSTM Method

To predict the behavior of electric vehicles, we first need a learning model. In addition to an accurate model, appropriate data are of great importance in this section. After collecting the data to be used to teach the model, the data outside the limiting intervals of the system must be removed. In addition, incomplete, distorted, and unreliable data should be deleted to ensure that the model is properly trained.

14.10 Conclusion

The presence of modern consumers, various applications, and the restriction of conventional resources have made it necessary for our distribution system's evolutionary adaption. Therefore, the concept of smart cities is being expanded. Among various factors, electric vehicles are receiving massive attention from experts. As mentioned in this chapter, they can boost the stability of the power grid, moderate power consumption, and develop transportation. Nevertheless, data collection, controlling, and monitoring of all sections' behavior and guaranteeing the network's security are valued highly. For instance, normalization and standardization of data are used for preprocessing. The network learns the pattern and weight coefficients of the network using the training data, and the results are compared with the real data or the test data, which is calculated by comparing the network result with the desired error result. The issue of optimal use of electric vehicles in the power system to reduce costs in the presence of the smart city, as a result of the smart grid and IoT platform, is a challenge that is examined in this chapter, and the use of this platform is a good basis for predicting vehicle behavior and therefore optimal Their company in the power network will be strengthened. For this reason, this chapter defines a smart city, smart grid, IoT, and an introduction to methods such as deep learning and neural networks, with the existence of electric vehicles and related components in distribution networks.

Exercise

- Introduce renewable sources in the model and consider charging these vehicles through renewable sources in the system.
- Investigate the presence of electric vehicles with probabilistic methods and compare it with the results of this chapter.
- Considering the parameters of the distribution network (frequency, voltage, pulse transformer, etc.) in the objective function of the model presented in the chapter.
- Consider the cost of reducing the car battery life in the costs related to the electric car.

References

1 Delsing, J. (2021). Smart city solution engineering. *Smart Cities* 4 (2): 643–661.

2 Zand, M., Nasab, M.A., Hatami, A. et al. (2020). Using adaptive fuzzy logic for intelligent energy management in hybrid vehicles. In: *2020 28th Iranian Conference on Electrical Engineering (ICEE)*, 1–7. IEEE. https://doi.org/10.1109/ICEE50131.2020.9260941.IEEE Index.

3 Ahmadi-Nezamabad, H., Zand, M., Alizadeh, A. et al. (2019). Multi-objective optimization based robust scheduling of electric vehicles aggregator. *Sustainable Cities and Society* 47: 101494.

4 Zand, M., Nasab, M.A., Sanjeevikumar, P. et al. (2020). Energy management strategy for solid-state transformer-based solar charging station for electric vehicles in smart grids. *IET Renewable Power Generation* 14 (18): 3843–3852. https://doi.org/10.1049/iet-rpg.2020.0399 IET Digital Library.

5 Ghasemi M, Akbari, E., Zand, M. et al. (2019). An efficient modified HPSO-TVAC-based dynamic economic dispatch of generating units, *Electric Power Components and Systems*, 47(19–20), 1826–1840. https://doi.org/10.1080/15325008.2020.1731876.

6 Nasri, S., Nowdeh, S.A., Davoudkhani, I.F. et al. (2021). Maximum Power point tracking of Photovoltaic Renewable Energy System using a New method based on turbulent flow of water-based optimization (TFWO) under Partial shading conditions. In: *Fundamentals and Innovations in Solar Energy*, 285–310. Singapore: Springer. ISBN: 978-981-336-456-1.

7 Rohani, A., Joorabian, M., Abasi, M., and Zand, M. (2019). Three-phase amplitude adaptive notch filter control design of DSTATCOM under unbalanced/distorted utility voltage conditions. *Journal of Intelligent & Fuzzy Systems* 37 (1): 847–865. https://doi.org/10.3233/JIFS-201667.

8 Zand, M., Nasab, M.A., Neghabi, O. et al. (2019). Fault locating transmission lines with thyristor-controlled series capacitors By fuzzy logic method. In: *2020 14th International Conference on Protection and Automation of Power Systems (IPAPS), Tehran, Iran*, 62–70. https://doi.org/10.1109/IPAPS49326.2019.9069389.

9 Zand, Z., Hayati, M., and Karimi, G. (2020). Short-channel effects improvement of carbon nanotube field effect transistors. In: *2020 28th Iranian Conference on Electrical Engineering (ICEE), Tabriz, Iran*, 1–6. https://doi.org/10.1109/ICEE50131.2020.9260850.

10 Tightiz, L., Nasab, M.A., Yang, H., and Addeh, A. (2020). An intelligent system based on optimized ANFIS and association rules for power transformer fault diagnosis. *ISA Transactions* 103: 63–74. ISSN 0019-0578. https://doi.org/10.1016/j.isatra.2020.03.022.

11 Zand, M., Neghabi, O., Nasab, M.A. et al. (2020). A hybrid scheme for fault locating in transmission lines compensated by the TCSC. In: *2020 15th International Conference on Protection and Automation of Power Systems (IPAPS)*, 130–135. IEEE. https://doi.org/10.1109/IPAPS52181.2020.9375626.

12 Zand, M., Azimi Nasab, M., Khoobani, M. et al. (2021). Robust speed control for induction motor drives using STSM control. In: *2021 12th Power Electronics, Drive Systems, and Technologies Conference (PEDSTC)*, 1–6. IEEE. https://doi.org/10.1109/PEDSTC52094.2021.9405912.

13 Padmanaban, S., Zand, M., Nasab, M.A. et al. (2021). Spider community optimization algorithm to determine UPFC optimal size and location for improve dynamic stability. In: *2021 IEEE 12th Energy Conversion Congress & Exposition – Asia (ECCE-Asia)*, 2318–2323. https://doi.org/10.1109/ECCE-Asia49820.2021.9479149.

14 Azimi Nasab, M., Zand, M., Eskandari, M. et al. (2021). Optimal planning of electrical appliance of residential units in a smart home network using cloud services. *Smart Cities* 4: 1173–1195. https://doi.org/10.3390/smartcities4030063.

15 Azimi Nasab, M., Zand, M., Padmanaban, S. et al. (2022). An efficient robust optimization model for the unit commitment considering of renewables uncertainty and pumped-storage hydropower. *Computers and Electrical Engineering* 100: 107846.

16 Nasab, A., Mortez, Z., Mohammad, P. et al. (2021). Simultaneous long-term planning of flexible electric vehicle photovoltaic charging stations in terms of load response and technical and economic indicators. *World Electric Vehicle Journal* 2021: 190. https://doi.org/10.3390/wevj12040190.

17 Zand, M., Nasab, M.A., Padmanaban, S., and Khoobani, M. (2022). Big data for SMART sensor and intelligent electronic devices–building application. In: *Smart Buildings Digitalization*, 11–28. CRC Press.

18 Panda, D.K. and Das, S. (2021). Smart grid architecture model for control, optimizationand data analytics of future power networks with more renewable energy. *Journal of Cleaner Production* 301: 126877.

19 Ali, Z.H. and Ali, H.A. (2021). Towards sustainable smart IoT applications architectural elements and design opportunities, challenges, and open directions. *The Journal of Supercomputing* 77: 5668–5725.

20 Asadi, A.H.K., Jahangiri, A., Zand, M. et al. (2022). Optimal design of high density HTS-SMES step-shaped cross-sectional solenoid to mechanical stress reduction. In: *2022 International Conference on Protection and Automation of Power Systems (IPAPS)*, vol. 16, 1–6. IEEE.

21 Kabir, M.H., Hasan, K.F., Hasan, M.K., and Ansari, K. (2022). Explainable artificial intelligence for smart city application: a secure and trusted platform. In: *Explainable Artificial Intelligence for Cyber Security*, 241–263. Cham: Springer.

22 Oladipo, I.D., AbdulRaheem, M., Awotunde, J.B. et al. (2022). Machine learning and deep learning algorithms for smart cities: a start-of-the-art review. *IoT and IoE Driven Smart Cities* 143–162.

23 White, R. (2021). *Global Harms and the Natural Environment*, The Palgrave Handbook of Social Harm, 89–114. Cham: Palgrave Macmillan.

24 Golla, Naresh Kumar, and Suresh Kumar Sudabattula. Impact of Plug-in electric vehicles on grid integration with distributed energy resources: A comprehensive review on methodology of power interaction and scheduling. *Materials Today: Proceedings* (2021).

25 Chen, Y., Peng, X., Xu, X., and Wu, H. (2021). Deep reinforcement learning based applications in smart power systems. *Journal of Physics: Conference Series* 1881 (2): 022051. IOP Publishing.

26 Lu, Q., Chen, Y., and Zhang, X. (2022). *Smart Power Systems and Smart Grids: Toward Multi-Objective Optimization in Dispatching*. Walter de Gruyter GmbH & Co KG.

27 Li, C. et al. (2021). Data-driven planning of electric vehicle charging infrastructure: a case study of sydney, australia. *IEEE Transactions on Smart Grid* 12: 3289–3304.

28 Sivaraman, P. and Sharmeela, C. (2021). Power quality problems associated with electric vehicle charging infrastructure. *Power Quality. Modern Power Systems* 151–161.

29 Alghamdi, T.G., Said, D., and Mouftah, H.T. (2021). Profit maximization for EVSEs-based renewable energy sources in smart cities with different arrival rate scenarios. *IEEE Access* 9: 58740–58754.

30 Dashtaki, M.A., Nafisi, H., Pouresmaeil, E., and Khorsandi, A. (2020). Virtual inertia implementation in dual two-level voltage source inverters. In: *2020 11th Power Electronics, Drive Systems, and Technologies Conference, PEDSTC 2020.*

15

Modeling and Simulation of Smart Power Systems Using HIL

Gunapriya Devarajan[1], Puspalatha Naveen Kumar[1], Muniraj Chinnusamy[2], Sabareeshwaran Kanagaraj[3], and Sharmeela Chenniappan[4]

[1]Department of EEE, Sri Eshwar College of Engineering, Coimbatore, India
[2]Department of EEE, Knowledge Institute of Technology, Salem, India
[3]Department of EEE, Karpagam Institute of Technology, Coimbatore, India
[4]Department of Electrical and Electronics Engineering, Anna University, Chennai, India

15.1 Introduction

The introduction of power electronic devices for various electrical applications has increased nowadays which leads to the increase in numerous tests conducted on them to ensure their functionality and reliability [1]. Conventional testing methodologies seem to be costly, time-consuming, and sometimes produce irrelevant output when connected to real-time systems. Hence in recent decades, to develop a high-quality controller, researchers have nowadays concentrated on replacing traditional testing methodology with the application of hardware-in-the-loop (HIL).

HIL is the testing methodology developed in the last decade to develop an efficient real-time embedded controller with quick response and less computation time. Simulation of sensors, actuators, and various mechanical components with their input and output can be done before integrating them into a real-time system. The components of HIL are shown in Figure 15.1.

The simulation part HIL can be split into two models, namely the plant model and the controller model [2]. The part that represents sensors and hardware components is called plant model and the part representing the controller is called controller model.

15.1.1 Classification of Hardware in the Loop

HIL is used to evaluate the performance of the equipments or system or equipments connected with actual system before the implementation at site/location. Based on the area and objective of the research, HIL can be classified into the following types:

☐ Signal HIL Model
☐ Power HIL Model
☐ Reduced-Scaled HIL Model

15.1.1.1 Signal HIL Model

In this model, the controller will be present in the hardware and the system model will be simulated in a real-time simulation environment. It is called a signal model because only control signals and measurement signals are managed here. Controllers developed for capacitor banks, electric vehicles (EVs) inverters, and standalone controllers are examples of signal HIL models. An example model of signal HIL is shown in Figure 15.2.

Artificial Intelligence-based Smart Power Systems, First Edition.
Edited by Sanjeevikumar Padmanaban, Sivaraman Palanisamy, Sharmeela Chenniappan, and Jens Bo Holm-Nielsen.
© 2023 The Institute of Electrical and Electronics Engineers, Inc. Published 2023 by John Wiley & Sons, Inc.

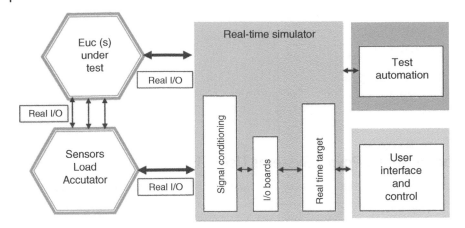

Figure 15.1 Components of HIL.

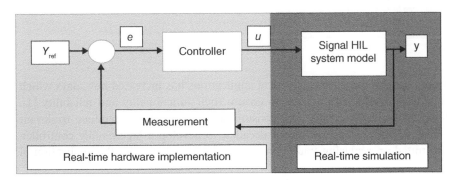

Figure 15.2 Signal HIL model.

15.1.1.2 Power HIL Model

In this model, whole model is split into two subcategories, namely tested objects and simulated ones. It is called a power model since both the system signal and control signal are managed. This model is highly preferred in renewable energy systems.

15.1.1.3 Reduced-Scaled HIL Model

It is similar to the power model but the equivalent subsystems with reduced power replace the tested objects. The characteristics of a tested object and the equivalent power system must be the same. Figure 15.3 shows the power and reduced scale HIL model.

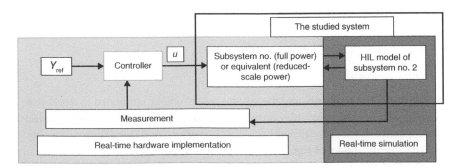

Figure 15.3 Power and reduced scale HIL model.

15.1.2 Points to Be Considered While Performing HIL Simulation

- Determine which real-time operating system is most suited for actual system needs. For real-time applications, there are a number of commercially available options. It is possible that all of them will work, but each one is unique. Bandwidth is essential.
- Make sure whether the system is actually modular. There is no use in having simply a system that can be used within a single supplier's operational framework. With conventional communication interfaces, such as DIL simulators, a system is necessary that helps in connecting with external systems.
- Establish use cases. A clear understanding of the use cases is usually beneficial. To understand HIL better, knowledge of hardware I/O and real-time simulations along with incorporating models from suppliers or customers is essential.
- Streamlining of the internal modeling architecture "Go" and "no-go" situations can be as basic as determining whether or not the two parties are compatible in real time. In the future, additional in-the-loop systems may be needed, so make sure it runs smoothly.
- Determine mobility whether essential or not before making a purchase to use a bench-top simulator for HIL setup. Or, check whether a system can be transported to a testing site.
- Sort simulations into different categories. With a simple classification system like fail-safe testing, subsystem interactions, parameter sweeps and human interactivity tests in place that will be able to identify which simulations will benefit from the addition of DIL and other simulation types to the process.

15.1.3 Applications of HIL

HIL simulation plays a significant role in the research of many fields, such as robotics, defense, aerospace, renewable energy, high voltage direct current (HVDC) [3, 4], flexible AC transmission systems (FACTSs), power system protection, control systems, automotive, and industrial automation. They also find applications in medical devices, machinery industries, semiconductors, etc. [5]. The evaluation of micro grids dominates the existing power system as they require detailed study tools and facilities for reliable testing. The nature of protection equipment, controlling devices, and power devices in micro grids cannot be predicted in real time. Micro grids have to supply reliable power to their consumers under various operating scenarios with constraints. Hence the secured operation of the micro grid can be verified by the test engineers using HIL. HIL helps to optimize the economic performance, reduces the risk of deployment of novel control, and ensures the safety of the system within the laboratory.

The real-time simulators can be connected to inverters, batteries, motors, loads, generators, and renewable energy sources. Hence HIL provides an opportunity to analyze the contingencies and difficulties of installation of renewable systems in the laboratory prior to real-time development. Simulation techniques seem to be the best way to reduce the complexity of electronic stability control (ESC) system which is used in automated vehicles. A validated multi-body system (MBS) model of vehicle can be used for creating calculation models required for real-time simulation. Dynamics of vehicles incorporated with ESC system can obtain an accurate level of simulation reliability by the application of HIL.

Operational adequacy measured from ECS control units can be evaluated using the test benches developed. Test benches also help in developing and debugging new algorithms for EVs. In aerospace and defense, HIL is used in a flight simulator and flight dynamic control as testing of real aircraft becomes complex. In industrial automation, HIL is used for controller-plant testing, because stopping the production in the middle to test control algorithms leads to heavy business loss.

Some of the HIL simulators presently available in the markets are: Real-time digital simulator (RTDS), dSPACE, OPAL-RT, TYPHOON, etc.

15.2 Why HIL Is Important?

HIL simulation is a technique that allows users to build the behavior of control schemes without having to use actual equipment at the site/location. HIL is a sort of real-time simulation in which a simulated real-time environment with a representation of physical process and an input/output controller is created [6]. This method is used to create and test the most sophisticated monitoring, protective, and tracking systems. HIL simulation enables users to evaluate their controllers on an authentic plant simulator that functions as a digital model of the entire system or its components, resulting in cheaper costs and improved functionality.

Because it was deemed expensive, HIL testing was formerly mainly employed in the automobile and aerospace industries. HIL systems are well renowned for offering quality visibility into just how an embedded system will perform, and expenditures in HIL systems can typically pay for themselves in as little as a few months. That is why HIL modeling is now widely used in a variety of industries and technologies, including power and energy, communications, semiconductors, and diagnostic implants.

15.2.1 Hardware-In-The-Loop Simulation

HIL simulation involves integrated code testing without having to use the actual system hardware. The practical equipment, detectors, and transducers are replaced by a computer model that runs in a simulator with appropriate inputs and outputs during HIL testing [7, 8]. All of these parts can communicate with process control and other equipment. The controller software developed from the controller model is stored in the controller hardware. Embedded computing is becoming more common in safety-critical systems, increasing the requirement for extensive testing. Clearly, system-level testing must be accelerated and automated as much as feasible within the restrictions of sample system cost and implementation effort. HIL simulation is a technique for undertaking extensive, cost-effective, and repeatable testing of the system of embedded systems. When embedded systems can't be tested readily, comprehensively, or consistently in their operating contexts, HIL simulation is frequently employed in their development and testing. A high-level perspective of an instance HIL simulation is shown in Figure 15.4.

The SUT – combination of sensor and actuator as shown in figure as having sensor and actuator interfaces an operator interface to display information and receives input commands. Operator directives are synthetically created by the simulation in this case to excite the SUT during testing. The use of synthetically produced operator commands allows test sequences to be automated and tests to be precisely repeated. It is also feasible to run the simulation in a mode where people see the operator displays and provides inputs. Although this mode is useful for user testing and process control, it eliminates the ability to accurately repeat test sequences.

Continuous improvements in computer hardware performance and price reductions have made it feasible to construct cost-effective HIL simulations for a much broader variety of goods. Another advantage of HIL modeling is that it greatly minimizes the likelihood of substantial faults being detected just after product has been put

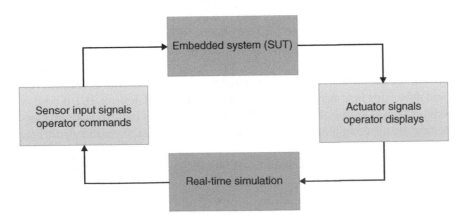

Figure 15.4 Block diagram for HIL simulation.

into production. HIL simulation is a useful tool for doing optimization design and hardware/software debugging throughout the product development process.

15.2.2 Simulation Verification and Validation

Before the HIL simulation can be utilized to generate valuable findings, it must first be proven that the SUT and its environment simulated adequately mimic the operating system and environment. Verification and validation are two essential phases in the process of verifying and documenting that the simulation's accuracy and fidelity are appropriate. Verification is the process of establishing that the HIL simulation comprises an exact implementation of the SUT and its environment's conceptual mathematical models [9, 10]. This stage can be carried out even before a prototype of the SUT has been created. Verification is usually done by comparing the results of the HIL simulation with the results of analytical calculations or results from separately generated SUT simulations.

Validation entails demonstrating that the HIL simulation accurately represents the embedded system and the real-world operational environment. The findings of system operational tests are often compared to simulation results as part of the validation process. This sort of validation test entails putting the embedded system in an HIL simulation through a test scenario that is equivalent to one that would be conducted by an actual system in a real-world setting. The two test results are compared, and any discrepancies are examined to see if they indicate a substantial difference between reality and the simulation.

If the functional test results do not match the HIL simulation results exactly, it may be important to improve the quality of the simulation modeling in specific areas. When comparing simulation findings to operational test results, flaws in the simulation hardware or software interfacing may become evident.

15.2.3 Simulation Computer Hardware

The simulation computer's real-time performance requirements are determined by the nature of the embedded device to be evaluated and its environmental situation, including: the SUT's I/O update rates and I/O data transfer speeds, the frequency band of the dynamic system made up of the SUT and its environment. The complexity of the SUT components and operating environment must be modeled in simulation software.

(i) I/O devices: In embedded systems, there are many distinct types of I/O devices.

An I/O device must be built in the simulation computer that connects to each SUT I/O port of interest in an HIL simulation. I/O interfaces for various signal kinds are accessible from a variety of sources, including:

- It's analog (DACs and ADCs)
- Digitally discrete (Ex, TTL or differential)
- Sequential (Ex, RS-232 or RS- 422)
- Data bus in real time
- Bus for instrumentation (IEEE-488)
- Create a network (Ethernet)
- Simulators for devices like Transducers and Thermocouples

A regular PC running a non-real-time operating system such as Windows NT may be capable of executing a valid and meaningful HIL simulation for an SUT has moderate I/O rates and a simulated environment that is not extremely sophisticated. A high-performance computing system is required for complicated, high I/O rate SUTs. The simulation computer must have system-level software that supports real-time computation and prevents code execution from being obstructed in any manner. Most general-purpose operating systems do not handle real-time features well enough to be usable in anything other than a low I/O rate HIL simulation. On the simulation computer, this scenario may entail the adoption of an RTOS or a specific real-time software environment.

15.2.4 Benefits of Using Hardware-In-The-Loop Simulation

Traditionally, control software has been examined on physical equipment using real/actual power. This procedure was extremely costly, inefficient, difficult to set up and operate, and possibly dangerous. Test engineers used HIL simulation instead of standard testing methods to provide high-quality controller software. This method has several advantages:

- HIL involves less commitment and administration than power lab testing, and there are no expenditures associated with potential equipment damage.
- Working in a virtual environment eliminates the possibility of ruining equipment or putting people in danger while conducting testing.
- **Fidelity**: Using a high-fidelity real-time emulation on an FPGA assures that the controller evaluated in HIL will function just as well when attached to real-world equipment.
- **Flexibility**: Evaluating a wide variety of circumstances, tweaking parameters on the fly, and testing defective circumstances may all be done in a virtual environment without the hazards involved with evaluating in a power facility.
- **Automation**: An endless number of automated test scripts may be written and run overnight or for a few days to gather information without the requirement for human intervention.
- **Speed**: An agile process of continuous testing enables for instant feedback on changes made to the control software, as well as immediate corrections in the event of inferior performance.
- **Time**: By saving time during the construction and use of the setups, the product may be released sooner.

HIL simulation, without a doubt, greatly enhances the process of evaluating control algorithms. This form of simulation allows for thorough testing of controllers while reducing risk, expense, and total testing time.

15.3 HIL for Renewable Energy Systems (RES)

15.3.1 Introduction

In recent years, due to the increase in demand, climatic conditions, and decrease in fossil fuels, various renewable energy resources, namely photovoltaic, fuel cells, wind turbines, and small hydro plants have gained importance all over the world. Among all the renewable sources, solar PV and wind technology acts as a promising source in the future [11]. In 2021, installed wind capacity is 743 GW which is rapidly increasing every year. The major issue with the wind farm is that unstable wind velocity causes a power imbalance between generated power and demanded power and also leads to frequency variation which affects grid stability.

Testing of wind turbines in the field becomes to be very expensive. The conventional methods for analyzing the integration of wind farms with the grid are software-based simulation, standalone hardware testing, and field demonstration. The drawback of the software simulation technique is that it cannot reproduce the output of the real-time simulation. Field demonstration is suitable for smaller systems and does not adopt for various power system configurations. Development of a wind turbine dynamometer acts as a better alternative to field testing. In recent years, computer technology and dynamometer testing are improved and joined together to form HIL testing. Rotor, hub, voltage, and frequency which cannot be measured in dynamometer testing can be measured by simulation. Real-time simulation of wind input and grid conditions can be simulated in HIL.

The integration of wind and solar generation in the grid produces an adverse effect on grid stability and reliability. Hence the solar plant operators and wind turbine manufacturers have to adapt to the change in grid code, change in rules and regulations, interconnection components, and so on. Under that circumstance, traditional testing cannot afford the low cost, reliable, and high-quality power supply. In order to overcome this, HIL testing schemes are proposed which combines real-time dynamic simulation controllers and powerful hardware to provide the capability to meet the reliable design consideration in a safe manner. Real-time-based HIL method is used in the verification of many control strategies as they provide high efficiency during the operation.

15.3.2 Hardware in the Loop

HIL simulation in wind turbines can be classified into two types, namely:

- Electrical HIL
- Mechanical HIL

15.3.2.1 Electrical Hardware in the Loop

The importance of power HIL in the dynamometer testing is analyzed by the introduction of missing components on the grid side. The greater scalability of power HIL can be modeled once it has both electrical strategy and mechanical strategy using which the entire wind system can be emulated. Using conventional methods, a single wind plant cannot be tested in isolation conditions from other parts of the power system. This drawback has been overcome by the application of power HIL which is capable of integrating the grid with the wind power.

Clemson University and the National Renewable Energy Laboratory (NREL) developed advanced power hardware in the loop (PHIL) to help wind power industries. Advanced PHIL combines a real time, multimegawatt grid simulator and variable frequency drive [12] and nontorque loading (NTL) systems with real-time dynamic simulation hardware (real-time digital simulator – Opal RT, NI PXI, NIcRIO) to create an environment suitable for testing in repeated conditions. The model exposed to perturbation coming from the grid side and variations from the wind side can be simulated using PHIL.

Wind power industries and other renewable energy sources [13, 14] can be tested using HIL grid simulators such as:

- Advanced testing tools are provided hence the reliability of the system is increased.
- Advanced validation tools are provided.
- Test can be performed as per national and international standards, including grid codes and interconnection requirements.
- Wind power compliance and auxiliary components testing at various megawatt levels.

PHIL test setup involves the following components, namely a dynamometer with test article, grid interface, and connecting auxiliaries in case of additional power hardware as shown in Figure 15.5.

To analyze wind active power control and low voltage ride-through, various tests were conducted by NREL using PHIL. For active power control, NREL is conducting tests on a new controller design to test whether they provide automatic generation control and frequency control. The impact of various grid codes on the turbine is necessary to evaluate LVRT control. A wind power system consisting of generators, converter, wind turbine step-up transformer, transmission lines, and load can be simulated in real time which provides voltage set points to grid interface. Same waveform is fed back to RTDS to promote closed loop simulation as shown in Figure 15.6.

15.3.2.2 Mechanical Hardware in the Loop

For a better understanding of mechanical HIL, both Clemson and NREL dynamometers are considered for studies. In the testing point of the HIL scenario, along with the gearbox, prime mover, and NTL, the load applied to the hub point can also be calculated in real time [15]. HIL is useful when the device under the test is influenced by the variation of load and varying environmental conditions. The real-time simulation is done to calculate the output of the device and the test equipment is checked for the replication of results at full scale as shown in Figure 15.7.

In the wind turbine, the load experienced by the hub depends on both the input and control strategy of the turbine. The input to the turbine will be wind and the control action depends upon the pitch, yaw, and generator torque. Wind Turbine Generator (WTG) comprises of mechanical and electrical controllers. The Nacelle accommodate the drive train and generator and fed to the Down Tower Assembly (DTA). Hence HIL testing system

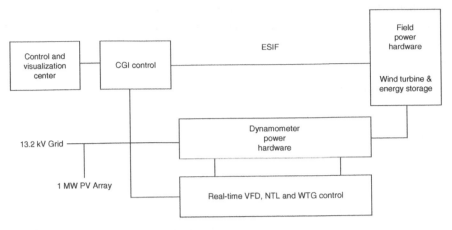

Figure 15.5 Power hardware in the loop setup.

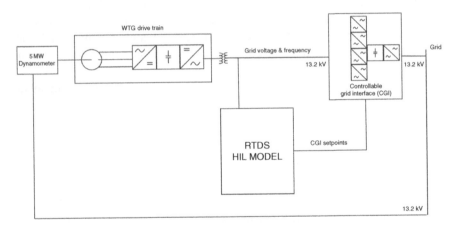

Figure 15.6 Example of power hardware in the loop setup.

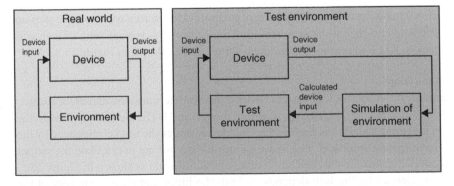

Figure 15.7 Real-world environment and HIL test environment interaction structure.

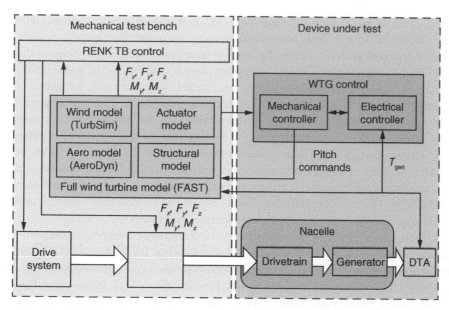

Figure 15.8 An interaction exists between the device under test, the computer simulation, and the test hardware.

considers both the input and control action from the nacelle to provide the loading conditions. In both Clemson and NREL test bench, the nacelle is either fitted to the floor or mounted on the short tower from the ground. For testing, a hub of the nacelle is removed and it is connected to the test bench through the main shaft to alter the mechanical properties.

Energy Systems Integration Facility (ESIF) is a tool to monitor the performance of the system. Variable Speed Drives (VFD) is accounted in the system to drive the motor with variable voltage and frequency. The Controllable Grid Interface (CGI) will monitor the efficiency of the operation of the wind turbine. To account for the effect of missing and to compensate for the structural properties of the wind turbine, the following HIL configuration for the wind turbine is considered as shown in the figure. With his configuration, the blade pitch mechanism and yaw mechanism are intercepted and given to the real-time model. The model makes use of this control signal and along with the wind input, and computes the load at the nacelle as shown in Figure 15.8.

The benefits of mechanical HIL in the testing of the wind turbine are listed below.

- By including nacelle control system in the physical testing, the wind turbine testing system can be used in physical and structural testing.
- Though nacelle is present in the field, HIL allows the occurrence of both electrical and mechanical transients for a few seconds to provide a mechanical boundary for the testing article.
- The component incorporated with the nacelle can be tested using the nacelle controller, thereby offering a fully integrated test.

The wind turbine developers get a unique opportunity by the arrival of modern wind turbine dynamometer testing coupled with faster computing capability since they reduce the testing time of the prototype. Dynamometer HIL data helps to upgrade and validate the design models as it is easier to collect the data from HIL than from the field.

15.4 HIL for HVDC and FACTS

15.4.1 Introduction

Environmental constraint plays a vital role in the development of a power transmission system. The integration of the renewable system into grids becomes highly complex when the reserve capacity is unavailable and the connecting AC links are weak [16, 17]. Thus the stability and flexibility of the power system can be enhanced by the application of HVDC and FACTS. HVDC system finds more advantageous compared to an existing system, as it provides AC power flow control in power grids, is suitable for long-distance transmission, and helps in the implementation of the super grid [18, 19]. Similarly, the FACTS controller plays a vital role in the transmission system by offering various grid services, including real and reactive power control, voltage control, mitigation of harmonics, power quality improvement, inter and intra area oscillation damping, and transient stability improvement [20]. Since the HVDC system consists of many controllers and protective systems, testing them becomes more expensive in real-time applications. RTDS was introduced in the 1980s to solve these issues, by offering a flexible solution for closed-loop testing for not only HVDC system but also for FACTS devices. Nowadays, real-time simulators are widely used in the field of power electronics, cyber security, distribution automation, and wide area protection. In the recent trends, the performance of grid modernization can be evaluated by closed-loop testing which is also said to be HI testing. The interoperability of various power equipment, their protocol, and control can be examined using HIL which also provides simultaneous testing of many devices.

The most attractive VSC-HVDC technology preferred nowadays is HVDC transmission-based modular multilevel converter (MMC) [21] because it has a high range of modularity and scalability. HVDC-based MMC overcomes the drawback of conventional two-level and three-level converters such as producing lower order harmonics at the AC side of the converter, reducing the losses, and reducing the switching frequency [22]. Many researchers are working on the control strategies and the modulation mode of MMC. To plan a primary system and to develop the test models, physical verification is necessary. Thousand of SMs are required to develop an MMC, such that it produces output with high redundancy, extreme power capacity, and decreased total harmonic distortion. The increase in the computational time and the complexity in the simulation of parallel systems becomes a major issue in the systems with more subsystems.

MMC even up to 200 levels can be developed to obtain perfect physical verification but their complexity lies in transmitting the huge capacitive voltages. Hence by considering the drawbacks of realization of MMC system and controller, HIL is established between real-time simulator and physical MMC controller. The step size of the system should be kept very low to reduce the time delay between the triggering of IGBT used in the converters and the output current simulated in the real-time simulator. In this chapter, application of HIL in HVDC-based MMC is discussed based on the results obtained from references.

15.4.2 Modular Multi Level Converter

Conventional HVDC and FACTS applications consist of line commutated converters (LCCs). The limitation of these converters is the production of lower-order harmonics and high reactive power consumption. The fourth generation of VSC is MMC, which was first introduced by Lesnicar and Marquardt in 2001 as given in Figure 15.9. Each converter leg has two arms, namely an upper arm and a lower arm. Each arm has many sub-modules (SMs). In each SM two IGBT switches, two anti parallel diodes, and a capacitor will be present as shown in Figure 15.10. The inductor and the resistor are provided in the circuit to minimize the voltage difference between the arms. For high-power/High voltage and medium power/high-voltage applications, MMC is considered the appropriate technology because of its modularity and scalability features [23]. The main component involved in the designing of HVDC-based MMC systems is converters. Various types of MMC are available, namely half-bridge MMC, full-bridge MMC, and alternate arm controller (AAC) [24]. Among them, full-bridge MMC and AAC can block the current flowing through the converter, when a fault occurs on the DC side. Hence they are called fault blocking converters and are highly preferred in HVDC grids using cables. Half-bridge MMC are preferred in submarine cables [25]. The block diagram of MMC-HVDC topology is given in Figure 15.11.

The advantage of VSC-MMC-HVDC technology over conventional LC topology is given below:

o Due to disturbance in the AC network commutation failure occurs, which can be eliminated in VSC-MMC-HVDC technology.
o Independent control of active and reactive power is generated.

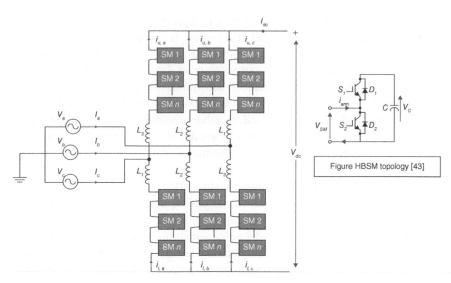

Figure 15.9   Three-phase MMC-HVDC topology.

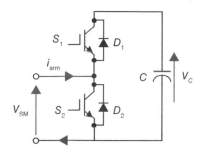

Figure 15.10   Submodule topology.

Figure 15.11   Block diagram of MMC-HVDC topology.

o Weak AC networks can be connected through VSC-MMC-HVDC technology.
o Harmonic distortion is low.
o Power quality can be improved and the requirement of the filter can be eliminated.
o The transformer is not necessary to assist the commutation process.
o Current and power flow directions can be changed without reversing the polarity.

15.5   HIL for Electric Vehicles

15.5.1   Introduction

As a go green initiative and advancement in reusable, renewable technologies. The electric car contributes to the growth of nation with zero emission; environmentally friendly, economic, maintenance-free, noiseless sustainable e-mobility is encouraged by both public and government. Testing and analysis of such vehicles was carried out by two approaches before, i.e. by doing typical system tests with medium to high power ratings: either on real hardware or via pure software simulation. A third method, dubbed HIL testing, is currently being investigated (HIL). This possibility is a combination of the previous two [26]. A straightforward approach would be to construct the entire system in real hardware components and then conduct various tests on the actual system. In terms of precision, this is by far the best solution. Other factors, such as financial constraints and physical space, are also relevant and, in many cases, render this technique ineffective due to the high costs and effort connected with implementation, as well as the high risk associated with the test itself. Another argument against this is the lack of flexibility provided by a pure hardware experiment. Typically, small or significant components of the system cannot be updated in order to examine new methodologies. In general, one may argue that a pure hardware test is impractical in a large number of situations.

Historically, the alternative to hardware testing has always been software simulation. This has significant advantages over pure hardware simulations in that it is more customizable and often costs a fraction of the money. Generally, one can divide a system's components into two categories: those that are easy to model or for which good and accurate models exist; and those that are difficult to model or for which no good and accurate models exist. The quality of the simulation findings is highly dependent on the tester's comprehension of the system and his or her ability to simulate it accurately. In other words, the more accurate the model, the more accurate the simulation results. If a single component of the whole system belongs to the second group or is not precise enough, the entire simulation will be inaccurate. In the end, the applicability of a pure software simulation is highly dependent on the user's knowledge of the system. It is possible that scenarios arise in which software simulations cannot be used due to a lack of understanding of the system or component.

15.5.2 EV Simulation Using MATLAB, Simulink

In order to speed up the development of EVs, engineers can utilize MATLAB®, Simulink®, and Simscape™ to make use of data and models. Simulink [27] and MATLAB are used to simulate EVs. In order to develop and optimize complicated EV structures, use MBSE (model-based system engineering).

- Design and simulate batteries and battery management systems (BMSs)
- Model fuel cell systems (FCSs) and create control systems for fuel cell control systems (FCCSs)
- Engineer traction motors and design motor control units (MCUs)
- Deploy control algorithms, integrate them, and test them
- Develop EVs using data-driven processes and AI

15.5.2.1 Model-Based System Engineering (MBSE)

Management of system complexity, improvement of communication, and optimization of systems are some of the goals of MBSE implementations. It is impossible to produce clear system descriptions without first synthesizing the requirements of key stakeholders into architecture models. In addition, MATLAB analytics can be utilized to optimize system architectures by capturing architecture metadata and directly linking to MATLAB for domain-specific trade studies. View the components of interest in a model in a simplified way by creating bespoke views for distinct engineering needs. Use simulation-based tests to validate requirements and validate the design of the system. Model-based design in Simulink can be used to translate and improve requirements into architectures with components ready for simulation and implementation.

15.5.2.2 Model Batteries and Develop BMS

Electric drive trains powered by batteries are becoming increasingly popular across a wide range of sectors [28]. Battery packs power everything from electric automobiles to electric aircraft to e-bikes and self-driving cars. It is critical to verify safety features and the robustness of algorithms for the state of charge (SOC) and state of health (SOH) at an early stage in order to save both costs and development time for battery pack management

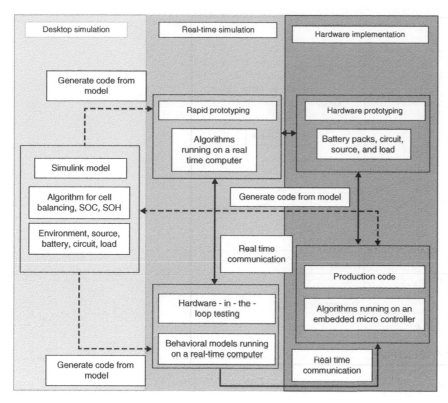

Figure 15.12 Battery management system.

systems. Engineers can detect design faults in the early stages of the BMS development by using automated HIL testing and a model-based design methodology. In order to do automated HIL testing, battery cell emulators utilizing a variety of industrial protocols and interfaces are used. It is possible to test important scenarios in a safe and automated manner with the help of these robots. For large-scale automated production testing, the same methodology can be used. This software is used for modeling, simulations, and development of BMSs. It is used to simulate non-linearities, thermal effects, SOC/SOH and deterioration of batteries, model battery-equivalent circuits and add fidelity with elaborate circuit topologies. BMS functions are monitoring of voltage and temperature as well as protection from overcharging as well as cell balance and isolation. In EVs, lithium-ion battery packs are the most common kind of energy storage. It is depicted in Figure 15.12 below, according to Mathworks White Paper [29].

15.5.2.3 Model Fuel Cell Systems (FCS) and Develop Fuel Cell Control Systems (FCCS)
Polymer electrolyte membrane (PEM) modeling is used to frontload the development of FCS and FCCS across a wide range of operating circumstances and operating environments.

- Modeling and simulation of FCS as well as the development of FCCS are both possible with it.
- First principles based on electro-chemistry or experimental data are used to model PEM fuel cells.
- Assist in the development of fuel cell electric vehicle (FCEV) performance and efficiency models (FCEVs).
- FCCS development, encompassing control logic, automatic code generation, and closed-loop validation, with support for AUTOSAR and certification procedures, is provided.
- Monitor and control power, voltage, and current, as well as temperature.
- FCS and control systems (FCCS) can be modeled and developed according to Math works White paper [29] as shown in Figure 15.13.

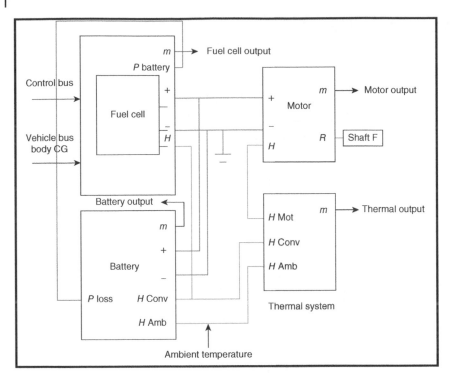

Figure 15.13 Model fuel cell systems (FCS) and fuel cell control systems (FCCS).

15.5.2.4 Model Inverters, Traction Motors, and Develop Motor Control Software

Before hardware testing, accurate motor modeling is used to speed up the development of motors and MCUs. An end-user can do the following with the help of the MATLAB, Simulink, and Simscape tools:

- Design, simulate, and validate power conversion systems using model libraries of power semiconductors and machines such as the permanent magnet synchronous motor (PMSM) and the induction motor (IM).
- Automate PID controller tuning, automatic code creation, and validation in closed-loop simulations, including HIL with support for AUTOSAR and certification operations, to make MCU development possible.
- According to Math works White Paper [29], the Figure 15.14 explains how to model inverters, traction motors, and motor control software.

15.5.2.5 Deploy, Integrate, and Test Control Algorithms

EV developers increasingly need to comply with safety standards. With MATLAB and Simulink, the user can:

- Automatically generate optimized C and HDL code
- Trace requirements, measure quality of code/models, and generate test cases automatically
- Comply with an ISO 26262 reference workflow to meet functional safety requirements
- Use tools that are pre-qualified for ISO 26262
- Leverage AUTOSAR Blockset (classic and adaptive) to model AUTOSAR software components, simulate compositions, and import/export ARXML files
- Integrate with CI/CD/CT pipelines, generate code, package for deployment, and automate regression testing

Figure 15.15 shows Deploy, Integrate, and Test Control Algorithms as per Math works White paper.

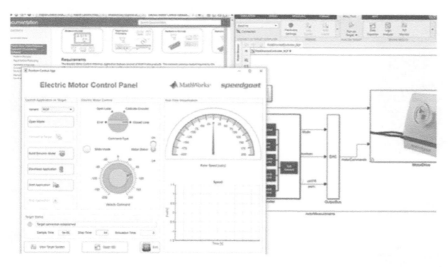

Figure 15.14 Model inverters, traction motors, and motor control software.

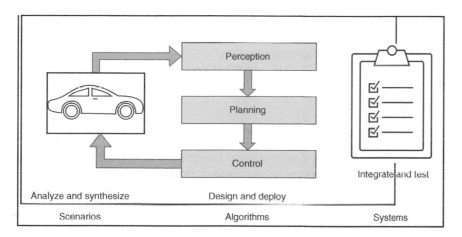

Figure 15.15 Deploy, integrate, and test control algorithms.

15.5.2.6 Data-Driven Workflows and AI in EV Development

Using test data and real-world driving data, the user can make design decisions, build reduced-order models that speed simulations, and develop maintenance services as shown in Figure 15.16. With MATLAB and Simulink, the user can:

- Leverage the complete AI workflow: Data preparation, AI modeling, simulation and test, and deployment on embedded hardware, edge devices, cloud, or enterprise servers
- Start with pre-built algorithms, models, and reference examples for AI modeling
- Access data from databases, cloud sources, binary files such as MDF and more
- Train models with point-and-click apps for machine learning and deep learning
- Import models from the broader AI community for transfer learning and deployment
- Integrate AI into system-wide models, and simulate and verify before moving to hardware
- Use AI capabilities for predicting remaining useful life, predictive maintenance, building digital twins, and bringing AI in Simulink

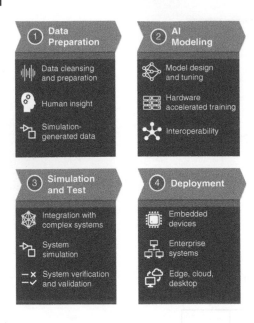

Figure 15.16 Data-driven workflows and AI in EV development.

The sustainability is a key factor HIL technology plays vital role in framing future transportation systems with enhanced features and cost efficient products to the development of nation.

15.6 HIL for Other Applications

A condition monitoring and fault diagnosis is the critical part in the operations of the electrical apparatus connected in the electrical girds. If any faults occurred due to natural phenomena or system failure, it can alter the grid running characteristics so that process interference occurred. Nowadays, the fault diagnosis and condition monitoring has been performed in online either destructive or non-destructive methods. Predicting the fault occurrence and accuracy are important to avoid complete plant shut down and rectifying the fault with in its tolerance clearing interval has to be carried out. A real-time fault generation and amplitude of the fault signature is the essential part of the condition pre-assessment system and it may be complex and to some extent that cannot be generated due to safety issues. In the recent time, hardware-based fault simulator is focused to address such issues. HIL simulation is used for the improvement and checking out of developed control structures that can be used for the operation of complex machines control and structures. With HIL simulation the physical part of a system or machine under control is not changed and is only simulated. In this section, it is described about how the HIL has been implemented to test and diagnose the electrical drive motor and converter faults.

15.6.1 Electrical Motor Faults

In electrical motors, the common mechanical faults are bearing fault, rotor damage, air gap eccentricity, and shaft bending. Faults associated with eccentricity occurs frequently during manufacturing and hence needs unique attention. The common electric faults are insulation failure, short circuit fault, and voltage imbalance. The embedded controller plays a significant role in the development of high-speed HIL system. FPGA has been implemented widely for devolvement of hardware real-time simulation. Normally, numerical simulation software such as MATLAB/Simulink, LabView, ETP, etc. are used to develop real-time hardware model. Figure 15.17

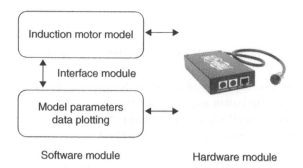

Figure 15.17 HIL structure for fault testing. Source: Tripp.Lite by Eaton.

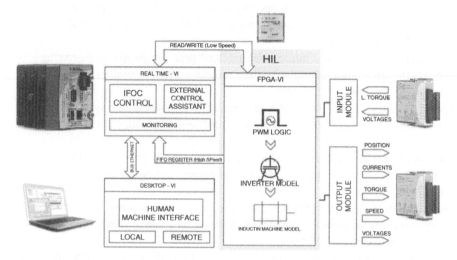

Figure 15.18   HIL implementation blocks. Source: Aiello et al. [30]/with permission of Springer Nature; NATIONAL INSTRUMENTS CORP.

Shows the structure of HIL system that is used to test the bearing, sensor, rotor and stator faults of induction motor. Real time involves the Indirect Field Oriented Control (IFOC) and External Control Assistant. The Pulse Width Modulation (PWM) technique is used to control the inverter of the induction machine. A dc motor set with the soft sliders and other components has been used for generating motor faults and variable resistor introduced in Tacho generator are used to generate the faults. The induction motor modeling programs are written in the MATLAB/SIMULINK programming language and can be compiled and downloaded to a dSPACE DSP board.

PMSM drive inter-turn fault is tested in HIL platform. Figure 15.18 shows the implementation blocks of the fault testing system. Finite element model is used to develop emulator model and FPGA controller used for testing the faults conditions.

15.7   Conclusion

Thus it is can be concluded that HIL testing is an excellent solution to increase the speed of simulation and testing. Also researches found it useful for automation process based on communication protocols. The various studies on power system allow its flexibility and scalability. It can also be used for applications in situation awareness and for protection and control. It is a good platform to cooperate with utilities, academics, and industries for knowledge sharing and updating as well as professional training.

References

1 Basu, K.P. (2009). Stability enhancement of power system by controlling HVDC power flow through the same AC transmission line. *Proceedings of ISIEA* 2: 663–668.

2 Ramesh, M. and Laxmi, A.J. (2012). Stabilty of power transmission capability of HVDC system using facts controllers. In: *2012 International Conference on Computer Communication and Informatics*, 1–7.

3 Allebrod, S., Hamerski, R., and Marquardt, R. (2008). New transformerless, scalable Modular Multilevel Converters for HVDC-transmission. In: *2008 IEEE Power Electronics Specialists Conference*, 174–179. https://doi.org/10.1109/PESC.2008.4591920.

4 Bakas, P., Okazaki, Y., Shukla, A. et al. (2020). Review of hybrid multi-level converter topologies utilizing thyristors for HVDC applications. *IEEE Transactions on Power Electronics* 36 (1): 174–190.

5 Paramalingam, J., Nakamura, F., Matsuda, A. et al. (2018). Application of FACTS devices for a dynamic power system within the USA. In: *2018 International Power Electronics Conference (IPEC-Niigata 2018 -ECCE Asia)*, 2329–2334. https://doi.org/10.23919/IPEC.2018.8507559.

6 Sidewall, K. and Zakonsjek, J. (2021). Experiences, and future potential: Real-time simulation for de-risking converter-connected generation, HVDC interconnections, and modern protection. In: *Proceedings of the CIGRE SEERC Conference*, Vienna, Austria, 30 November 2021.

7 Jeon, J.H., Kim, J.Y., Kim, H.M. et al. (2010). Development of hardware in-the-loop simulation system for testing operation and control functions of microgrid. *IEEE Transactions on Power Electronics* 25 (12): 2919–2929.

8 Yoo, C.H., Cho, W.J., Chung, I.Y. et al. (2012). Hardware-in-the-loop simulation of DC microgrid with multi-agent system for emergency demand response. In: *IEEE Power and Energy Society General Meeting*, 1–6. https://doi.org/10.1109/PESGM.2012.6345678.

9 Das, A., Nademi, H., and Norum, L. (2011). A pulse width modulation technique for reducing switching frequency for modular multilevel converter. In: *India International Conference on Power Electronics 2010 (IICPE2010)*, 1–6. https://doi.org/10.1109/IICPE.2011.5728082.

10 Rohner, S., Bernet, S., Hiller, M., and Sommer, R. (2010). Modulation, loss, and semiconductor requirements of modular multilevel converters. *IEEE Transactions on Industrial Electronics* 57 (8): 2633–2732.

11 Nguyen, V.H., Tran, Q.T., Guillo-Sansano, E. et al. (2020). Hardware-in-the-loop assessment methods. In: *European Guide to Power System Testing* (ed. T.I. Strasser, E.C.W. de Jong and M. Sosnina), 51–66. Cham: Springer https://doi.org/10.1007/978-3-030-42274-5_4.

12 García-Martínez, E., Sanz, J.F., Munoz-Cruzado, J., and Perié, J.M. (2020). A review of PHIL testing for smart grids – selection guide, classification and online database analysis. *Electronics* 9 (3): 382. https://doi.org/10.3390/electronics9030382.

13 Gardiner, J.D. (2008). Real-l. In: *2008 DoD HPCMP Users Group Conference*, 379–381. https://doi.org/10.1109/DoD.HPCMP.UGC.2008.43.

14 Electric vehicle. MathWorks. (n.d.). Retrieved 21 March 2022. https://in.mathworks.com/solutions/automotive/electric-vehicle.html

15 Saleem, A., Issa, R., and Tutunji, T. (2010). Hardware-in-the-loop for on-line identification and control of three-phase squirrel cage induction motors. *Simulation Modelling Practice and Theory* 18 (3): 277–290.

16 Merlin, M.M.C., Green, T.C., Mitcheson, P.D. et al. (2014). The alternate arm converter: a new hybrid multi-level converter with DC-fault blocking capability. *IEEE Transactions on Power Delivery* 29: 310–317.

17 De Jong, E., De Graff, R., and Crolla, P. et al. (2012). European white book on real-time power hardware in the loop testing. Technical Report No. R- 005.0, DERlabConsoritum.

18 Huang, S. and Tan, K.K. (2012). Fault simulator based on a hardware-in-the-loop technique. *IEEE Transactions on Systems, Man, and Cybernetics Part C: Applications and Reviews* 42 (6): 1135–1139. https://doi.org/10.1109/TSMCC.2012.2182992.

19 Scelba, G., Scarcella, G., Cacciato, M., and Aiello, G. (2016). Hardware in the loop for failure analysis in AC motor drives. In: *2016 ELEKTRO*, 364–369. https://doi.org/10.1109/ELEKTRO.2016.7512098.

20 Alvarez-Gonzalez, F., Griffo, A., Sen, B., and Wang, J. (2017). Real-time hardware-in-the-loop simulation of permanent magnet synchronous motor drives under stator faults. *IEEE Transactions on Industrial Electronics* 64 (9): 6960–6969. https://doi.org/10.1109/TIE.2017.2688969.

21 Faruque, M.D.O., Strasser, T., Lauss, G. et al. (2015). Real-time simulation technologies for power systems design, testing, and analysis. *IEEE Power and Energy Technology Systems Journal* 2 (2): 63–73.

22 Jandaghi, B. and Dinavahi, V. (2019). Real-time HIL emulation of faulted electric machines based on nonlinear MEC model. *IEEE Transactions on Energy Conversion* 34 (3): 1190–1199. https://doi.org/10.1109/TEC.2019.2891560.

23 Dorn, J., Huang, H., and Retzmann, D. (August). A new multilevel voltage-sourced converter topology for HVDC applications. In: *CIGRE session*, B4. Paris, France: CIGRE.

24 Montoya, J., Brandl, R., Vishwanath, K. et al. (2020). Advanced laboratory testing methods using real-time simulation and hardware-in-the-loop techniques: a survey of smart grid international research facility network activities. *Energies* 13: 3267. https://doi.org/10.3390/en13123267.

25 Mejia-Barron, A., Tapia-Tinoco, G., Razo-Hernandez, J.R. et al. (2021). A neural network-based model for MCSA of inter-turn short-circuit faults in induction motors and its power hardware in the loop simulation. *Computers and Electrical Engineering* 93: 107234.

26 National Instruments. website address: http://www.ni.com. (accessed on 27 September 2020).

27 Palladino, A., Fiengo, G., and Lanzo, D. (2012). A portable hardware-in the-loop (HIL) device for automotive diagnostic control systems. *ISA Transaction* 51 (1): 229–236.

28 Röck, S. (2011). Hardware in the loop simulation of production systems dynamics. *Production Engineering* 5: 329–337. http://dx.doi.org/10.1007/s11740-011-0302-5.

29 Witt, M. and Klimant, P. (2015). Hardware-in-the-loop machine simulation for modular machine tools. *Procedia CIRP* 31: 76–81. http://dx.doi.org/10.1016/j.procir.2015.03.015.

30 Aiello, G., Cacciato, M., Scarcella, G., and Scelba, G. (2017). Failure analysis of AC motor drives via FPGA-based hardware-in-the-loop simulations. *Electrical Engineering* 99 (4): 1337–1347. https://doi.org/10.1007/s00202-017-0630-3.

16

Distribution Phasor Measurement Units (PMUs) in Smart Power Systems

Geethanjali Muthiah[1], Meenakshi Devi Manivannan[1], Hemavathi Ramadoss[1], and Sharmeela Chenniappan[2]

[1]*Department of Electrical and Electronics Engineering, Thiagarajar College of Engineering, Madurai, India*
[2]*Department of Electrical and Electronics Engineering, Anna University, Chennai, India*

16.1 Introduction

Globally, the need for electrical grids is enlarging. Especially, the distribution systems are steadily expanding [1]. Due to the depletion of conventional energy sources, expenditure, and other factors, renewable energy resources (RES) are getting more attraction. The inclusion of renewable energy sources may cause uncertainties in the electrical grid [2]. For example, in over-capacity generation managing a large-scale renewable energy source is crucial. In addition to the renewable energy sources, electric vehicles (EVs), demand response loads are also increasing.

The enduring progression of power distribution systems specified by the accomplishment of the smart grid structure is initiating novel challenges to operations and planning action [3]. Focusing on these questions requires new advances and technologies to prolong reliability and safe supply to end consumers. High penetration levels of intermittent distributed generation (DG) can direct to major impacts on device deployments and operations [4]. For illustration, variable output from solar photovoltaic and wind plants can produce voltage and power deviations in the feeders. Furthermore, as penetration stages raise, concerns considering stability and interfaces among DG units are turning into additional importance. Uncontrolled plug-in electric vehicle (PEV) charging can result in distribution device overloading and violations of voltage limits. Moreover, harmonic insertion from power electronics elements like drives, LEDs, computers, and distribution energy resources and PEV inverters can enlarge total harmonic distortion (THD) intensity on distribution feeders and adapt the conventional models of voltage and current signals [4–6].

Hence, a smart grid that accommodates both real-time data and consumption analysis is essential. In this aspect, wide area monitoring systems (WAMSs) facilitate the employment of huge electrical energy storage systems in distribution systems [5]. The objectives of real-time measurements are equipped with phasor measurement units (PMUs). PMUs within the framework of wide-area monitoring, protection, and control (WAMPAC) are to be familiar with an important aid in making sure the protected operation and stability of transmission systems. As a result, as conventional radial supply feeders develop into vastly complex, meshed, dynamic, and active systems it becomes evident that PMUs can afford related functionalities and profits to make certain reliable and secure supply [6].

The deployment of PMUs at optimal feeder sites can raise the real-time monitoring, investigation, and synthesis abilities of functions. PMUs can give the necessary functionality and accuracy to attain time-synchronized phasors that can be applied to advance distribution circuit characterization and widen monitoring instruments to assess and moderate the impact of intermittent DG [7]. Despite the obvious advantages, the function of PMUs in distribution systems is still un-explored; this is generally due to financial factors. Though the drivers and needs for execution are quickly rising, there is increasing attention to relating PMUs to distribution systems.

Artificial Intelligence-based Smart Power Systems, First Edition.
Edited by Sanjeevikumar Padmanaban, Sivaraman Palanisamy, Sharmeela Chenniappan, and Jens Bo Holm-Nielsen.

In the structure of WAMPAC, PMUs are renowned for the stability of the power system and to perform a secure operation. PMUs are capable of functioning in a complex and meshed network by ensuring reliability [8, 9]. In particular, with the employment of the PMU devices at appropriate bus and feeder locations, the PMUs could execute an operation even more precisely. This advances the monitoring applications of the devices and alleviates the consequences of an intermittent DG. Due to financial aspects, the purpose of PMUs in distribution systems is still in the implementation stage. Though the requirement for the PMU devices is enlarging, the accomplishment of the devices is highly increasing and improves the interest of PMUs in distribution systems [10]. The complexity of the power system requires analysis, control, and operation of distribution networks. The voltage and angle of the power system buses change according to the state of the operation. In a steady-state, for continuous data transfer PMUs require medium bandwidth. While in dynamic state applications the PMUs need a high range of bandwidth [11]. With reference to North American synchrophasor initiative reports, PMU applications are classified in terms of automation, planning, reliability, and operation.

In an AC power network, the evaluation of the synchronized magnitudes of voltage and current can be effectively done using distribution PMUs [11, 12]. This synchrophasor technology gathers time-tagged data from diversified locations by means of specific global positioning system (GPS) signals in a high-speed duration. The PMUs are capable of capturing all the basic electrical parameters that are required to estimate the power flow [13]. The voltage phasor values are measured from the node of the system, whereas the current and power flow is taken from the branch of the network. The distribution PMUs are reasonably different from the PMUs deployed in a transmission system [14, 15]. In the transmission system, the PMU data are used by comparing the values from different substations. Nowadays, the PMU data are vital for the stability and monitoring of the power grid [15]. In contrast, the distribution synchrophasor values are considered from substations at a range of locations and also from various points on the identical distribution circuit. The major applications of PMUs in the distributed system shall be listed as:

(i) With the time-tagged phasor values.
(ii) PMUs can be used for situational awareness and state estimation.
(iii) To sense the fault occurrence and to determine the fault location.
(iv) PMUs can be connected with flexible AC transmission systems (FACTSs) devices to balance the sensitive loads.
(v) For balancing load and generation during islanding PMUs are used.

Still, many more applications are explored. This chapter intends to know the potential applications of PMUs in distribution systems and to present an imminent theory behind their prospective implementation.

16.2 Comparison of PMUs and SCADA

Conventionally, supervisory control and data acquisition (SCADA) systems have occupied an immense function in power systems [16]. They are well-organized and consistent in operating a power system. At present, owing to the fast-developing communication technologies and the need for rapid data transfer, SCADA devices are facing challenges.

Furthermore, the traditional devices have lesser accessibility to the latest technologies and updated devices such as intelligent electronic devices (IEDs), high-voltage direct currents (HVDCs), and FACTSs. The SCADA devices lack an interactive grid system between the power producers and the consumers [17]. The periodicity of SCADA is time-consuming for about two to four samples/cycle and is deficient in time synchronization. Consequently, the range of coverage is for a shorter distance when compared to advanced devices. The salient features of SCADA and PMU devices are listed in Figure 16.1.

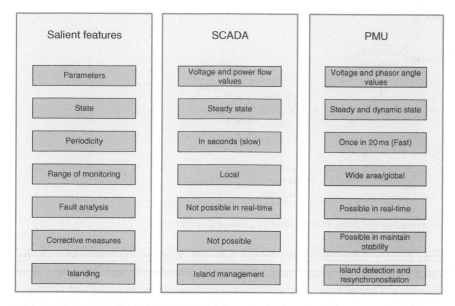

Figure 16.1 Comparison of SCADA and PMU devices.

For this reason, the synchrophasor technology turns out to be a resolution. Synchrophasor technology transfers a set of time-tagged data faster (once in 20 ms) and more precisely covers a wide area [18]. Moreover, they are capable of operating in the dynamic condition of a power system. In transmission systems, the PMUs are appropriate for wide-area situational awareness in monitoring the voltage, frequency, and oscillations, whereas in distribution systems, due to the sudden changes in active loads and renewable energy sources PMUs are highly effective. Since it adds the cost of installation, the system operators are restricting the device placements in the distribution systems [18, 19]. Deciding on the minimum order of PMUs using optimization techniques is the cost-effective solution for this issue.

16.3 Basic Structure of Phasor Measurement Units

PMUs measure the instantaneous and phasor values of electrical parameters such as voltage, current, frequency, and angle. PMU devices deal with time-tagged data for monitoring and controlling purposes. It consists of a data transmission unit, GPS receiver, CT, phase-locked oscillator, converter, and microprocessors. With these units, PMUs are capable of transferring data with an accuracy of 1 μs [19]. To describe the performance of the power grid, the finest way to express the sinusoidal waves is said to be the phasors. The sinusoidal function $x(t)$ is expressed as

$$x(t) = X_m \cos(\omega t + \phi) \tag{16.1}$$

The phasor representation of the sinusoidal wave is the magnitude and angle of $x(t)$ as follows:

$$X = \left(X_m / \sqrt{2}\right) e^{j\phi} \tag{16.2}$$

The phasor representation is expressed with angular frequency ω. While comparing the phasors, the frequency must be the same throughout the power network. The nominal frequency must be maintained, else the change in frequency leads to a change in phase [19]. In other words, if the frequency is below the nominal value, then the phase angle will reduce. While on the contrary, the phase angle increases, when the frequency gets higher than the nominal value. The calculated parameters from the multiple locations are presided by the power-flow

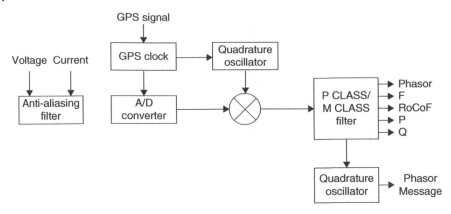

Figure 16.2 Block diagram of phasor measurement units. ROCOF, Rate Of Change Of Frequency.

equations [19, 20]. These equations comply with Kirchhoff's circuit law and describe the particulars of the parameters involved. The real power flow relies on voltage angle differences whereas the reactive power flow depends on voltage magnitude differences. The voltages of the distribution side should be around 5% of 120 V_{rms}. Thus, to maintain the flat voltage profile, the reactive power should be fed to the necessary locations of the distributed system [20]. The variation in the frequency also denotes the amount of load to be generated to meet the demand. Any fluctuations in these electrical parameters may lead to power system instability. This will lead to huge power loss and increases the project cost.

The word "synchronous" means that the abovementioned parameters should be acquired from multiple locations at the same time in seconds. As shown in Figure 16.2, this clock synchronization will be done by the GPS [21]. This technology supplies a 5 V pulse each second with an accuracy of 1 μs. These time-tagged GPS data are called synchrophasor data.

The anti-aliasing filter removes the signals that breach the Nyquist rate [22]. The voltage and current phasors at the input node are digitized through an analog-digital converter. The work of an oscillator is to convert the one pulse per second signal into a series of high-speed timing pulses. This in turn is used in the sampling process of a waveform. The phasor calculations are executed in the microprocessor and the phasors are converted into positive sequence measurements [23]. These values are time-stamped using a clock. The time-stamped data are transferred to a data center by using a modem.

The merits of PMUs are:

- Broad coverage, low latency, low cost
- Flexibility, reliability, and accuracy
- Adaptability with 5G technology
- Wireless communication

16.4 PMU Deployment in Distribution Networks

Observability of a complex distribution system is essential to function steadily and efficiently to amplify the precision and consistency of the system. The deployment of PMUs in a distributed network should be done in order to achieve complete observability [23]. Complete observability is when the parameters such as voltage and current phasor values are calculated. With the determined values of the voltage and current phasors of the installed buses, the line parameters of the other buses can be resolved using ohm's law. This brings all the buses under coverage and thus complete observability is achieved [23].

Several heuristics and meta-heuristic methods for the optimal placement of phasor measurement units (OPPs) are discussed for transmission lines. But, the PMU deployment techniques for distribution networks are an upcoming research focus [24].

Usually, the OPP techniques deal with four steps. They are as follows:

(i) Detection of appropriate locations for PMUs
(ii) Investigation for the minimal number of PMU devices
(iii) Checking for redundancy and
(iv) The execution time for the OPP algorithm

By this process, the radial network of distribution systems is evenly rationalized to the optimization. Thus, a technique with minimal PMU device locations and minimum execution time will be the best algorithm to find out the solution for OPP.

16.5 PMU Placement Algorithms

PMUs are placed at the substations and gather time-tagged data sequences of all buses in a power system. These data are stored in a data storage device and is available at remote locations for future investigation purpose. The bad data are rejected by the phasor data concentrators (PDCs) that are placed next to the PMUs [25–27].

For the suitable locations of PMUs, optimization techniques for PMU placements are used. To sort out the multi-dimensional problems, optimization algorithms are categorized as the deterministic method, heuristic method, and meta-heuristic method. This minimizes the cost of the devices and the complexity of the power system.

In the paper [9], the authors discuss a global search algorithm to optimize the locations of μPMUs in distribution networks. This objective is done to achieve complete observability with minimal computational time. In the article [10], in distribution networks, radial topology is selected and downsized. Then exhaustive search method is performed. Several research articles discuss least square estimation for the PMU placement in distribution networks. In this, algorithms differ in objectives such as state estimation, short circuit fault identification, and complete observability. Manuscript [11] illustrates a combined monitoring procedure with traditional flow measurement devices and μPMUs. The algorithm aims to reduce installation costs and complete observability. The method discussed in [12] deals with the nonlinear programming algorithm to attain the cost optimization using the combination of μPMUs and power flow measurement devices in distributed networks. The optimization technique using greedy search is used in the article [13]. The algorithm considers node selection strategy and information evaluation. In taking the cost as a constraint, the smart meters and pseudo-measurement devices are used in a combined manner to reduce the cost of the device. The objective is carried out to achieve both incomplete and complete observability. The topological process is executed with an enhanced information evaluation and node selection strategy.

16.6 Need/Significance of PMUs in Distribution System

It is critical to validate the envelope of design aspects such as peak loads or fault currents in a conventional power system network with radial power distribution and one-line power flow. But with the recent advancements and growth in DERs like hybrid EVs, renewable energy generation, and demand response programs more short periods and unexpected fluctuations and distributions are introduced [28]. Hence, the entire working state of the power system should be monitored continuously.

16.6.1 Significance of PMUs – Concerning Power System Stability

The applications, as well as the advantages of utilizing the PMU data, are useful to improve the system's reliability and make the project cost-effective. By using PMUs in DG systems, the possibility of incorporating renewable sources also improves. By improving the resilience, the fault location is determined rapidly and the time needed for the restoration process is comparatively less [28, 29]. This includes the fast reclosing of lines, forensic analysis, and fast resynchronization of islanded connections, oscillation detection, and synchronization of power generation. This in case reduces the fault interruptions and improves the power system stability. By recognizing the device failures, further consequences are avoided. The PMUs are capable of measuring the real and imaginary values along with the phase angle values. By this, the generator synchronization and reclosing are done. Whenever a fault or any interruption occurs in a distribution network, the PMU can examine the events and frequently informs the operator in the control room. The evaluations of parameters done by PMU are more precise than the data gathered by conventional devices. The model validation of the distribution network can be done by PMUs. The identification of fault and fault location is executed at the moment of the occurrence and the speedier action prevents the power outage. The time length of unintentional outages could be reduced.

16.6.2 Significance of PMUs in Terms of Expenditure

The expenditure of PMUs includes the installation cost, maintenance cost, communication links, and labor charges. The PMU cost is high when compared to traditional devices like SCADA [1]. To overcome this shortcoming, the optimization technique and congestion reduction methods are the solution. Despite the expenditure involved in the incorporation of PMUs in distribution systems, it is done in modern power system networks. Because while integrating RES the cost of fossil fuel consumption is mitigated [30]. Hence, the total fuel cost will be reduced. Also, with the faster recovery from fault and isolation, the power outages could be prevented. This benefits both the operators and the consumers. With the fault localization techniques, the fault buses and their locations can be accurately determined. This reduces the labor time in search of fault lines. Summarily PMUs in distribution systems will be justifiable and become essential in the present-day situation.

16.6.3 Significance of PMUs in Wide Area Monitoring Applications

The real-time observations of power network stability analysis are fulfilled by the PMU measurements. Generally, real-time observing is the precondition to react on time.

As referred to in Figure 16.3, time-stamped phasor measurements provide a perspective of the dynamic process that is significantly more valuable than normal RTU measurements of RMS values. Synchrophasors, in particular, provide the three key aspects that conventional measurement techniques do not:

- Messages per second rates ranging from 1 to 60
- High accurate time-synchronized measurements from all localities
- Voltage, current, and digital measurements with high precision (status and alarms)

Major applications of D-PMUs in distribution networks are listed as follows:

- Angle/frequency observation
- Root cause analysis
- Voltage stability observation
- Improvised stability monitoring
- DG/independent power producer (IPP) application
- Power system re-establishment

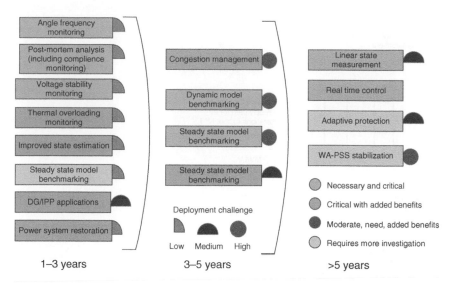

Figure 16.3 Applications of wide area monitoring systems (WAMS). WA-PSS, Wide Area-Power System Stabilizer.

Of all the above, in this chapter focus is made on a few of the following major applications concerning distribution networks.

- Automation of system reconfiguration to manage power restoration
- Planning for high DER interconnection (penetration)
- Voltage fluctuations and voltage ride-through of the DER
- Controlled islanding
- Fault-induced delayed voltage recovery (FIDVR) detection

16.7 Applications of PMUs in Distribution Systems

PMU technology for transmission is currently available and was used many years ago, but distribution is still in the early stage [2]. In future, it is expected that PMU will find a wide range of applications like system reconfiguration, planning for DER penetration, fault location, islanding, and protection at the distribution level.

16.7.1 System Reconfiguration Automation to Manage Power Restoration

Generally, the normal operation of distribution systems is in radial topology, but with the particularity to allow transfer of loads interrupted during an event to other circuits using circuit breakers or re-closures strategically located at different points of the grid. If this reconfiguration is done in an automated mode then restoration will be done as quickly as possible and effective power management will be established.

16.7.1.1 Case Study
Automation of System Reconfiguration Figure 16.4 depicts an example of a typical system configuration. System reconfiguration could be utilized to handle restoration after a distribution circuit fault.

Reconfiguration During Fault Conditions For instance, in Figure 16.4 if a fault arises at Substation N° 1, then the protection scheme will usually trip the breaker at the HV/MV substation [3]. Hence, all the customers on this feeder will be disconnected. Because of this action, the protection scheme makes the system safe. But, because of a single fault, all the customers connected to the whole feeder are getting interruption of power.

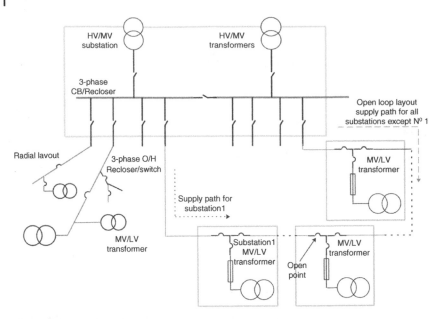

Figure 16.4 Typical MV/LV distribution system.

However, if an automation system is introduced it can reconfigure power flow with the minimum number of customers disconnection in such kind of situation. Consequently, the number of customers without power and the duration of disconnection will be considerably reduced.

By closing the connection of the open point and by opening the main circuit switch of substation N° 1, a major portion of the system shall be restored to the normal operation [4].

Reconfiguration During Normal Operating Conditions System reconfiguration is also needed during normal operating conditions for the following purposes:

- To balance load through the transformation of customers aided from one feeder to another
- To manage voltage
- To improve power delivery efficacy

Pennsylvania Power and Light (PP&L) gathered load data based on time-of-day, on six feeders in its Lehigh Division at 15-minute intervals in 1984 and 1985. They investigated reconfiguration with Westinghouse and discovered that losses might be condensed by 14%.

The current (losses) reduction was mostly accomplished through the distribution lines, with loads being moved to the feeder's lightly loaded part. While applying this strategy across the 26 feeders in the whole division, approximately 2500 MegaWatt-hours (MWh) of power was reduced annually. PP&L was able to boost efficiency from 0.2% to 98.95% by employing these strategies [1].

Usually, automated configuration has the following drawbacks:

(1) Lack of monitoring about the state of equipment placed among the substation and the customer
(2) Exact deficiency of information about phase angle to enable load transfer (with the massive introduction of DERs in distribution level, this check becomes critical).
(3) Delays in communications

To overcome the above-mentioned drawbacks, PMU with adequate communication support can be used effectively.

16.7.2 Planning for High DER Interconnection (Penetration)

Usually, in the conventional distribution systems, there will be no active sources within the system. Hence, planning has been done straightforward due to the oneline power flow. Customer and utility adoption of DER (such as wind, solar PV, geothermal, and energy storage) within distribution systems add complexities to the system [18]. As a result, a complete simulation model must augment conventional planning practices. PMUs provide the required measured information to develop distribution planning model calibration and validation.

The distribution level connects the majority of DG and IPPs. Most of these energy sources are renewable like wind, solar, etc. These resources are dynamic in nature. To do power flow analysis and stability analysis in such kinds of dynamic systems, PMU measurements make significant contributions. PMU measurements can also be employed for distribution-level control of generation units [19, 20].

16.7.2.1 Case Study

Synchrophasors Usage for Energy Accounting A sample of PMU measurements being used in a distribution network for energy accounting is provided and explored [23]. In Figure 16.5, the project "RegModHarz" [23] is deployed to a region "Harz" in Germany.

The region under study has the following power generation sources:

- 12 MW thermal power stations
- 80 MW pumped storage
- 250 MW wind power

The controllable load includes 10 MW industrial loads and 0.5 MW residential loads. The aim is to consolidate all native decentralized power production into a single virtual power plant and to control loads wherever potential so that the region can manage and operate without any external power flow.

The PMUs are installed at the regional grid's border connections with the associated power network. They send their data to the central component "Virtual Power Plant," which streamlines generation and controllable loads. This kind of virtual power plant project implementation is still in the pipeline in Germany [23].

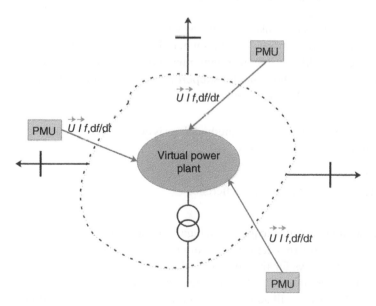

Figure 16.5 System configuration in project "RegModHarz."

In this project the major aids obtained are:

- Energy accounting with high precision
- Generation and consumption must be coordinated
- Decision-making with higher reliability
- Providing a database for services such as the energy market

The virtual power plant optimally coordinates the power balance among these components. The following three various operating modes are possible in this case:

- To obtain a maximum of renewable generation
- Cut-off load peaks
- To obtain maximum network stability

16.7.3 Voltage Fluctuations and Voltage Ride-Through Related to DER

Voltage fluctuations may happen all of a sudden that can be caused by changing system loads. Consequently, there will be noticeable changes in RMS voltage levels. Sometimes voltage sag is produced due to the large motor starting. Connecting DER (such as solar PV, wind and energy storage, etc.) to distribution-level feeders may result in unpredictability in feeder node voltages, particularly in fragile residential and rural networks [23, 24]. These fluctuations result in voltage instability, which frequently necessitates reactive power compensation or energy-storage provision (inverters in Volt/VAR control mode) to maintain a stable voltage profile. In such situations, PMUs provide invaluable information to monitor accurately the real performance of the distribution system, in particular during high DER penetration.

16.7.4 Operation of Islanded Distribution Systems

Islanding is a portion of the utility system that got isolated from the remaining utility system and continues to operate. This portion contains both load and generation [25].

When an islanding condition exists, the isolation point is most likely on the low voltage distribution line. However, islanding can occur on higher voltage distribution or transmission lines when there are a large number of microgrids and other DERs exist. PMUs can have a significant role in serving microgrids to manage frequency the same as the grid frequency during islanding conditions. They can also support the re-integration of islanded microgrids into the main grid as discussed earlier under system reconfiguration.

Case Study: Islanding Detection Islanding detection applying PMU-based methodology is developed by GE in 2012. The methodology is based on any one of the following three algorithms:

- First algorithm – Angle and Magnitude difference
- Second algorithm – The rate of change of angle difference and angle difference
- Third algorithm – THD difference

It is to be noted that all the three algorithms complement an adequate coverage of all operating circumstances to have a non-detection zone (NDZ) equal to zero. Figure 16.6 depicts the aim of the application "Island State Detection." The continuous frequency measurement allows for the immediate detection of network splits. For example, if two lines on the right side of Figure 16.6 were interrupted at the same time, two isolated areas would result. The SIGUARD Phasor Data Processor (PDP) system will automatically notify this via alarm indications. "SIGUARD PDP" is a kind of WAMS developed by Siemens. The IEEE C37.118 protocol is used to receive the synchrophasor data stream via the SIGUARD PDP server.

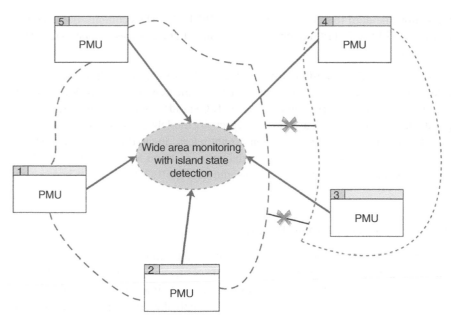

Figure 16.6 Island mode detection principle.

In this case, the measurements are distributed to the user interfaces by the server (UIs). Typically, UI is installed on separate computers. To ensure safe and continuous operation, the server computer is housed in a secure environment. The "SIGUARD PDP Com" component communicates with other WAMSs. During the interface, it behaves similarly to a PMU, using the same IEEE C37.118 protocol. Nonetheless, the SIGUARD system operator can choose which measurements to transmit and which not to transmit. For instance, voltage phasor and frequency values shall be transmitted and current phasor values shall not be transmitted. The elements Server, Archive, and Com form the PDC, which can function independently. In information relating to wide-area monitoring, the abbreviation "PDC" is frequently used. The control center rooms are where the user interfaces are located. The SIGUARD Engineer can be used for PMU, application, communication, calculation, and graphics configuration. The configuration must be changed if necessary. For example, the addition of one PMU is simple and can be completed instinctively. The user interface (Figure 16.6) is intended to provide a clear vision of the network dynamics. The power network status curve at the top of the screen is a fictitious value that shows the monitored system's distance to serious status. It is calculated by subtracting the distances of entire measurements from their respective limits. It performs the function of a traffic light, but it also indicates the system's state tendency. On 4 November 2006, a similar interruption occurred in the European transmission system. However, it was not immediately apparent to entire applicants. The obtainability of measurement streams from the PMUs is used for Island State Detection; however, currents and voltages phasor values are not required.

Actual Standardization Status In its current form, the extensively used standard IEEE C37.118 (2005) defines the communication interface for PMU data. It is presented in all currently available PMU products and WAMSs which is a useful solution [1]. New developments in standardization are taking place in two directions:

First Direction: IEEE C37.118 can be divided into two parts.

IEEE C37.118.1 as the performance standard for dynamic phasor and dynamic frequency measurement behavior of PMUs. This issue has been discovered to be critical for the accuracy of PMUs, which are designed to improve the network's dynamic measurements. Details for testing and calibrating PMUs will also be included.

IEEE C37.118.2 is the communication standard that replaces the former IEEE C37.118's content. Because it is intended to be compatible with current IEEE C37.118 implementations, only slight changes can be made. The time limit for this will be a release in 2011.

Second Direction: The development of an IEC standard for synchrophasors. IEC61850-90-5 will be the name of this standard. The IEC61850 standard will be used for communication, while the new IEEE C37.118.1 standard will be included in this standard via the "dual logo" principle. This new IEC standard will also cover new applications like adaptive relaying and wide-area controls. It also discusses security models. It is demonstrated how to transition from IEEE C37.118 (2005) to an IEC61850-90-5 solution. Nonetheless, due to the large installed base of the IEEE protocol, PMUs and PDCs will almost certainly offer both communication options (IEEE C37.118.2 and/or IEC61850-90-5) after the release of this standard. It will take some time for IEC61850-90-5 to become an International Standard (IS).

16.7.5 Fault-Induced Delayed Voltage Recovery (FIDVR) Detection

Mostly FIDVR is developed in systems with a modest amount of single-phase induction motor (IM) loads [25]. After huge trouble such as fault, switching surge, etc., these motors, which are connected to mechanical loads with constant torque, will come to a stop condition. Also, they may draw approximately 6 times their nominal current. This situation may cause the system voltage to drop for several seconds. To mitigate the FIDVR phenomenon, two types of solutions have been discussed in the literature:

- Supply-side methods (injection of dynamic VARs via SVC, etc.) and
- Demand-side methods (disconnection of loads using measurements, etc.).

Utilities prefer to use the supply-side solution, which involves calculating the amount and location of SVCs and STATCOMs during the planning phase [26]. These methods rely on contingency sets and possible operating cases, as well as extensive time-domain simulations, to confirm that the installed devices can reduce FIDVR in a variety of scenarios [27]. Meanwhile, because the planning phase cannot account for all rare and severe events, a measurement-based technique that can supervise FIDVR will assist the grid in dealing with unexpected/rare contingencies [28].

The widespread use of PMUs by utilities in the current era has led to the development of ways to quantify the intensity of FIDVR in real time to conduct appropriate control steps to avoid IM loads from stalling further [29].

16.8 Conclusion

To summarize, the PMUs are used in power system networks for monitoring both the transmission lines and distribution networks. PMU technology for transmission is currently available, but in distribution it is still in the early emerging stage. The devices check the stability and observability of the power system. The devices placed at wide-area locations are synchronized in terms of time, thus, the data transfer is done in a very faster manner.

There is upward attention in the industry to operating PMUs in distribution system applications. This importance has been primarily provoked by the rising penetration stages of new causes and loads, mainly of photovoltaic DG. Though financial issues still bound the prevalent accomplishment of PMUs, further distribution equipment is initially including this ability also as an optional or standard aspect. Additional challenges for placement consist of the choice of ample communications and information technologies for conducting enormous volumes of information presented by PMUs, and the particular data storage, processing, and analysis limitations. To perform the operations described in Sections 16.7.1–16.7.5, it is necessary to require and process the data provided by PMUs. It is important to note that in some applications, such as security, this must be done in real-time, although in others, such as load modeling, the goal is to use previous data and give distribution engineers expensive knowledge rather than massive volumes of undefined data. As a result, specific data processing methods need to be extended to eliminate the relevant data for each application. Another important consideration is the most advantageous sample rate range, as well as the classification of the minimum number and location of PMUs in distribution feeders. This will minimize overburdening communication and data storage systems while yet accomplishing the goals. All of these are important topics that will require more research in the future to develop the foundations for PMU execution in distribution systems.

Major concluding remarks:

- In the presence of fluctuation and uncertainty, high-resolution voltage phase angle measurement offers a new option for precise and flexible distribution network monitoring and control.
- A wide variety of specific diagnostic and control applications will depend on improved visibility and simplicity of the distribution system, meaning better facts of the system state in real time.
- High-precision monitoring systems will be required to provide clear, precise, and thorough observation of varying system behavior for the effective management of distribution networks with high renewable penetration, demand reactivity, and distributed control. μPMUs are a strong candidate for economically creating this functionality.
- To perform any of the exact applications it will be needed to observe what phenomena can be detected at the declaration of the μPMU (D-PMU), and what can be reliably concluded from those observations. The need for hierarchical, layered, distributed control of aggregated distributed resources and loads – especially clusters capable of islanding – must be studied in connection to voltage phase angle to pursue the potential for active coordinated control based on PMU data.
- DERs may seamlessly shift between linked and isolated modes, as well as provide local power quality, dependability, and support to the core grid.
- Security and infrastructure resiliency considerations may help to enable the spread of such initiatives in the future. Increased visibility and precise measurement will be important features of the new infrastructure, regardless of how distribution networks develop.

References

1 IEEE C37.118-2005 (2005). *Standard for Synchrophasors for Power Systems*. IEEE C37.118-200.
2 Liu, Y., Wu, L., and Li, J. (2020). D-PMU based applications for emerging active distribution systems: a review. *Electric Power Systems Research* 179: 106063.
3 Shahsavari, A., Farajollahi, M., Stewart, E. et al. (2019). Situational awareness in distribution grid using micro-PMU data: a machine learning approach. *IEEE Transactions on Smart Grid* 10: 6167–6177.
4 Kim, J., Kim, H., and Choi, S. (2019). Performance criterion of phasor measurement units for distribution system state estimation. *IEEE Access* 7: 106372–106384.
5 Gholami, M., Abbaspour, A., Moeini-Aghtaie, M. et al. (2020). Detecting the location of short-circuit faults in active distribution network using PMU based state estimation. *IEEE Transactions on Smart Grid* 11 (2): 1396–1406.
6 Meenakshi Devi, M. and Geethanjali, M. (2018). Fault localization for transmission lines with optimal phasor measurement units. *Computers and Electrical Engineering*, Elsevier 70: 163–178.
7 Meenakshi Devi, M. and Geethanjali, M. (2020). Hybrid of genetic algorithm and minimum spanning tree method for optimal PMU placements. *Measurement*, Elsevier 154: 107476.
8 Tahabilder, A., Ghosh, P.K., Chatterjee, S., and Rahman, N. (2017). Distribution system monitoring by using micro-PMU in graph-theoretic way. In: *Proceedings of the 4th International Conference on Advances in Electrical Engineering (ICAEE)*, Dhaka, Bangladesh (28–30 September 2017), pp. 159–163.

9 Chen, X., Chen, T., Tseng, K.J., and et al. (2016). Hybrid approach based on global search algorithm for optimal placement of *m*PMU in distribution networks. In: *Proceedings of the IEEE Innovative Smart Grid Technologies-Asia (ISGT-Asia)*, Melbourne, VIC, Australia (28 November–1 December 2016), pp. 559–563.

10 Chen, X., Chen, T., Tseng, K.J., and et al. (2016). Customized optimal *m*PMU placement method for distribution networks. In: *Proceedings of the IEEE PES Asia-Pacific Power and Energy Engineering Conference (APPEEC)*, Xi'an, China (25–28 October 2016), pp. 135–140.

11 Dusabimana, E., Nishimwe, H.L.F., and Yoon, S. (2019). Optimal placement of micro-phasor measurement units and power flow measurements to monitor distribution network. In: *Proceedings of the International Council on Electrical Engineering (ICEE)*, Hong Kong, China (2–6 July 2019). http://sgcl.ssu.ac.kr/wp-content/uploads/2019/05/ICEE19J-178-FP-2.pdf.

12 Dusabimana, E. and Yoon, S. (2020). A survey on the micro-phasor measurement unit in distribution networks. *Electronics* 9: 305.

13 Wu, Z., Du, X., Gu, W. et al. (2018). Optimal micro-PMU placement using mutual information theory in distribution networks. *Energies* 11: 1917.

14 Von Meier, A. and Arghandeh, R. (2014). Every moment counts. In: *Renewable Energy Integration*, 429–438. Crossref, https://doi.org/10.1016/b978-0-12-407910-6.00034-x.

15 Wache, M. (2011). *Benefits of Synchrophasor Solutions for Distribution Networks*. CIRED.

16 Wang, W. et al. (2020). Advanced synchrophasor-based application for potential distributed energy resources management: key technology, challenge and vision. In: *2020 IEEE/IAS Industrial and Commercial Power System Asia (I&CPS Asia)*, pp. 1120–1124. https://doi.org/10.1109/ICPSAsia48933.2020.9208606.

17 Hataway, G., Flerchinger, B., and Moxley, R. (2013). Synchrophasors for distribution applications. In: *Power and Energy Automation Conference,*Spokane, Washington (26–28 March 2013). https://cms-cdn.selinc.com/assets/Literature/Publications/Technical%20Papers/6561_SynchrophasorsDistribution_RM_20121030_Web.pdf?v=20191011-220136.

18 Cardenas, J. and Patynowski, D. (2020). Phasor measurement units (PMU) in distribution systems and distributed energy resources (DER). In: *GCC* (November 2020).

19 Sánchez-Ayala, G., Agüerc, J.R., Elizondo, D., and M. Lelic, (2013) Current trends on applications of PMUs in distribution systems. In: *2013 IEEE PES Innovative Smart Grid Technologies Conference (ISGT)*, pp. 1–6. https://doi.org/10.1109/ISGT.2013.6497923.

20 Eto, J.H., Stewart, E., and Smith, T. et al. (2015). Scoping study on research and development priorities for distribution-system phasor measurement units. Several Lab Research.

21 Roop, M. and Bower, W. (2002). *Evaluation of Islanding Detection Methods for Photovoltaic Utility Interactive Power Systems*. IEA PVPS.

22 Cardenas, J., Mikhael, G., Kaminski, J., and Voloh, I. (2014). Islanding detection with Phasor Measurement Units. In: *67th Annual Conference for Protective Relay Engineers*, pp. 229–241. https://doi.org/10.1109/CPRE.2014.6799003.

23 Patynowski, D., Cardenas, J., and Menendez, D. et al. (2015). Fault Locator approach for high-impedance grounded or ungrounded distribution systems using synchrophasors. In: *2015 68th Annual Conference for Protective Relay Engineers*, pp. 302–310.

24 Intelligent Monitoring System (ILMS). GE Grid Automation.

25 Wang, D., Takoudis, C., Wilson, D., and Murphy, G. (2015). Quantifying benefit of angle constraint active management on 33 kV distribution networks. In: *GE and SP, UK. CIRED*. https://www.semanticscholar.org/paper/QUANTIFYING-BENEFIT-OF-ANGLE-CONSTRAINT-ACTIVE-ON-Wang-Takoudis/fda021e3a4bff23dff87c9a58c073df3ec5e2c65.

26 Kandel, A. (2016). *Benefits and Costs of PIER Research Enabling Synchrophasor Applications; Staff Report*. Sacramento, CA, USA: California Energy Commission.

27 Zafar, R., Mahmood, A., Razzaq, S. et al. (2018). Prosumer based energy management and sharing in smart grid. *Renewable and Sustainable Energy Reviews* 82: 1675–1684.

28 Usman, M.U. and Faruque, M.O. (2019). Applications of synchrophasor technologies in power systems. *Journal of Modern Power Systems and Clean Energy* 7: 211–226.

29 Popovski, P., Trillingsgaard, K., Osvaldo, S., and Durisi, G. (2018). 5G wireless network slicing for EMBB, URLLC, and MMTC: a communication-theoretic view. *IEEE Access* 6: 55765–55779.

30 Eissa, M.M. (2018). New protection principle for smart grid with renewable energy sources integration using WiMAX centralized scheduling technology. *International Journal of Electrical Power & Energy Systems* 97: 372–384.

17

Blockchain Technologies for Smart Power Systems

A. Gayathri[1], S. Saravanan[1], P. Pandiyan[2], and V. Rukkumani[3]

[1]*Department of EEE, Sri Krishna College of Technology, Coimbatore, Tamil Nadu, India*
[2]*Department of EEE, KPR Institute of Engineering and Technology, Coimbatore, Tamil Nadu, India*
[3]*Department of EIE, Sri Ramakrishna Engineering College, Coimbatore, Tamil Nadu, India*

17.1 Introduction

The development of smart energy systems in various applications raises a variety of concerns and opportunities [1, 2]. Applications like smart infrastructure systems, hybrid electric vehicles, and non-conventional energy sources [3, 4] are becoming more popular, which leads to an increase in the trading of carbon and energy issues [5, 6]. A global energy shortfall is resulting from natural gas, coal, and other energy supplies being insufficient to fulfil demand. Natural gas costs in Europe have risen by a factor of ten this year, while power shortages caused by a scarcity of coal have forced several Chinese firms to close. In the meantime, several of the UK's gas pumps have gone dry, leading the military to act to minimize chaos. In India's power plants and coal reserves are also critically low. It also manages common factors like energy management through demand response management [7]. In order to utilize the above-mentioned needs, it's important to realize smart energy systems and have to aim for technologies that allow us to accomplish those needs [8, 9]. Blockchain technology has been integrated into various applications such as the Internet of Things (IoT), Cloud computing, and Artificial Intelligence (AI) for monitoring and managing specific applications. Blockchain (BC) technology is used to make sure that communications between IoT devices are secure. It also provides a distributed, decentralized, and publicly viewable shared register for storing blocks of data and treating and proving them in an IoT network.

In recent days, interest in Blockchain technology has considerably increased. It is a digital-distributed ledger that can maintain and do updates on a decentralized system, also known as a peer-to-peer system, and operates to define certain protocols [10]. A concurrence of different technologies connected to the network, data, consent, similarity, and automation, are necessary for the successful creation and implementation of Blockchain [11]. Blockchain has several distinguishing characteristics, including decentralization, the establishment of a trustless network (in which nodes can resolve conflicts without the intervention of a centralized authority), tamper-proof data storage, fault tolerance, and suitability [12]. Technology has been chosen, which has a significant effect on Blockchain characteristics and its performance [13]. The use of Blockchain technology in smart energy systems is becoming more popular in research topics since this technology can benefit from integrating new and innovative technologies. Because the Blockchain is so unique, it can help more smart energy applications move forward [14].

Some Blockchain technology creates gaps, limiting its applicability to smart energy applications. A few gaps can be network management, data management, consensus management, identity management, automation management, and a lack of suitable implementation platforms. These gaps can be rectified by choosing appropriate Blockchain techniques. Significant new research in Blockchain technologies is still required to meet the diverse and often stringent latency, privacy, and security requirements of smart energy applications. The majority of Blockchain platforms with embedded combinations of Blockchain technology solutions are computing and resource intensive, and hence not entirely suitable for smart energy applications [15, 16].

This chapter specifically deals with the path of integrating Blockchain technology into smart energy systems by developing flexible Blockchain platforms. Modular Blockchain platforms, where embedded technology options can be changed at any time, would also be needed to help and speed up Blockchain integration in a wide range of smart energy applications.

17.2 Fundamentals of Blockchain Technologies

Blockchain is a software-oriented technology that is well known for securely enabling virtual transactions with full transparency without the interference of a third party. It is done with the participation of the sender and receiver [17, 18]. In general, the actual assumption of Blockchain contains a series of blocks with a sequential chain-like structure. Each block carries specific information related to the transactions [19]. The transactions that were made in the specific block were properly registered with the detailed structure where the information during and after the transactions will not be altered or taken out of the piece of the block. Blockchain technology is based on an internet protocol that supports a user interface where the data is represented, an Application Programming Interface (API) that fetches the data, database models for storing information, logic operations for validating and processing data, and platform functionality where the data is hosted. Even though the technology is hard-coded in several programming languages, Solidity looks like it will be the high-level object-oriented programming language where the latest improvements to Blockchain happen [20].

17.2.1 Terminology

While the structure of the blockchain is taken into consideration, there are more features inside the blockchain that make this technology more reliable. Figure 17.1 interprets the actual design of the Blockchain; Make a sharp focus on a specific blockchain block. The block consists of the information related to the transaction. The first ever block in the chain is known as the foundation block, also called the Genesis Block. Only after adding the genesis block to the chain will the subsequent block connection take place. The main identity for the blocks in the chain is the hash address. Each block has its hash address. Each block in the series will be connected based on the hash address as they contain the previous block hash address and the current block hash address. If there

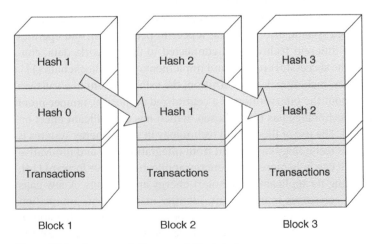

Figure 17.1 Conceptual diagram of Blockchain.

is any alteration in the block connection, then there will be a change in the hash address [21]. Once there is a change in the hash address in a particular block, the subsequent blocks' hash addresses don't change. Therefore, the following blocks fail in validation, which results in the termination of the block. This feature makes blockchain technology more secure in recent times because each block change can be traced and easily found [22]. This makes blockchain technology more secure.

17.2.2 Process of Operation

The steps of the blockchain transaction process are illustrated in Figure 17.2. To initiate a transaction on the blockchain, there are a few procedures to be followed. The first of its kind is requesting the transaction in the block. The requisition can be made by any user. Once the transaction has been requested, this information will be passed over to all the users in the network. Following that, there is a verification process where all the nodes will take part in the process of verification through the hashes [23]. After the successful completion of verification, the transaction takes place, resulting in the addition of a block with the previous block to the chain with the transaction information contained in the block and the hash address of the previous block and current block [19]. But the problem raised by the attacker with 51% of the attack is that, with the help of superfast computers, they can be able to alter the blockchain and re-evaluate the hash address of the other blocks within a short time. This is one of the most aggressive drawbacks related to blockchain trends.

To overcome the downside of the technology, several analyses were made to define the specific algorithm known as Consensus [23]. Consensus is simply the validation of all nodes in the network, which prevents the duplication and manipulation of transactions added to the chain network. This acts as a pre-validation process before connecting with the nodes. The process occurs at certain time intervals. The various intervals involved in the process are initiation, validation, and addition. Validation intervals are determined by some factors, including the size of the block, the information contained within the block, and the time of the process. As per the concerns, certain algorithms were created and are employable in certain practical scenarios. The various approaches were discussed in the following topic [24].

17.2.2.1 Proof of Work (PoW)

Proof of work (PoW) is a mechanism for adding new blocks of transactions to a cryptocurrency's blockchain. The work in this scenario entails creating a long string of characters called a hash that resembles the target hash of the current block. The crypto miner now can add that block to the blockchain and receive rewards [25].

17.2.2.2 Proof of Stake (PoS)

A type of consensus mechanism where there are no miners, instead they create their validators for the transactions. First, the stakes must be placed for the validation during the transactions, and then the validator node verifies the

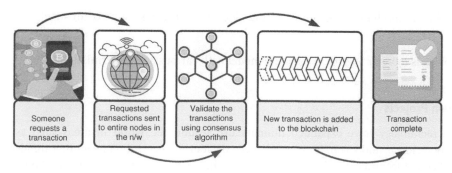

Figure 17.2 Blockchain transaction steps.

accurate things. If there is any inaccuracy present during validation, it leads to a loss of stake. If the information is valid, it will be added to the chain network. This one is more efficient and less time-consuming than the Proof of Work [25].

17.2.2.3 Proof of Authority (PoA)

As the name implies, this algorithm will be used by authorized parties, i.e., the concerned users and acceptable accounts, to perform network transactions. This approach will be more secure and concentrate on the hotspots [25].

17.2.2.4 Practical Byzantine Fault Tolerance (PBFT)

The approach is more specific for the recurring analysis, which occurs twice. Normally, it consists of primary and secondary control where the validation is first done by the primary control. If the validation passes over the primary, then the secondary will catch up with the validation done by the primary. Then there will be the green light for adding the blocks to the blockchain [25, 26].

Apart from the above concerns, Blockchain may have certain access criteria where the access permission will be provided to the users as per the access criteria. The various access criteria are listed below. They are

(i) Public: As the name indicates, this will be accessible to all the users. Anyone can perform transactions, visualization of data, and the addition of blocks to the chain network.
(ii) Private: In this case, only the users who are allowed to do so can use the different functions in the Blockchain.
(iii) Clan Users: This is a group that contains the required members who are eligible for the operations in the network, like viewing, verifying, and inclusion of blocks in the network.

17.2.3 Unique Features of Blockchain

Blockchain may have some characteristics that distinguish it from other trends. The features are listed below [19]:

(a) **Decentralized Network Users:** The whole functionality of the network can be performed by all the node users who are taking part in the transaction.
(b) **Resilience:** Due to the decentralization capability, the entire network becomes robust.
(c) **Time reduction:** Transactions on the nodes occur in a short time without the intervention of a third party, which makes it more attractive over the top of other technologies.
(d) **Reliability:** One of the best aspects of Blockchain is its immutable functionality; once something is recorded, it is permanent. No other changes will take place.
(e) **Transparency:** The operations that are happening across the network will be visible to all the users on the network.
(f) **Security:** Despite the all-defensive mechanism built around the Blockchain, the possibility of an attack is a tedious thing.

The entire blockchain technology is more or less like a database model with various access permissions.

17.2.4 Energy with Blockchain Projects

The capabilities of Blockchain technology have expanded dramatically around the world in areas such as E-Commerce, the IoT, Health and Science, and so on. It can work well considering its position over other trends and its ability to work with embedded technologies, but it is still a work in progress in many industries. Some use-cases which employ Blockchain technology were mentioned in the following discussions [17].

17.2.4.1 Bitcoin Cryptocurrency

The cryptocurrency rose to prominence as a result of Blockchain technology. In the year 2008, a man named Satoshi Nakamoto [27] floats a coin into the virtual world named "Bitcoin," which is not owned by any government authorities. At that time, the transaction that is happening will be visible to all the users who are taking part in it. Anyone can perform actions on the network. The possibility of cryptocurrency transactions is only via blockchain technology. This technology gradually advanced and became well-known at a later stage. Other cryptocurrencies, in addition to Bitcoin, such as Ripple, Litecoin, and Ethereum, played an important role in Blockchain technology [28].

17.2.4.2 Dubai: Blockchain Strategy

The Dubai government launched the Blockchain strategy in their market in 2016. This open-source platform serves as the foundation for other countries around the world to participate in this technology. Dubai established Blockchain on three pillars: government productivity, pioneering technology sustainability, and industry growth [29].

17.2.4.3 Humanitarian Aid Utilization of Blockchain

In 2017, the United Nations launched a project called Humanitarian Aid to use Blockchain to transfer funds for beneficiaries in Pakistan's rural areas. This is employed for the clear visibility of information regarding all types of transactions performed over here to all the members and for security purposes [30].

17.3 Blockchain Technologies for Smart Power Systems

Furthermore, in the preceding cases, the goal is to understand the architecture, features, and working principles of blockchain technology, as well as to see some concepts related to cryptocurrency transactions. The new beginning of the topic brings the employment of Blockchain in electrical power networks. In recent trends, everything is being digitalized, and the updating of the normal electrical grid to the smart grid paves the way for a new way of working. By making several analyses, the characteristics and workings of blockchain move coherently with the electrical network systems. The various features that bring hands up to the electrical system are enhanced security, decentralization, robustness, higher visibility of information, higher computational analysis, and lower transaction time. With these, the more advanced version of Blockchain technology comes into the picture by accompanying the electrical networks [31].

17.3.1 Blockchain as a Cyber Layer

The establishment of the Smart Grid with the Blockchain application makes the entire electrical network more reliable by protecting the system with a defensive cyber layer. The sustainable growth and efficient working of the Smart Grid act in each step of the smart grid's operation, such as trading energy from the side that makes it to the side that distributes it and protecting data through a secure cyber layer. As clearly shown in Figure 17.3, trading occurs between the generating power plants and the distribution end. The things that come into the picture during this trade are the very first thing is the generation side, which automatically updates the data to the Blockchain as per the production and microgrid involved in the transmission control, like transmission to different substations and e-vehicle charging stations. Those things will be recorded on separate Blockchain networks respectively. All these trading activities during the period of transactions must be well secured by the Cyber-Physical Security Blockchain as the defensive mechanism for protecting the data against malicious attacks. All the actions under the network will be interconnected with one another for the coherent action and accuracy of the information that is transformed. The Digital World furnishes a lot where data is the core performer in the entire plot, paving the way for Blockchain technology, which raises the performance of the smart grid in an efficient way [32].

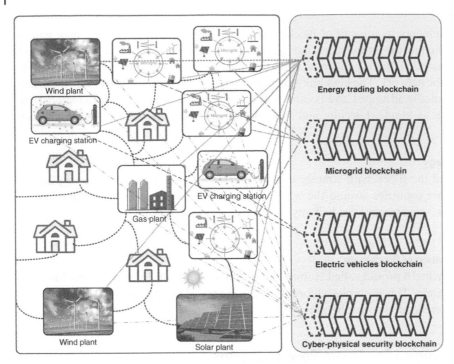

Figure 17.3 Smart Grid—New cyber layer using Blockchain. Source: Adapted from [31]/with permission of IEEE; Wajan/Adobe stock; ekazansk/Adobe stock; ryanking999/Adobe stock.

17.3.2 Agent/Aggregator Based Microgrid Architecture

When the blockchain enters the commercial application of the smart grid, the data flow becomes massive in terms of trading. As things are slowly getting updated to the trend within the electrical network, the computational data might become a complex process to handle in terms of the capacity of storage and the way for the analyses. This may result in severe complexity regarding the control and performance of the microgrid. Figure 17.4 illustrates the futuristic approach to the scenario discussed above. Despite the approach of dumping data to a single network, this is the mechanism brought about by having a separate chain network for the various components in the smart grid, i.e., for measuring the metering devices such as Smart Meters, Wind Energy Meters, and Solar Energy Meters, the approach is to aggregate the meter to the production and to have a separate chain network for all the meters. This, along with collecting the data separately, makes it easy to keep track of operations, make analyses, and make decisions. The Blockchain capability increases the usage of this technology in a wide range and may be adaptable to various industries as per the application related [31].

17.3.3 Limitations and Drawbacks

There are also some downsides related to technology. Taking a closer look at the negative aids in the development of solutions and the careful operation of the chain network [31]. The various limitations are,

(i) **Smaller ledger:** A ledger is nothing more than a node that is linked to the chain. The ledger system's capacity is determined by the network's size and the energy trading body's size. As a result, not all ledgers are small. A secure ledger system is always used in conjunction with blockchain technology. In some applications, it might be smaller in size, which could bring risks in security, alteration of information, and volume of data storage in the network.

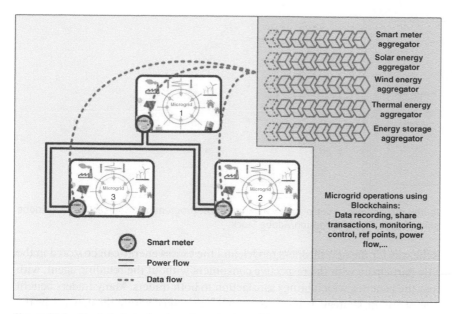

Figure 17.4 Blockchain technology-based microgrid automation [31].

(ii) **Delayed transaction:** Even without the intervention of a third party, the time consumed for processing the transaction is very high.

(iii) **Transaction cost and network range:** First, the cost of a single transaction is quite high, and second, the blockchain takes a long time to respond, so it's not a good choice for transactions that need to happen quickly.

17.3.4 Peer to Peer Energy Trading

The world is getting ready for digital evolution with the growth of the internet market everywhere [33]. Playing around with the data requires a step above all the technologies to get some stable play. Though it has some disadvantages, the process of upgrading the way of operation can improve the performance wisely and bring back more efficient technologies with the unique features of data visualization to all nodes, information security, and validation for network additions. The visualization market was booming in the commercial and electrical industries.

The technology is making it possible for people in the same P2P energy network to trade energy with each other. In order to have a conversation with the end–end parties, the intermediate one is not necessary. In the concept of energy trading, the person who produces can utilize the surplus energy as per their needs and then trade with the rest of the things left. This initiates a new way of the market in selecting the source of trading and getting the goods needed as per the requirements. This new transformation may enhance the industry, which results in a direct impact on the investment of small-scale energy producers, like renewable energy producers.

The main scope of blockchain in the peer–peer concept is to experiment with the transition from the traditional approach of a centralized market to a decentralized one, which is the independence of participation mostly reliant on microgrid controls. The peer–peer concept has gained interest recently in the area of electricity and commercial markets [34] due to the adaption of roof–top photovoltaic panel energy generation [35], integration of renewable energy resources [31, 36], and prosumer models that may be traded within the network or injected into the grid [37]. As a benefit of the awareness given in society, individuals are taking part in the energy generation of their own. With these energy generations, they can know about the actual quantity of energy generated and many conservative methods of energy savings. They will mostly use hybrid connectivity in this area of generation. The energy can be utilized as per their requirement, and they can also avail of the option of trading the electricity with

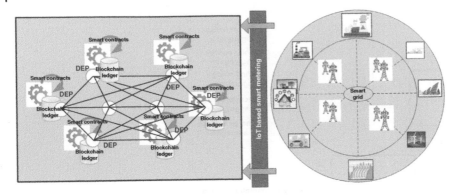

Figure 17.5 Blockchain-based distributed management, control, and validation of DR programmes. Source: Djvstock/Adobe stock; Thruer/Adobe stock; Happypictures/Adobe Stock; Ylivdesign/Adobe Stock.

the other consumers, which paves the way for the new business model, and the excess energy can be stored in the battery. The trading can also directly participate with the respective consumers without the retailing agent, with the price that needs to be sold as per the benefits, which brings satisfaction to both traders. Many traders benefit from Blockchain technology in terms of trading strategy. Transactions can be performed within the network by the enablement of protection. In this way, the demand, supply, actual generation, and price of the unit can be discussed within the network. As this technology became more popular, many countries tried to adopt it. However, the idea of peer-to-peer energy trading is still being researched and improved [38].

Figure 17.5 shows how Blockchain-based distributed management, control, and validation of demand response (DR) programs in low/medium voltage smart grids work. These programs work in a decentralized and high-reliability way by using energy flexibility transactions that can be tracked and can't be changed and by validating DR in almost real-time. In this approach, the grid is represented as a graph of peer nodes that can communicate with one another using a blockchain-based architecture to provide fully decentralized energy demand and generation matching while maintaining grid stability.

The smart grid is responsible for the construction and management of a distributed ledger based on Blockchain technology. Each "distributed energy prosumer" (DEP) has IoT-based energy metering devices and registers that keep track of the data in blocks about how much energy is being produced or used. So, a DEP is shown as a node in the peer-to-peer distributed energy network. It can keep a copy of the ledger that is automatically updated whenever new energy data is recorded.

When it comes to DEP participation in DR initiatives, one of the most significant barriers is data security and privacy, but the distributed ledger technology in use here provides a unique solution to these issues. There is still research to be done on how to record energy transactions in a way that can't be changed in the centralized approaches, even though a great deal of effort has been invested into ensuring that smart meters protect consumers' personal information.

Decentralization and autonomy can be achieved through demand-side management for smart energy networks as described in [39]. In this architecture, the Blockchain is used to create a decentralized, secure, and automated energy network, which allows all of its nodes to operate independently without the need for centralized supervision and control by a distributed system operator. As an additional benefit of using it, it may be used to store energy consumption information in a secure and tamper-proof manner in the blocks that are acquired from smart meters. The smart contract shown in Figure 17.6 provides decentralized control, calculates rewards or penalties, validates demand response agreements, and applies rules for balancing energy demand and output on the power grid. Finally, this concept is confirmed and evaluated by creating an Ethereum Blockchain prototype using energy use and production data from UK buildings. The findings show that this model is capable of adjusting demand in near real-time by executing energy flexibility levels as well as validating all demand response agreements. But it's

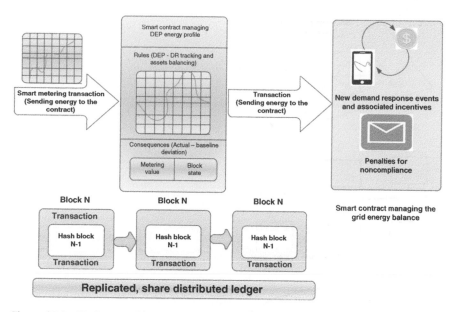

Figure 17.6 Six layer architecture of smart contracting for energy applications.

not clear how the energy profile's privacy has been protected on this public blockchain. The identity of a user can be deduced from publicly available transaction data.

17.3.5 Blockchain for Transactive Energy

Transactive energy helps in the flow and control of power within the power system network and provides communication for the energy transmission over all the grids for economic production and distribution management. Unlike traditional networks, this transactive energy system will interact with all levels of the energy network, and it is critical to building the transitive energy Blockchain network to make all financial transactions during energy trading visible to all participants [40]. The Micro-source Controller (MC) can be used to set up this system in a microgrid. The MC is the central point of communication, neutralizes demand, manages the generation and load-side sources, and helps make decisions during exchange operations. On the other hand, Distributed Signal Operator (DSO) acts as a localized energy trader and communicates (sub-level of Micro-source Controller) with the local transactive energy network for the current market situation of trading energy. Given the level of complexity in trading, the question of access and protection of microgrid from malicious attacks arises, because otherwise it would lead to misleading network system activities and, ultimately, change market situations. Information exposure in the network should have some limitations, i.e., critical information regarding the generation and operation should be avoided by providing information about the generation capacity and expense to regulate the normal operation of the entire transactive energy system [41].

To increase visibility and trustable factors in the decentralized network, here is the design for the transactive energy system approach shown in Figure 17.7.

As per the approach, the system is constructed with three functional layers. The top of the stack, known as the market layer, constitutes the financial network agents for the trading prospective for peer–peer transactions. The bottom of the stack, known as the Physical Layer, which constitutes the Distributed Energy Network, links the microgrid network over all the areas of generation, transmission, distribution, and storage. The layer that lies between the market and the physical layer is known as the "Cyber Layer. It represents the point of communication for the two layers: the defensive layer for data protection, decision-making computing capability, and automated

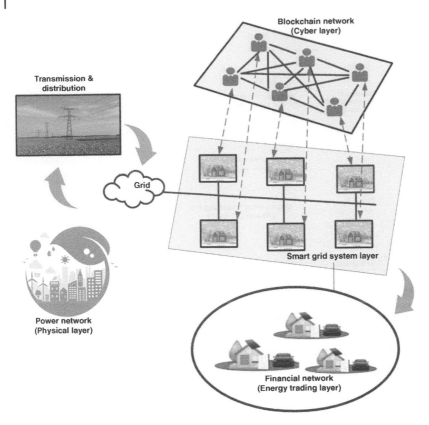

Figure 17.7 Design of transactive energy systems. Source: Luctheo/Pixabay; Barks/Adobe Stock.

control and flow process. The Cyber Layer plays a vital role in the network as they are the ones that control the entire electrical and market transactions by collecting real-time information for processing the decisions of the physical layer and energy trading in the market layer. The Cyber Layer investigates the simultaneous operation of analyzing the actual generation and distribution in the physical layer, thereby processing the energy trading across the nodes of the agents. In the Cyber Layer of the Blockchain network, the Micro-source Controller and Distributed Signal Operator perform the controlling operations and enable secure data storage for the sustainable network for validation of nodes between the Market Layer and the Physical Layer. For efficient operation at the cyber layer, there should be a requirement for high data storage capability. Due to the increase in complexity level of transactive energy storage systems, in order to enhance the protection of data, brings the scope of cybersecurity into the blockchain network, which increases the width of the protection layer and thereby increases the security and trustability factor among the participants in the network [42].

17.4 Blockchain for Smart Contracts

A Smart Contract is nothing but an executable program that is created using different programming languages such as C++, JavaScript, Python, Go, Solidity, PHP, C#, Simplicity, Ruby, Rust, SQL, Erlang, Rholang, and CX. The question that arises is normal regarding the relationship between the Blockchain and smart contracts [43]. By creating executable code, transactions such as releasing funds, registering products, and updating notifications in the Blockchain can be done with an automated flow [44]. The code will trigger the process when specific conditions are met [45]. This includes the advantages of low processing fees without the intervention of a third party

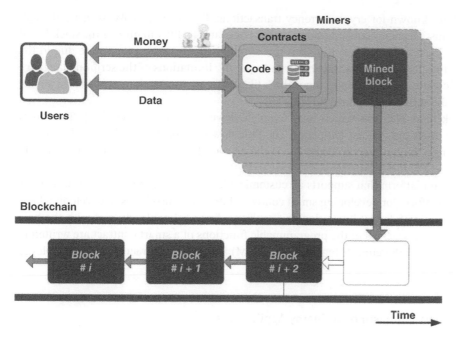

Figure 17.8 Working on smart contract system.

and minimal processing time compared to traditional transactions. Initially, the ideology of smart contracts was introduced by Szabo in 1994, but at that time it was just an idea [46]. These smart contracts were supposed to be much needed after the boom of blockchain technology. But this smart contract provides the hand for the transactions on the blockchain. It depends upon the business process, and legal and market capability. The entire working model of smart contracts is illustrated in Figure 17.8 and will be followed in further discussions.

The Smart Contract consists of features like account balance viewing, executable programs, and a storage facility. Miners in the blockchain include the Smart Contract and Miner Block, where the entire validation process takes place before the transaction node is added to the chain network. As per the working process concerns, the users first initiate the request for the transactions and then it will first process the smart contracts with the passing of all the conditions and consensus mechanism from the miner block. Then it proceeds further to add the node to the blockchain. The smart contract will be automatically upgraded after each transaction. Information visibility will be passed over to all the nodes in the network. The contracts can maintain the accounts by sending or receiving money from the users and then creating a new contract for any further transactions. The brighter side of the smart contract includes speed processing, maintaining accurate data, transparency, less time consumption, and, finally, defensive protection. Generally, smart contracts will be categorized into two different types [47]. They are deterministic and non-deterministic smart contracts. The deterministic contract doesn't require any external information for the processing of data. But a non-deterministic contract needed an external party for the information to process the transactions. An Oracle database or other data sources are examples of sources of information that come from outside the company.

17.4.1 The Platform for Smart Contracts

Various blockchain platforms have their smart contracts where they are developed and employed as per the needs of the respective blockchain. Some of the blockchains require high-level programing languages for the deployment of smart contracts; they are designed as per the requirements. In this section, some public platforms are discussed.

Bitcoin is a popular blockchain known for cryptocurrency transactions. Bitcoin uses PoW, so it needs high computing power as well as exhaustive energy consumption [26, 48]. For example, Bitcoin utilizes the stack-based scripting language with simple logic for processes. But with this simple logic, it takes a long time due to the involvement of multiple authorities for a single transaction [28]. As per the limitations of the scripting language side [45, 49], the possibility of creating some complex algorithms like loops for transactions tends to be a risky one as per the limitations from the scripting language side [45, 49].

NXT is another public blockchain platform which uses template-driven smart contracts [50]. The smart contracts were developed using those pre-defined templates. The little downside of this platform is that it does not allow any customizable changes in the smart contract due to the non-availability of Turing Completeness in the scripting languages.

Ethereum is another blockchain platform that supports all customizable and advanced smart contracts and is intended to serve as a common platform for developing smart contracts [49, 51]. It also helps in developing logic with a higher complexity level like withdrawal limits, looping concepts, financial contracts, etc. with the help of the Turing-Complete scripting language. All of the programmable functions of a smart contract are written in stack-based byte code and tested in the Ethereum Virtual Machine (EVM). The smart contract can also be deployed in several high-level languages.

17.4.2 The Architecture of Smart Contracting for Energy Applications

A multi-layer architecture is proposed to describe the flow of information that begins with agent input. As seen in Figure 17.9, smart contract operations include six different layers of information transmission that must be passed through before the information can be acted on by the physical assets. The six layers identified are as follows:

- Input from agents, devices, and the grid
- Algorithms for energy management
- Native smart contracting capabilities
- Blockchain applications
- Computational methods and procedures
- Communication layer

Each layer and its functions are described in Figure 17.9. Data from an agent, device, and/or the grid is required for Layer 1. Bids and offers from peer-to-peer energy trading agents, available signals from a smart charging electric vehicle, and grid voltage levels to initiate an automatic demand-side action are examples.

This data is then given to energy management algorithms built by energy researchers. This layer typically involves optimization approaches as per the literature. For example, an enhanced efficient settlement algorithm might be used to rectify the imbalance between delivered and contracted energy. This layer would handle any intelligent decision, such as negotiations, control schemes, and so on. These calculations could be done outside of the smart contract to save money on extra computing costs.

The third layer involves contract programming, which is typically done in a standalone smart contract language. Different languages are used to code Layers 2 and 3 in several ways. As a result, they are represented as distinct layers. This layer is where agents and devices are registered, as well as any type of financial transaction and gas usage computation. An outcome is a digital contract consisting of code, which is the final product.

Integrating the smart contract onto a blockchain block is Layer 4. This brings together the things that need to be checked, encrypted, etc. A well-known example is the Proof of Work algorithm that is utilized for Bitcoin.

Layer 5 is where implementation and computation happen, and it includes interacting with virtual machines like the EVM.

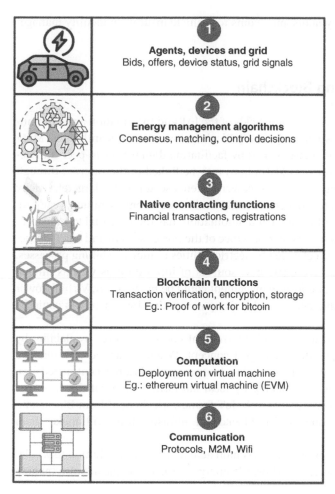

Figure 17.9 Six layer architecture of smart contracting for energy applications. Source: [52]/with permission of Elsevier; photo3idea/Adobe Stock; bsd studio/Adobe Stock; sripfoto/Adobe Stock; Anton Shaparenko/Adobe Stock.

Finally, the data is transmitted through communication protocols. This could include cable and/or wireless machine-to-machine (M2M) connectivity. For example, the smart contract might instruct the smart meter to submit data to a piece of software.

17.4.3 Smart Contract Applications

In real-world scenarios, there are numerous applications in which smart contracts play a major role in blockchain platforms. The notable ones are

Internet of Things [53]: The IoT is known for the exchanging of information among different things over the internet. The function of a smart contract in this case is to exchange the information with defined security access.

Management of Music Rights [54]: In this scenario, the smart contract will first register the owner of the music and then proceed with the transactions happening once the music goes up on the commercial markets.

Aside from the aforementioned use cases, several others dominate the global market, such as E-Commerce for cryptocurrency transactions between traders and product sellers via e-commerce applications; E-Voting ensures the accuracy of polled votes and verifies the identity to avoid duplicate castings; motor insurance to enable notifications for validation period and tracking of payments and claims for vehicle insurance; identity management; and digital R&D.

17.5 Digitize and Decentralization Using Blockchain

Potential blockchain solutions can also be found in the developing fields of the IoT and Information and Communication Technology (ICT) development, including in smart homes [17]. Blockchains can significantly assist M2M data transfer and communication between connected smart devices by facilitating digital peer-to-peer transactions. By 2020 [55], an increasing number of smart devices (20.8 billion) may be linked to the Internet through connectivity. ICT and Smart Meters are rapidly being used in the power systems sector. For a large electricity provider, the number of smart meter readings is estimated to increase to 220 million/day from 24 million/year [56]. This approach, when integrated with the capabilities of big data and automation, has the potential to transform the energy value chain. Data-driven insights can enhance the performance of the power system and asset inspections, resulting in cost savings. Digitalization provides a chance for electric utilities to increase billing processes, network efficiency, and supply chain efficiency and investigate new sources of innovation as well as unique business models [55]. Data-driven demand collection and load management could be made better, Virtual Power Plants could be made easier, and the use of renewable energy and active participation by consumers could be increased [56].

Smart connected devices can manage and respond to cost and intermittent energy resource availability by automatically regulating their electricity usage as per the smart grid concept. When expanded to a massive number of sensors sending enormous amounts of data at a specific frequency, traditional centralized systems become inefficient. Distributed control and decision-making by local authorities help cut down on the amount of computing power needed to run future power plants well. IoT frameworks benefit from Blockchain technology, and an open-source, accessible, and collaborative Blockchain infrastructure will make them compatible [17].

Decentralized energy trade has drawn the greatest amount of blockchain activity to date. In addition to wholesale energy trading systems that give end users full access to wholesale energy markets, people are also making peer-to-peer energy trading systems.

Blockchain can cut transaction costs in wholesale energy trading while also offering open, accessible information that can be accessed by a variety of stakeholders, especially agencies that can ensure compliance requirements. At the same time, blockchain could cut out the intermediaries, lowering trading costs, and perhaps expanding trading volumes, allowing small-scale customers to engage in the electricity sector [17]. The speed and scale of transactions that a blockchain network can support are constraints in this field. Moreover, a key concern is an open access to all competitors of trade-sensitive documents. Platforms that give end-users access to energy markets can help the grid unlock new flexibility services. Furthermore, such programs can raise consumer knowledge and provide them with more options when it comes to their energy supply, thereby leading to speedier switching and increased competition. Another problem in this field is that it can be hard to meet all of the regulatory and legal requirements.

17.6 Challenges in Implementing Blockchain Techniques

As per the inference gained from all the things discussed above regarding the construction and operation of blockchain in the electrical sector, it is not a simple one that can be easily embedded into the electrical network. It takes a large number of professional and innovative technicians to integrate the technology into a real-world scenario. From a technical point of view, the implementation process needs to take into account scalability, the accuracy of the information, transaction costs, an algorithm for validating transactions, and security. Since security is a major concern, there were many loopholes where attackers could easily get into the spot and crack it. Another issue to consider is user flexibility and privacy in network data administration [42].

On the other side of the electrical network, they were centralized on one network, and due to the network's high complexity, it is difficult to trade independently. One of the important discussions going around the globe is energy consumption. The volume of energy consumed by the blockchain network is quite high. This may result in it increasing exponentially over the upcoming years. As this technology has been booming in recent years, there should be research around the technology to optimize its consumption and come up with a new solution for its successful implementation. Though it has a lot of advantages, one major downside makes the entire technology a downfall. In a recent case study from the Life Cycle Assessment Expert Anders S.G. Andrae, it was documented in 2015 that ICT is responsible for 1757 TWh of world energy consumption [57]. In the worst case scenario, it could quadruple in the next 10 years. As the public becomes more aware of blockchain, it is vital to conduct the necessary research in order to optimize it and make it more efficient.

Blockchain technology is in its early stages of development, and various technological gaps might limit its use in smart energy systems. This section will discuss it.

17.6.1 Network Management

It is necessary to use appropriate protocols and algorithms in order to manage a blockchain network effectively. All these protocols are essential for forwarding transactions, spreading data, discovering nodes, and keeping track of bad ones. This means that the performance of these protocols directly influences overall delay, bandwidth, and the rate at which transactions are processed. In this context, it is necessary to design network management protocols that are security-aware, delay-aware, scalable, and privacy-aware in order to integrate Blockchain technology into smart energy systems in the future. Furthermore, protocols must be able to accommodate a wide range of compromises in order to meet the delay and bandwidth needs of smart energy applications, which can be achieved through the use of adjustable variables.

17.6.2 Data Management

It is more difficult to perform off-chain data management, which is often required for resource-constrained nodes in smart energy systems. This is because it requires the synchronization and availability of conventional databases, which is difficult to achieve in most cases. Determining the right quantity of data to keep on-chain and off-chain for diverse purposes is critical. It's also difficult to store off-chain data securely. Additionally, data formats and database structures can change between enterprises or applications. Additionally, novel ways for dealing with numerous types of database schema, query processing, and data models on the blockchain are required to be developed.

17.6.3 Consensus Management

The quickest consensus-management technique is the Proof-of-Authority (PoA) algorithm. However, some applications have an extremely strict delay and bandwidth requirements, which PoA may not be able to meet. Consensus-management strategies for smart energy applications need to be improved. For example, the use of implicit consensus, which is suggested in [16], could be examined.

17.6.4 Identity Management

Due to Know Your Customer (KYC) regulations imposed by regulators, a decentralized, trustworthy identification method must be employed in various smart energy applications. When compared to a more private self-sovereign identity-management system, this has fewer advantages. In some smart energy systems, it can also be hard to get back stolen identities, especially for nodes with important or private data.

17.6.5 Automation Management

In order to prevent hacking or invocation of smart contracts under various situations other than those intended by the original programmer, smart contracts must be written and secured properly. If a smart contract is not properly written and secured, it may be hacked or invoked under circumstances that are not consistent with the original programer's intentions. The handling of non-deterministic smart contracts poses an even greater security risk. Software and templates are needed for the establishment of secure and well-written contracts in the context of smart energy applications involving vital data and industrial infrastructure. Smart contract execution frequently necessitates sequential processing, which might cause transaction verification to be delayed. In order to meet the high-performance needs of different applications, the right parallel processing sharing strategies must be developed.

17.6.6 Lack of Suitable Implementation Platforms

A lot of prominent blockchain platforms aren't modular and don't have smart energy technologies embedded. Among the shortcomings of the platforms include their lack of support for off-chain data management and non-deterministic smart contracts, both of which are typically necessary for nodes with limited resources. In order to support a wide range of smart energy applications, the development of open-source and flexible blockchain platforms with appropriate embedded technologies is essential.

17.7 Solutions and Future Scope

The development of blockchain-based energy trade is still being developed. The development of Blockchain technology 3.0 could result from the advancement of new technologies such as Artificial Intelligence, Data Science, Big Data analytics, and IoT.

Several peer-to-peer energy trading companies, like Verv, Greeneum, and Drift [58–60], have combined Blockchain with modern tools like machine learning and artificial intelligence. These technological advancements are being researched in order to ensure functionality and security. For instance, collected sensor data stored on a Blockchain ledger may be used for analytics by artificial intelligence. Big data analytics may increase data security and reliability, and machine learning may be used to identify abnormal and fraudulent actions in Blockchain technology.

Smart contracts typically have a very basic purpose. They're rule-based applications that can do a limited sequence of tasks using pre-defined logic. Smart contracts may be able to self-adapt and work in a semi-autonomous manner with the fusion of artificial intelligence and machine learning, considerably broadening the range of their utility. As a result, smart contracts, including machine learning and artificial intelligence, might enable node-to-node discussions on the tokenized value of a unit of power, as well as node-to-node cooperation to optimize household energy use. In P2P energy trading, there is an interesting synergy between new technologies and blockchain that needs more investigation.

Future studies, on the other hand, may go deeper into the technical challenges surrounding blockchain technology. The RENeW Nexus pilot project used a dynamic pricing mechanism [61]. This concept has ample space for more research, such as looking at the usage of a dynamic pricing model for peer-to-peer energy trading.

17.8   Application of Blockchain for Flexible Services

Due to its role-play with data and data protection, blockchain can change the concept of entirely different phenomena. According to the various Blockchain concerns expressed, the integration of Blockchain technology with the smart grid paves the way for the integration of renewable resources, energy storage devices, and e-vehicles [62, 63]. The various Blockchain applications associated with the power system are as follows [64]:

(a) **Power Generation:** From the perspective of generating units produced, Blockchain helps the traders to understand the actual generation and operating status of the power grid in order to increase the profit across the market by deciding the trading as per the generation.
(b) **Power Transmission and Distribution:** Based on the idea of decentralization, Blockchain makes it possible to control and protect the grid with the help of a Micro-Source Controller and a Digital Signal Operator. This solves the problem that traditional power systems have.
(c) **Power Consumption:** Blockchain is helpful in the case of power consumption too by maintaining and encouraging the trading performed by the prosumers to increase the profit zone from both the seller and buyer sides. By the way, it also manages the energy storage systems.
(d) **Trading in Energy:** Hahn et al. [65] experiment with intermediary-free blockchain trading, which is profit-making and benefits the traders. Luo et al. [66] employed the concept of a multi-layer blockchain trading concept wherein one layer consists of different market agents and the other layer is a blockchain-based platform for the transactions [67]. This will enhance the market for the integration of renewable energy resources.
(e) **Electric Vehicles (EVs):** In today's scenario, electric vehicle use is slowly rising, and people are adopting to using electric vehicles [68]. To increase the use of e-vehicles and raise public awareness about the use of green energy. Governmental authorities need to facilitate the maximum number of e-vehicle charging stations, which are mostly connected to the smart grid. Due to the connection across the grid, the grid is facing enormous fluctuation, which leads to instability over the entire distribution network [69, 70]. To maintain stability, the blockchain comes into the picture to recognize the consumers' the best nearest charging stations that can deliver the supply without any fluctuations, at an affordable price, and with the most secure functionalities [71, 72].
(f) **Microgrid Management:** The operation of distributed energy resources is entirely dependent on microgrid functionality because all demand-side control is performed here [73, 74]. So, Blockchain is employed for the Distributed Energy Resources scheduling mechanism with the help of highly computed ability and supervision of data in a secured manner. Munsing et al. [75] experimented with the system for optimizing distributed energy resources in decentralized optimal power flow. With the help of the approach, they can reduce the peak-to-average ratio of load on the demand side [76]. A new set of discussions on the decentralization of medium-voltage direct current link control through blockchain makes the grid operators so viable across the energy system network [77].

17.9   Conclusion

Blockchain technology has acquired significant attention in recent years, primarily due to the growing business of digital currencies. The numerous and distinct characteristics of Blockchain technology have prompted numerous researchers to start investing in it. At the moment, the view of Blockchain as a technology is far too positive, and the promises for the technology's value are far too exorbitant, much like the perspective and predictions at the inception of the internet. As with other major industries, the electric grid is using an impressive amount of blockchain technology. This chapter reviewed the most recent applications and utilization of Blockchain technology in the smart grid sector, both from a research and industrial standpoint. Additionally, it showcased the numerous benefits of Blockchain technology in the electrical power system, as well as some frameworks for the smart grid. However, considerable hurdles must be overcome in future years to successfully incorporate the Blockchain into the power grid.

References

1 Liserre, M., Sauter, T., and Hung, J.Y. (2010). Future energy systems: Integrating renewable energy sources into the smart power grid through industrial electronics. *IEEE Industrial Electronics Magazine* 4 (1): 18–37.

2 Farhangi, H. (2014). A road map to integration: perspectives on smart grid development. *IEEE Power and Energy Magazine* 12 (3): 52–66.

3 Strasser, T., Siano, P., and Ding, Y. (2018). Methods and systems for a smart energy city. *IEEE Transactions on Industrial Electronics* 66 (2): 1363–1367.

4 Hafez, O. and Bhattacharya, K. (2018). Integrating EV charging stations as smart loads for demand response provisions in distribution systems. *IEEE Transactions on Smart Grid* 9 (2): 1096–1106.

5 Sousa, T., Soares, T., Pinson, P. et al. (2019). Peer-to-peer and community-based markets: a comprehensive review. *Renewable and Sustainable Energy Reviews* 104: 367–378.

6 Tushar, W., Yuen, C., Mohsenian-Rad, H. et al. (2018). Transforming energy networks via peer to peer energy trading: potential of game theoretic approaches. *IEEE Signal Processing Magazine* 35: 90–111.

7 Hustveit, M., Frogner, J.S., and Fleten, S.-E. (2017). Tradable green certificates for renewable support: the role of expectations and uncertainty. *Energy* 141: 1717–1727.

8 Haider, H.T., See, O.H., and Elmenreich, W. (2016). A review of residential demand response of smart grid. *Renewable and Sustainable Energy Reviews* 59: 166–178.

9 Ma, K., Yu, Y., Yang, B., and Yang, J. (2019). Demand-side energy management considering price oscillations for residential building heating and ventilation systems. *IEEE Transactions on Industrial Informatics* 15 (8): 4742–4752.

10 Tschorsch, F. and Scheuermann, B. (2016). Bitcoin and beyond: a technical survey on decentralized digital currencies. *IEEE Communication Surveys and Tutorials* 18 (3): 2084–2123.

11 Narayanan, A., Bonneau, J., Felten, E. et al. (2016). *Bitcoin and Cryptocurrency Technologies: A Comprehensive Introduction*. Princeton University Press.

12 Vukolić, M. (2017). Rethinking permissioned blockchains. In: *ACM Workshop on BlockChain, Cryptocurrencies and Contracts*, 3–7.

13 Xu, X., Weber, I., Staples, M. et al. (2017). A taxonomy of blockchain-based systems for architecture design. In: *IEEE International Conference on Software Architecture (ICSA)*, 243–252.

14 Dinh, T.T.A., Liu, R., Zhang, M. et al. (2018). Untangling blockchain: a data processing view of blockchain systems. *IEEE Transactions on Knowledge and Data Engineering* 30 (7): 1366–1385.

15 Dunphy, P. and Petitcolas, F.A. (2018). A first look at identity management schemes on the blockchain. *IEEE Security and Privacy* 16 (4): 20–29.

16 Hassan, N.U., Yuen, C., and Niyato, D. (2019). Blockchain technologies for smart energy systems: fundamentals, challenges, and solutions. *IEEE Industrial Electronics Magazine* 13 (4): 106–118.

17 Andoni, M., Robu, V., Flynn, D. et al. (2019). Blockchain technology in the energy sector: a systematic review of challenges and opportunities. *Renewable and Sustainable Energy Reviews* 100: 143–174.

18 Mattila, J., Seppälä, T., Naucler, C., Stahl, R., Tikkanen, M., Bådenlid, A., and Seppälä, J. (2016). Industrial blockchain platforms: an exercise in use case development in the energy industry (No. 43). ETLA Working Papers.

19 Bronski, P., Creyts, J., Crowdis, M. et al. (2015). *The Economics of Load Defection: how Grid-Connected Solar-Plus-Battery Systems Will Compete with Traditional Electric Service–Why it Matters, and Possible Paths Forward*. Rocky Mountain Institute (RMI), Homer Energy.

20 Corea, F. (2017). The convergence of AI and Blockchain: what's the deal. Retrieved October, 6, 2020.

21 Hasse, F., von Perfall, A., Hillebrand, T. et al. (2016). Blockchain–an opportunity for energy producers and consumers. *PwC Global Power & Utilities* 1–45.

22 Ioannidou, M. (2018). Effective paths for consumer empowerment and protection in retail energy markets. *Journal of Consumer Policy* 41 (2): 135–157.

23 Smith, M. and Ton, D. (2013). Key connections: The us department of energy? s microgrid initiative. *IEEE Power and Energy Magazine* 11 (4): 22–27.

24 Bamberger, Y., Baptista, J., Belmans, R., and Buchholz, B.M. (2006). Vision and strategy for europe's electricity networks of the future. In: *European technology Platform SmartGrids* (ed. M. Chebbo, J.L.D.V. Doblado and M. Wasiluk-Hassa).

25 Competition and Markets Authority (2016). Energy Market Investigation: Summary of Final Report.

26 Stifter, N., Judmayer, A., and Weippl, E. (2019). Revisiting practical byzantine fault tolerance through blockchain technologies. In: *Security and Quality in Cyber-Physical Systems Engineering*, 471–495. Cham: Springer.

27 Nakamoto, S. (2008). Bitcoin: a peer-to-peer electronic cash system. *Decentralized Business Review* 21260.

28 Babbitt, D. and Dietz, J. (2014). Crypto-economic design: a proposed agent-based modeling effort. In: *English. Conference Talk*. Notre Dame, USA: University of Notre Dame.

29 Triantafyllidis, N.P. and van Deventer, T.O. (2016). *Developing an Ethereum Blockchain Application*, 2015–2016. University of Amsterdam System and Network Engineering, MSc.

30 Peyrott, S. (2017). *An Introduction to Ethereum and Smart Contracts*. Bellevue, Washington: Auth0 Inc.

31 Musleh, A.S., Yao, G., and Muyeen, S.M. (2019). Blockchain applications in smart grid–review and frameworks. *IEEE Access* 7: 86746–86757.

32 Mylrea, M. and Gourisetti, S.N.G. (2017). Blockchain: A path to grid modernization and cyber resiliency. In: *Proc. North Amer. Power Symp. (NAPS)*, 1–5.

33 Kakavand, H., Kost De Sevres, N., and Chilton, B. (2017). The Blockchain revolution: An analysis of regulation and technology related to distributed ledger technologies. Available at SSRN 2849251.

34 Tushar, W., Saha, T.K., Yuen, C. et al. (2020). Peer-to-peer trading in electricity networks: an overview. *IEEE Transactions on Smart Grid* 11 (4): 3185–3200.

35 Felder, F.A. and Athawale, R. (2014). The life and death of the utility death spiral. *The Electricity Journal* 27 (6): 9 16.

36 Guerrero, J., Chapman, A.C., and Verbič, G. (2019). Local energy markets in LV networks: community based and decentralized P2P approaches. In: *2019 IEEE Milan PowerTech*, 1–6. IEEE.

37 Azim, M.I., Tushar, W., and Saha, T.K. (2020). Investigating the impact of P2P trading on power losses in grid-connected networks with prosumers. *Applied Energy* 263: 114687.

38 Mylrea, M. and Gourisetti, S.N.G. (2017). Blockchain for smart grid resilience: Exchanging distributed energy at speed, scale and security. In: *In 2017 Resilience Week (RWS)*, 18–23. IEEE.

39 Pop, C., Cioara, T., Antal, M. et al. (2018). Blockchain based decentralized management of demand response programs in smart energy grids. *Sensors* 18 (1): 162.

40 Li, Z., Bahramirad, S., Paaso, A. et al. (2019). Blockchain for decentralized transactive energy management system in networked microgrids. *The Electricity Journal* 32 (4): 58–72.

41 2019. *LO3 Energy: The Future of Energy, Blockchain, Transactive Grids, Microgrids, Energy Trading*, https://lo3energy.com/.

42 Di Silvestre, M.L., Gallo, P., Guerrero, J.M. et al. (2020). Blockchain for power systems: current trends and future applications. *Renewable and Sustainable Energy Reviews* 119: 109585.

43 Alharby, M. and Van Moorsel, A. (2017). Blockchain-based smart contracts: A systematic mapping study. *arXiv* preprint arXiv: 1710.06372.

44 Stark, J. (2018). Making sense of ethereum's layer 2 scaling solutions: state channels, plasma, and truebit. https://medium.com/l4-media/making-sense-of-ethereums-layer-2-scaling-solutions-state-channels-plasma-and-truebit-22cb40dcc2f4 (accessed 23 July 2018).

45 Buterin, V. (2014). A next-generation smart contract and decentralized application platform. white paper, 3.

46 Szabo, N. (1997). Formalizing and securing relationships on public networks. First monday.

47 Morabito, V. (2017). Smart contracts and licensing. In: *Business Innovation Through Blockchain*, 101–124. Cham: Springer.

48 Memon, M., Hussain, S.S., Bajwa, U.A., and Ikhlas, A. (2018). Blockchain beyond bitcoin: blockchain technology challenges and real-world applications. In: *In 2018 International Conference on Computing, Electronics & Communications Engineering (iCCECE)*, 29–34. IEEE.

49 Dos Santos, R.P. (2017). On the philosophy of Bitcoin/Blockchain technology: is it a chaotic, complex system? *Metaphilosophy* 48 (5): 620–633.

50 Lewis, A. (2016). A gentle introduction to smart contracts. *Bits on Blocks* 1.

51 Wood, G. (2014). Ethereum: a secure decentralised generalised transaction ledger. *Ethereum Project Yellow Paper* 151 (2014): 1–32.

52 Kirli, D., Couraud, B., Robu, V. et al. (2022). Smart contracts in energy systems: a systematic review of fundamental approaches and implementations. *Renewable and Sustainable Energy Reviews* 158: 112013.

53 Christidis, K. and Devetsikiotis, M. (2016). Blockchains and smart contracts for the internet of things. *IEEE Access* 4: 2292–2303.

54 Egbertsen, W., Hardeman, G., van den Hoven, M. et al. (2016). Replacing paper contracts with Ethereum smart contracts. *Semantic Scholar* 35: 1–35.

55 Burger, C., Kuhlmann, A., Richard, P., & Weinmann, J. (2016). Blockchain in the energy transition. A survey among decision-makers in the German energy industry. *DENA German Energy Agency*, 60.

56 Sagiroglu, S., Terzi, R., Canbay, Y., and Colak, I. (2016). Big data issues in smart grid systems. In: *In 2016 IEEE International Conference on Renewable Energy Research and Applications (ICRERA)*, 1007–1012. IEEE.

57 Andrae, A. (2017). Total consumer power consumption forecast. *Nordic Digital Business Summit* 10: 69.

58 Verv's AI-Based Energy Insights & Predictive Maintenance Solution for Air Conditioning, Heat Pumps and Home/Office Appliances, https://verv.energy/.

59 Greeneum is a decentralized marketplace for management and trading of clean and sustainable energy, https://www.greeneum.net/.

60 Drift Trader, formerly Drift Marketplace provides a real-time green energy trading platform that provides businesses and consumers an economic option of trading clean energy directly from the generating units, https://www.drifttrader.com/.

61 Wongthongtham, P., Marrable, D., Abu-Salih, B. et al. (2021). Blockchain-enabled Peer-to-Peer energy trading. *Computers and Electrical Engineering* 94: 107299.

62 Ipakchi, A. and Albuyeh, F. (2009). Grid of the future. *IEEE Power and Energy Magazine* 7 (2): 52–62.

63 Farhangi, H. (2010). The path of the smart grid. *IEEE Power and Energy Magazine* 8 (1): 18–28.

64 Berntzen, L., Meng, Q., Vesin, B., Johannessen, M. R., Brekke, T., and Laur, I. (2021). Blockchain for smart grid flexibility-handling settlements between the aggregator and prosumers.

65 Hahn, A., Singh, R., Liu, C.C., and Chen, S. (2017). Smart contract-based campus demonstration of decentralized transactive energy auctions. In: *2017 IEEE Power & Energy Society Innovative Smart Grid Technologies Conference (ISGT)*, 1–5. IEEE.

66 Luo, F., Dong, Z.Y., Liang, G. et al. (2018). A distributed electricity trading system in active distribution networks based on multi-agent coalition and blockchain. *IEEE Transactions on Power Systems* 34 (5): 4097–4108.

67 Pipattanasomporn, M., Kuzlu, M., and Rahman, S. (2018). A blockchain-based platform for exchange of solar energy: laboratory-scale implementation. In: *2018 International Conference and Utility Exhibition on Green Energy for Sustainable Development (ICUE)*, 1–9. IEEE.

68 Yilmaz, M. and Krein, P.T. (2012). Review of the impact of vehicle-to-grid technologies on distribution systems and utility interfaces. *IEEE Transactions on Power Electronics* 28 (12): 5673–5689.

69 Mukherjee, J.C. and Gupta, A. (2014). A review of charge scheduling of electric vehicles in smart grid. *IEEE Systems Journal* 9 (4): 1541–1553.

70 Liu, C., Chai, K.K., Zhang, X. et al. (2018). Adaptive blockchain-based electric vehicle participation scheme in smart grid platform. *IEEE Access* 6: 25657–25665.

71 Kang, J., Yu, R., Huang, X. et al. (2017). Enabling localized peer-to-peer electricity trading among plug-in hybrid electric vehicles using consortium blockchains. *IEEE Transactions on Industrial Informatics* 13 (6): 3154–3164.

72 Hou, Y., Chen, Y., Jiao, Y. et al. (2017). A resolution of sharing private charging piles based on smart contract. In: *2017 13Th International Conference on Natural Computation, Fuzzy Systems and Knowledge Discovery (Icnc-Fskd)*, 3004–3008. IEEE.

73 Samad, T. and Annaswamy, A.M. (2017). Controls for smart grids: architectures and applications. *Proceedings of the IEEE* 105 (11): 2244–2261.

74 Danzi, P., Angjelichinoski, M., Stefanović, Č., and Popovski, P. (2017). Distributed proportional-fairness control in microgrids via blockchain smart contracts. In: *2017 IEEE International Conference on Smart Grid Communications (SmartGridComm)*, 45–51. IEEE.

75 Münsing, E., Mather, J., and Moura, S. (2017). Blockchains for decentralized optimization of energy resources in microgrid networks. In: *2017 IEEE Conference on Control Technology and Applications (CCTA)*, 2164–2171. IEEE.

76 Noor, S., Yang, W., Guo, M. et al. (2018). Energy demand side management within micro-grid networks enhanced by blockchain. *Applied Energy* 228: 1385–1398.

77 Thomas, L., Zhou, Y., Long, C. et al. (2019). A general form of smart contract for decentralized energy systems management. *Nature Energy* 4 (2): 140–149.

18

Power and Energy Management in Smart Power Systems
Subrat Sahoo

Hitachi Energy Research, Vasteras, Sweden

18.1 Introduction

Smart power system is a relatively emerging concept that holds different meanings to people, communities, organizations, and institutions. Often, even the smartness it has to offer is widely misconstrued. Therefore, this chapter aims towards addressing this ambiguity and bring more clarity to its readers by setting up appropriate boundaries. Power and energy management are quite interchangeably addressed due to the slimline between their function to the end-user. However, in a broad sense, the power system comprises the whole network of infrastructure, its operation, and the underlying principles of making the electricity available at the walls of a domestic consumer or a factory floor of an industry, right from its production. The energy system, on the other hand, refers to the accessibility, usability, and functional assessment of the end product of the power systems, i.e. electricity. Energy being the time integral of power attaches the time duration to the power equation for its definition. Therefore, power must continue to flow and effectively gets consumed by an endpoint, if it has already been produced; whereas energy is a commodity that can be traded even before its production and can even hold negative value, depending on whether the power finds an end consumer or not. The energy can be conserved or stored from the produced power, with the help of several storage techniques. The definitions are relatively philosophical and therefore, the author hopes that readers will find the interchangeable usage of the terms quite acceptable.

Smart power systems by definition are expected to be futuristic, in terms of their operation and ability to seamlessly integrate different forms of energy production, consumption, and storage, based on a given set of criteria, driving economy, scalability, and sustainability of energy. It also embodies a lot of technological transformations that have happened elsewhere but finds its transition into the power and energy sector, by virtue of their respective appropriateness and usefulness in the power industry. These transformations are essential to understand and test these cross-domain technologies and platforms and their corresponding implications.

The last couple of years has seen a sudden change in some of the energy scenario dynamics that was neither anticipated nor planned for. Few contributors to these modified scenarios are discussed below. Notwithstanding the recent developments, the transformation in the power industry must go on, to keep pace with other trends driving the world, namely Industrial internet of things, digital transformation, ultrafast communication technology in wired and wireless mediums, and big data analytics. The power industry is, thus, a driver and an actor at the same time.

18.1.1 Geopolitical Situation

Following the Russian invasion of Ukraine [1], a lot has changed in the global development roadmap. Ukraine being a natural resource-rich, agriculture centric, industrial country, and geopolitically very relevant to Europe, the

Artificial Intelligence-based Smart Power Systems, First Edition.
Edited by Sanjeevikumar Padmanaban, Sivaraman Palanisamy, Sharmeela Chenniappan, and Jens Bo Holm-Nielsen.

war is slated to bring in mid- to long-term uncertainties not just for Ukraine but also for the whole of Europe, and the entire world for that matter; due to the latter's growing dependence on Oil and Gas (O&G) from Russia, as well as the agricultural imports from Ukraine [2]. As a repercussion, the price of energy and its ancillary services have already started shooting up. At the same time, the western world is now making efforts to minimize this dependency gradually, by phasing in the O&G embargo and becoming sustainable with renewable energy sources (RESs).

On the technology front, it is worthwhile to mention Ukraine has an installed base of around 54 GW of electricity, which is one of the largest in Europe, with a bulk of production coming from nuclear power. Following the recent war situation, it shredded the grid-tie with Russia on 24 of February 2022. The system operator Ukrenergo formally joined the continental Europe transmission system operator (TSO) – European network of transmission system operators of electricity (ENTSO-E) on the 16 of March 2022, after running locally, briefly after disconnecting from Russian grid [3]. This emergency request was handled by earnest efforts from both ends, although it was long-time due since 2017. This was a huge step for Ukrenergo on dissociating from Russian frequency regulation, but at the same time exposing the Ukrainian grid to transparency and the European wholesale market pricing competition [4]. Nonetheless, the cyber vulnerabilities this integration brings into ENTSO-E are yet to be thoroughly assessed and countered for, which remains a huge security risk for Europe.

18.1.2 Covid-19 Impacts

There was an 8% reduction in the global energy demand in the year 2020 [5], following the global pandemic and subsequent downturn that wiped out trillions of dollars of economy, following 6.2 million deaths [6] and counting, and lockdown of several energy intensive industries. This has completely regressed the global energy transition roadmaps due to a huge disruption in supply chain management. It also led to a large setback in technological development for the power electronics engineering for converter, valve systems, high voltage direct current (HVDC), power quality (PQ) management devices, silicon carbide (SiC) related developments, chip shortage across several verticals of industrial production, leading to perennial delay on the software interfaces, panels, database management systems, and the likes.

On the technology front, this unusual situation shifted the proportion of energy demands between industrial (decline) and residential (steady rise, due to stay-at-home almost globally applied, following multiple lockdown guidelines) establishments. The cyclic pattern of weekdays and weekend load profile seemed to vanish, requiring a steady energy demand (albeit several percent lower, due to reduced commercial operations) that had a very different energy mix, needed from the grid operators [7]. The consumer mix and changing weather patterns required even curtailment of bulk of RES to address the grid resilience in the modified scenario of reduced market demand. The day ahead pricing even went negative for countries like Germany [8] with O&G prices reaching a historic low during 2020 (US$ 12 a barrel for crude oil import, for example).

18.1.3 Climate Challenges

Reports suggest that the Arctic and Antarctic circles are experiencing global warming 60% faster than the regions farther from the north and south poles [9]. This in turn has resulted in a 1 °C rise in temperature just in the last decade. Whether the rise in temperature is totally attributed to the consistent Greenhouse gas (GHG) emission or not is still debatable, but the containment of GHG emissions can certainly alleviate the global climatic situation. The climate action force is creating several task forces and many developed countries are pushing the boundaries in their national decarbonization efforts, as well as making the technology and funds accessible, for other developing countries to help negate the global warming limits [10].

On the technology front, for power industries, it means more renewables getting integrated at various nodes, voltage levels, and geographical locations, of course where the access to these sources exists. However, the growing energy demand and systematic phasing out of fossil-based sources need these transitions rather earnestly, at the same time maintaining or augmenting the grid stability. This is primarily due to the increased volatility not only in the production but also in the consumption sectors, e.g. hundreds of kilowatts (kWs) of eV chargers, intermittently drawing power from the grid at any given time. These contributors result in reduced inertia, require large topological changes for the control and protection schemes and present enormous and serious PQ challenges in the system.

Owing to some of the above-mentioned reasons, there is a renewed gross domestic product (GDP) growth forecast by international monetary fund (IMF) [11] as of April 2022 that was further revised following the war-induced uncertainties and instabilities, including global supply chain disruption, refugee crisis, unrest, anxiety, and many unsettling issues. This is also creating impediments in the energy sector investments towards their decarbonization goals, casting a doubt on various commitments made by individual countries in conference of the parties (COP) 26, towards their carbon neutrality objectives.

This chapter aims to discuss a few resiliency topics which could technologically address some of the challenges outlined above.

18.2 Definition and Constituents of Smart Power Systems

The reason for building a smart power system is to handle the power in a more energy and technology efficient, reliable, versatile, and affordable manner. This has to be produced from clean sources and preferably from renewables so that the GHG emissions are contained- the energy sector is decarbonized. However, the progress in this domain is not encouraging enough, at least not globally nor in a scalable manner for various reasons.

Some tangible reasons are RES do not necessarily exist close to the consumers due to geographical as well as aesthetical constraints, such as in the proximity of mega cities or industries, where bulk of the power demand exists. Therefore, a large transmission corridor is needed to evacuate the power from the production site. This is a constraint that is almost globally present, without exception for the countries [12]. More importantly, the interconnection of various national grids across countries and even continents are becoming rather common [13, 14, 15], in order to facilitate bidirectional power flow and take advantage of mutual grid strengths.

Secondly, the level of intermittency of RES makes them less dependable, for a 24 hour and 365 days production cycle. This problem has to be addressed at varied complexity level, by mixing different energy sources, complementing the base load production from nonrenewable sources, addressing various congestions in the grid, and operating the grid sometimes, not only at maximum capacity, if desired, but also at overloading capacity, when needed, by doing a health trade-off of performing assets and their utilization factor. The stability and grid code compliance are further adhered to, by supplementing various control functions in voltage source converters (VSCs), namely, DC voltage control, active and reactive power control, AC voltage and current control, convert firing control, and tap changer control, etc. Momentary fluctuations for the frequency and reactive power compensation are handled by the grid level storage as well as series and shunt compensation devices, including static compensators (STATCOMs) [16].

On top of all these developments, the energy must be traded between the producers and the consumers and there are many intermediate traders, who take advantage of various predicted forecast in days, weeks, and months ahead and capitalize on the surplus or deficit offered on a given day and time. The rising speculation of energy production and demand gap makes the market much more interesting in recent times, where the excess productions even bid for negative pricing advantage consumers, failing which the surplus energy must be curtailed, than feeding to the grid. Therefore, beyond commercial interests, strategic technology advancement is imperative, which would pave way for storing or conserving this excess renewable energy which is otherwise wasted.

All these elements of operation essentially form the ingredients for the smart power system. This section briefly touches upon some aspects of these dynamics.

18.2.1 Applicable Industries

Electrification across industries and domestic consumption is gaining new momentum in recent times, with the forward trajectory observed, especially in the last decade. The sectors that counted on other forms of energy sources, namely coal, oil, gas, or heat are now turning to electricity because of energy efficiency improvements, and arguably the generation coming from RES. This is due to competing technology and drop in operating prices that led to levelized cost of electricity (LCoE) reaching parity or faring better, in favor of RES. This kind of sector coupling is finding renewed demands in some particular sectors, namely electromobility for all modes of transportations and data centers.

The European Union (EU) alone expects 30 million electric vehicles (EVs) in all forms of road transport, not just personal cars but also cargo and public transports by the year 2030 and the numbers are expected to shoot up to around 220 million by 2050. The numbers currently stand at around six million according to Gartner [17]. These are the vehicles originally replaced by their internal combustion engine (ICE) counterparts, basically switching the fuels, thus putting enormous pressure on the grid to keep those batteries charged, on demand and thus aggravating the already existing challenges of peak management within the grid, since these charging come into a network in the most asynchronous manner, without any prior schedule. The European development of springing EV is indeed global, with a massive surge of two and three wheeled EVs, swarming in the developing economies including India, Thailand, Vietnam, and China and also Latin America and most parts of Africa.

The quest for clean energy is taking precedence equally in the aviation industry [18] as well as the maritime industry, both for short- and long-distance ferries [19]. There are 170 electric propulsion aircraft projects running by 2020 with commercial aspiration expected in as early as 2025 [20].

While the source for all these three modes of transport is considered for electrification, there is consideration for alternative fuels too, such as hydrogen powered, which also contributes to green energy (when sourced via a green way). Notwithstanding the alternative or hybrid solutions, where a combination of battery- powered by electricity and alternative fuel (Hydrogen, Methanol, biofuels, etc.), the electrification is still going to drive the race. This would require smart charging solutions, to best utilize the available resources and their operations. It also brings in another important aspect of utilizing the eVs as a mobile energy storage device, thus contributing to the grid decongestion by operating in Vehicle-to-grid (V2G) or vehicle-to-vehicle (V2V) mode, offering a host of grid flexibility opportunity.

The challenge that remains at the moment is the preparedness of the grid to accept the influx of the charging infrastructure, which only for eVs range from 2.3 kW to a lot of fast chargers (FCs) ranging up to a whopping 350 kW [21] and there are also discussions of ultra-fast chargers (UFCs) that could go up to a MW each [22], which is equivalent of supplying electricity to a thousand households at a given time, i.e. the size of a village. So, this much of drainage of power from the grid is unsolicited, at least in the current mechanism, and thus requires a complete shift in infrastructure planning and accommodation of such charging infrastructure, well supported with the PQ measurements induced at the point of common coupling (PCC). A massive roll out of standardization and awareness are needed, as the technology and the sales of eVs continues to push.

On a similar note, the increased dependence on video streaming/conferencing services (such as Netflix, Amazon Prime, HBO, Apple TV, Zoom, Skype, Teams, and Google Meet), social media platforms (Facebook, Instagram, LinkedIn, Google), cloud services (Google, Amazon, and Microsoft) have given rise to a completely new segment of industrial consumer of electricity. Just to put it in perspective, it costs 7 kWh of electricity to handle, transfer and store 1 GB of data [23]. Hence, the environmental impact of storing data in cloud is not that promising yet, till the whole world has switched to 100% renewables. The bulk of this power is used for maintaining the servers (~43%) and cooling and power provisioning (~43%) [24].

In 2020 global data centers consumed c. 250 TWh of electricity, i.e. 1% of the global electricity demand. This figure excludes the energy consumption from crypto mining in that year, which stood at c. 100 TWh; an equivalent of production from ten large nuclear power plants [25]. This consumption is equivalent of the yearly electricity consumption of Argentina or Norway [26]. Approximately 2500 numbers of installed data centers existed in Europe by the year 2021 alone, as presented in [27]. There are many other drivers, in varying capacity operating in this space. Information and communication technology (ICT) companies operate majorly in this smart environment that pretty much applies to the power systems field who want to take giant steps on acting for price volatility, reduced environmental impact, energy efficiency measures, using artificial intelligence (AI)-based techniques, to bring in positive changes in this field. Though, data centers tend to be energy efficient over the years, with annual growth rate pegged at 50% year-over-year, they are slated to surpass other sectors to be the fifth largest energy consumer in the world [28]. For records, major sectors, by far, consuming the bulk of the electricity are – manufacturing, mining, construction, agriculture, followed by transportation [29]. There is a lot of political and technology push to make the data centers green, by their energy consumption, by effective RES integration in its operation.

There are many other applicable industries which are thriving under the pursuit of more electrification possibilities, even in a sector coupling setup. With global climatic condition going extreme, heating, ventilation, and air conditioning (HVAC) are much in demand; thus, efficient energy transition forcing its essential inclusion into the bulk electricity consumer space. Therefore, conscious, and urgent efforts are needed to make this transition possible and viable by preparing the grid to handle this influx.

18.2.2 Evolution of Power Electronics-Based Solutions

Power electronics has become a cornerstone for various application domains within the power industry because of its versatile usage in both AC and DC applications. Figure 18.1 offers a schematic of various portfolios of power systems actors and the role played by power electronics-based solutions across the entire spectrum of generation, transmission, and distribution as well as consumer platforms. Power electronics, in a broad sense, is defined from the ultra-high voltage range (in thousands of kVs) to 12 V DC of consumer electronics, such as TVs or personal computers or even cellphones. Especially with the advent of VSCs (that use Insulated Gate bipolar transistors (IGBTs) over line commutated converters (LCCs), those use thyristors), there are dramatic improvements in firing angle control, active and reactive power controls of up to 100%. Both the methods are still preferred over cost considerations, modernization challenges, though VSC has a far technical superiority over the LCC one.

HVDC technology finds more traction in recent times for nonsynchronous interconnections which has varying frequency and voltage magnitude possibilities, as well as reduced power losses over bulk distances due to the absence of skin and proximity effects in an AC system that dominates with increasing voltage levels. Medium voltage direct current (MVDC) also uses similar technology where co-existence of several PCCs paves way for a DC

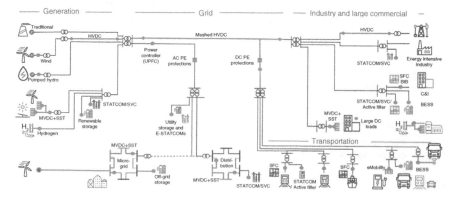

Figure 18.1 Evolution of power electronics in all spectrums of electricity utilization. Source: [30]/with permission of Hitachi Energy Ltd. UPFC = unified power flow control, PE = power electronics, C&I = commercial and industrial.

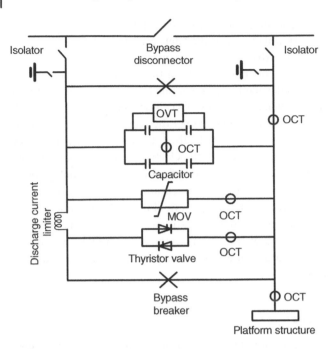

Figure 18.2 Critical components of a TCSC device. OVT = optical voltage transducer, OCT = optical current transducer, MOV = metal oxide varistor, TCSC = thyristor-controlled series capacitor.

connectivity, over its AC counterpart. Solid state transformers (SSTs) are another feather in the cap of the transformer technology which offers active voltage and current regulation and sometimes can do single to three-phase conversion. Operating at a higher frequency than the normal frequency (50 or 60 Hz) it offers superior efficiency performances as well [31].

On the other hand, flexible AC transmission systems (FACTSs) offers a host of benefits in the AC system via series and shunt compensation techniques. Series compensation facilitates increased power flow up to 50% by capturing the power oscillations and dynamic disturbances. An illustration of a thyristor-controlled series compensation (TCSC) is shown in Figure 18.2 below. Such a compensation helps effectively boosting the mean transmitted power, as shown in Figure 18.3.

FACTS devices help achieve major PQ problems on the consuming devices thus offering increased application in industrial furnaces, utilities, static frequency converters (SFCs) for railway application, extensive eV charging infrastructure, where flickers can be reduced by up to 75%, voltage quality can be regulated [32], saving the grid

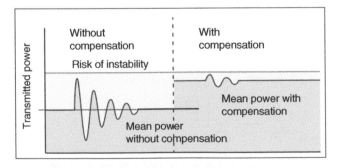

Figure 18.3 Illustration of mean power compensation of transmitted power following a TCSC compensation.

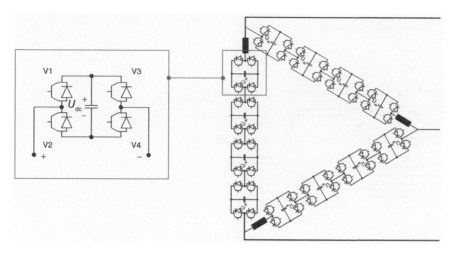

Figure 18.4 An illustration of a multilevel VSC converter technology.

from black outs that costs as much as US$ 1 billion per hour. The range of FACTS solutions extends to PQ solvers in the AC network, besides the reactors, STATCOMs and in a modified environment, even enhanced STATCOMs (e-STATCOMs). It bridges the complexity of weak grids in the AC network and finds a wide usage in the SFC applications as well, that operates at different operating frequency than the standard 50 or 60 Hz for obvious reasons of being decoupled from mainstream grid supply.

STATCOMs further employ modular multilevel VSC converter (MMC) topology to attain a smoother wave shape that offers less harmonics, compared to two-three level topology. MMC topology are widely deployed in present times for its obvious benefits of reduced footprint, modular structure, low switching losses, due to lower switching frequency, etc. Of course, it is more expensive compared to its peers. An illustration of an MMC topology in a static var compensator (SVC) environment is presented in Figure 18.4.

The constant shift towards increased power electronics driven converters in both generation and consumption will impact the system stability over a broader frequency range of converter operation. This would mean that the assessment of the utilization of these converters has to be holistic, aggregating and in real-time, to capture the broad frequency operation phenomenon and then propose appropriate mitigation techniques [30].

18.2.3 Operation of the Power System

In February 2021, following a severe snowstorm, the mighty state of Texas in the US was left plunging into a systematic blackout, that lasted for 17 days and impacted 25 counties, five million households, US$ 195B worth of property damage leading to more than 100 casualties. It left the entire public out of electricity, heating, water, and gas. This was a catastrophe waiting to happen, when repeated warnings of grid vulnerability to the impending storm were ignored by the authorities and the locals alike [33]. The grid personnel were not used to the winter storm of this magnitude and the residents did not heed to the request to decrease the electricity consumption, in the absence of enough power to keep the supply-demand in balance. The peak demand started rising 20 GW additional, when the 26 GW generation already fell apart. The combined failure of gas, electricity and water caused havoc to maintain the frequency, despite a whopping 28.4 GW of reserve capacity.

Being almost an independent system operator in the region, Electric Reliability Council of Texas (ERCOT) had little assistance available from its interconnections, given the already pressing demands in the neighboring grid. As shown in Figure 18.5 below, the grid frequency dropped from 60 to 59.4 Hz for approximately 4.5 minutes, dropping nearly half of its capacity (52.2 GW out of 107.5 GW). The grid could have resulted in a complete collapse, had the situation lasted for nine minutes [34].

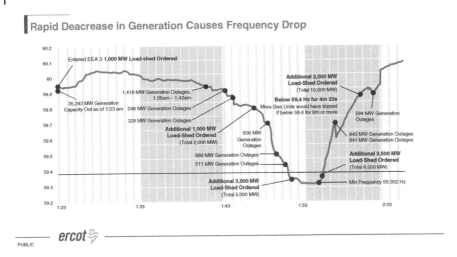

Figure 18.5 RoCoF illustration of ERCOT network on the 15 February 2021.

Similar black out situation occurred in UK on 9 August 2019, where there was a systematic outage of windfarms, following two lightning strikes. The impact left one million households plunged into darkness, while 2 GWs of power were wiped out of the system [35]. The frequency dropped to 48.8 Hz (50 Hz system) before the restoration, despite attempts to arrest it through primary and secondary response deployments and sooner low frequency demand disconnection (LFDD). These are not standalone incidents and nor happening in a weak grid scenario and thus rings alarm all around the world, about the physical vulnerabilities brought by multiple actors.

In 2015, there was a successful stage-1 intrusion made into the Ukrainian grid to begin reconnaissance for a possible cyber-attack, allegedly by the Russian attackers [36]. A stage-2 attack was launched on 23rd December 2015, where all circuit breakers were remotely opened for a total of 23 substations at 35 kV level and seven substations at 110 kV levels. The attack was further amplified when the UPS power of the utility control center was disabled, firmware downloaded to the serial/ethernet communication devices to take them out of service, telephone dial was denied to the call center and operator workstation hard disk was erased. This was a virtual war that impacted 230 000 customers for a period of one to six hours.

All the above examples paint a very grim picture of the level of preparedness of the utilities towards natural or manmade disasters. This is not withstanding the fact that the UK incident was a clear case of an outdated protection scheme that was not designed for all the RES integrations, enforced in the system. This calls for a change in perspective, both on the cyber and physical space, to ensure a healthy operation of the power system. Some mitigation mechanisms to guarantee an efficient operation and upkeep of the entire infrastructure, while making room for significant modernization investments must be carefully thought through.

18.3 Challenges Faced by Utilities and Their Way Towards Becoming Smart

Utilities around the globe are grappled with multiple challenges from reduced investment funds towards modernization, rapid prosumerism, lack of experienced workforce, quick change in topology and technology that makes the existing workforce without training, less skilled, or rather unfit to adapt to the transition, etc. On top of these problems, the utilities in the transmission and distribution (T&D) sector deal with a huge number of aged assets such as transformer, switchgear, and high voltage (HV) equipment, as well as millions of kilometers of transmission line and the support infrastructure. All around the globe, majority of these infrastructure are in dire need of modernization, end-of-life replacement, or serious maintenance. With all the distributed generation placements

Figure 18.6 Risk, cost, and performance optimization challenges of utility grids under the changing scenarios.

at possible voltage levels, these infrastructures are further confronted with all the modern grid requirements of overloading, reverse power flow scenarios [37], momentary generation curtailment and more dynamic operational regime, where their ability to operate are tested to the limit. The above situations are captured in Figure 18.6, where the utilities challenges can be summarized as a complex interaction of risk, cost, and performance optimization, in the presence of the above-mentioned constraints.

The problems can only be solved with innovative technology, collaborative network of workforce, use of ICT, use of rule-based as well data-driven digital twin infrastructure, and an effective reliability analysis that eventually decides an optimal routine of maintenance regime, with due knowledge of the health of the assets.

One notable way of managing these contradicting situations, and thereby prolonging the life of the assets and at least postpone larger investments are by adopting to condition monitoring (CM) techniques, where a close watch on the status and health of the equipment are kept. Such a philosophy helps planning maintenance of the assets, based on historical progression of failure modes, still overlooking the transient faults that happens instantly. Monitoring the individual assets and understanding them in silos, does not answer the mutual interaction of the assets due to systemic conditions, modified operational behavior, as well as contributions from assets from existing or neighboring grids or substation. An example of sympathetic inrush event is presented in Figure 18.7 below, where transformer T8 in station 2 was energized at a given time. This energization resulted in a lateral impact of sympathetic inrush, experienced not only by transformer T7, which is in the same feeder but also on the neighboring stations (Station#1 and #3) which are 4.5 and 12 km apart, respectively [38].

The phase voltage on the 130 kV side, magnitude of current and per phase power output of T7 were plotted in Figure 18.8, at the time of energization (t1), and subsequent perturbation for around 20 seconds, where a loss of symmetry was observed, due to the participation of transformer T7, in the operation of T8, even leading to a

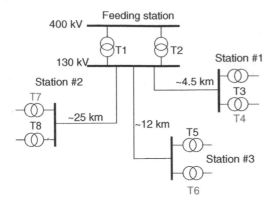

Figure 18.7 Sympathetic inrush event experienced by transformers in the neighboring feeder, due to energization of T8 transformer. Source: [38]/with permission of IEEE.

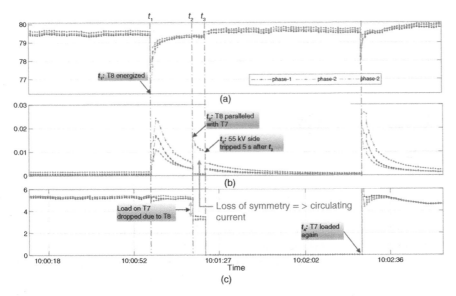

Figure 18.8 Illustration of sympathetic behavior in the grid for an existing transformer T7, following energization of an incoming transformer T8. (a) 130 kV side phase voltage (kV), (b) magnitude of current T7 (kA), (c) per phase power out of T7 (MVA).

circulating current that seems to have nullified at time t4 after c. one minute, when T8 starts to share the load. The sequence of entire operation is explained in detail in [38], including the notable behavior in transformers T4 and T6 of Figure 18.7.

It is crucial to understand the genesis of systemic events in order not to isolate certain behavior of an asset due to its own fault. This kind of cross coupling becomes more prevalent in recent times due to widespread growth of distributed generations and thus ubiquitous PCCs in the entire network.

Because of the broader collaborative grid network right from distribution level and upwards, there is a constant need for understanding the holistic view of different assets in a substation [39], or even an extended fleet of resources, that is working under a larger feeder network. Consolidation of all these data from different PCC and the basis for understanding even the single or multiterminal integration of MVDC and HVDC networks becomes more important in recent times. The following Section 18.3.1 discusses some possible approaches in reaching the smart power system objective by the utilities.

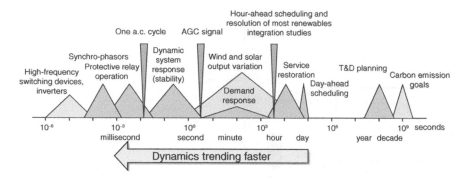

Figure 18.9 Time scale of different responses in a power systems world. Source: With permission of von Meier.

18.3.1 Digitalization of Power Industry

Following the changing generation and consumption scenario, a self-evolving grid is expected, so it responds fast to the volatilities offered on either side. This applies to the operation, control and protection which happens in shorter-time scale versus the maintenance planning and long-term healthy upkeeping procedures, those happen in a relative longer time horizon. Figure 18.9 elaborates on various response time of different operation of power system components that operates even in microseconds. Figure 18.9 also includes some of the operation planning, maintenance objective as well strategic decisions that happens in time scale of years. It is interesting to note that the modern power system is expected to cover this broad horizon in terms of various decision making and accordingly classify them under the array of control and protection, monitoring as well as long-term strategy. This is possible due to significant advancement in the digital infrastructure, embedded multicore processors and micro controllers, with fast computing abilities, low power consumption, extreme ruggedness, and hostile environmental operations. Critical operations employ multiple serial and parallel redundancy for various switchover principles to avoid availability issues, following single component failure. Extensive self-supervision is also adhered to, so that they fail, only in extremely unusual conditions.

The role of digital substations is just getting intensified, thanks to its inherent ability to bring safety into the control room by effectively digitizing the voltage and current signals in the switch yard and transporting all those data over IEC 61850-9-2 process bus communication protocols. Some of the recent studies and interviews suggest that in the event of a natural disaster, such as hurricane Sandy in 2012 in New York, the control and measurement copper wires were severely corroded due to the saltwater contamination, that would have been very hard to get rid of with kilometers long of communication wires. Digitized communication offers tremendously high-speed data at 4 kHz or even higher sampled rate, transported with help of optical fiber cables, so that faster decision making can be accomplished for operation of the protection relays. It also simultaneously reduces 80% of cabling that used to happen over point-to-point copper wire connections. The advantage in the modified digital substation is it brings in a lot of functional consolidation of the past generation electro-mechanical relays, with digital relays that cross talk and even offer redundancy to each other and communicate data flow upwards to the network control centers (NCCs), in an interoperable, IEC 61850-8-1 station bus environment. Such a philosophy, by far, helps the utilities to choose their vendors at will and the digital interfaces are standardized to work seamlessly across the original equipment manufacturers (OEMs). A schematic representation of a typical substation components is depicted in Figure 18.10. The station bus handles Manufacturing Message Specification (MMS) and Generic object-oriented system-wide event (GOOSE) signals and process bus handles sampled value (SV) signals and GOOSE for communication into the switchgear downstream. The bay level intelligent electronic devices (IEDs) acts with a precision time protocol (PTP) clock for time synchronization. There is an optional station bus communication provisioned to access the MMS signals, from the breaker IEDs. The GOOSE signal is also communicated among different IEDs

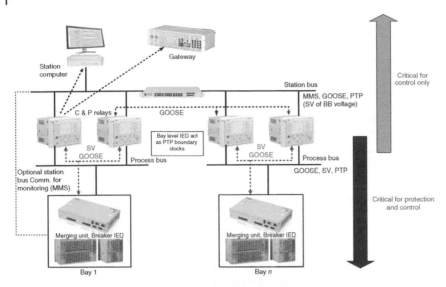

Figure 18.10 Schematic representation of various components in a digital substation. C&P = control & protection, BB = bus bar.

within the station bus. The digital substation offers a host of benefits of reducing the physical footprint by at least 50%, project cycle times by 40%, improving the safety of operating personnel [40].

In addition, it offers a host of customer values, which are core elements of smart grid functions. To name a few – it deals with accurate real time phasor measurements, thermal loading of the line, short-term emergency handling provisions due to continuous monitoring of asset status, including dynamic loading possibilities, adaptive protection, flexibility for reconfiguration and parametrization, and power oscillation monitoring. It also works as an essential enabler for the fast and automated response of the alternating grid status, thus participating in grid stability [41].

There are many pilot projects taking shape in different countries since the evolution of the concept in the last two and half decades. Landsnet, the Icelandic system operator however seems to push the boundaries in these implementations, after having five installations in the last two years with a very ambitious roadmap chalked ahead for the next five years [42], as shown in Figure 18.11.

18.3.2 Storage Possibilities and Integration into Grid

With increased volatility of the grid and the rotational inertia virtually disappearing from the electric power system, there arises urgent need of stabilizing the grid from momentary fluctuations. The fluctuations can be defined as very sudden if that disturbs the normal notion of defined rate of change of frequency (RoCoF). In the event of a sudden loss of hundreds of MWs of power from the grid, the frequency could go for a free fall and thereby engaging the RoCoF protection support, wherein various forms of fast frequency support kicks in. Depending on the duration of these engagement they are labeled as primary, secondary, and tertiary controlled storage solutions, as illustrated in Figure 18.12. When the support is even more urgent, before the primary control, such situations are referred to as loss of inertia in the system and that must be replenished synthetically. These demands are super critical where the form of storage can come from high power density devices, and therefore ultracapacitors may be employed to deal with such a situation.

While the primary services help in effective frequency balancing, decongestion of the grid and voltage control, the ancillary services offered by the storage solutions, mainly are meant to offer flexibility solutions. They also participate in trimming the demand curve stabilization. Medium scale battery energy storage system (BESS)

Digital Substation overview

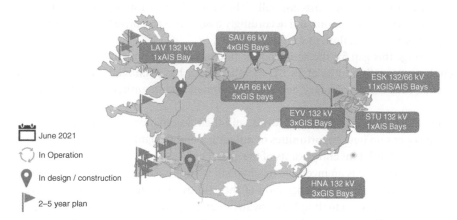

Figure 18.11 Digital substation current and future roadmaps for Icelandic system operator, Landsnet. Source: [42]/with permission of Birkir Heimisson.

Figure 18.12 Typical example of various forms of fast frequency inertia beyond the inertia.

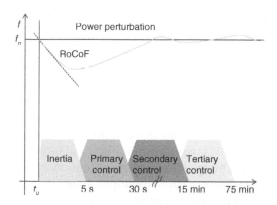

solution deals with primary reserve with fast reaction time and can offer higher power output compared to the energy density. In this sense, this can be considered to offer virtual inertia, synonymous to the spinning reserve that reacts to small fluctuations in the frequency. Several pilot projects are undertaken within ENTSO-E to test different efficacy of energy storage solutions, together with other integrated or isolated environments, such as optimal system mix of flexibility solution for European electricity (OSMOSE), COORDINET, cross border renewable energy (CROSSBOW), enable ancillary services by renewable energy sources (EASY-RES), Flexitranstore, graphical islands flexibility (GIFT), and grid solutions (GRIDSOL) [43].

Various energy storage technology commonly employed for grid storage application includes pumped hydro, flywheel solution, electric battery, flow battery, and ultra-capacitors. However, more unconventional methods are being tried recently. One such example is from a company called Energy vault that uses blocks of concrete lifted by excess electric energy and stored at a height of potential energy [44]. The round-trip efficiency of stacking and unstacking of these concrete blocks is claimed to be 85% (similar range as Lithium-ion batteries-90%). Green Hydrogen has been a strong contender for a while to be directly used as energy storage [45] or consuming Hydrogen in alternative platforms by producing ammonia [46], methane [47] or methanol [48], etc.

The availability and reliability of the grid are dictated by the duration and frequency of supply interruptions. The system average interruption duration index (SAIDI) and system average interruption frequency index (SAIFI) are the indicator of the resiliency offered by the grid that has significantly lowered in recent years in Europe, according to the World bank data. A comparison of these reliability indices for 2015 and 2020 for various member states of

EU is presented in [49]. The estimate suggests that during 2010–2015 the electricity disruptions experienced by the customers cost them in the range of 10–25 billion Euros annually. Therefore, there is a strong push for legislation, also driven by countries to classify the interruptions and accordingly associate a response time to address it, without causing penalty to the operator towards failure to supply electricity.

Storage plays an integral role to bridge this gap for the active power fluctuation requirements. The extent and source of electricity outage including generation units of 28 EU-member states has been studied in detail [50] to understand the disruption events, thus using voluntary demand curtailment and quantifying a value for the lost load.

With the widespread evolution of eVs in various shapes and forms, and their acceptance to be implemented in the grid for bidirectional power flow opens up new opportunities of utilizing a mobile bank of distributed energy storage solutions, ubiquitously available at different entry points of the grid. The V2G facilitates this concept once the legislation becomes more versatile and there is an incentive mechanism for more participation from the individual vehicle owners. This will largely alleviate the decongestion of the grid by allowing ICT driven legalized option of bidirectional power flow. Besides, even adopting legal means to employ those storage as a back up to the tertiary reserve, very similar to the LFDD scheme employed by the utility operators can facilitate/incentivize individual users. This in other words means, if an eV is connected to the grid (plugged-in) and has signed up for a bidirectional power flow, then the utility operator could in principle use these as contingency storage reserve, of course at the cost of incentivizing the participant.

18.3.3 Addressing Power Quality Concerns and Their Mitigation

PQ is an acknowledged concern in the power system for quite some time. However, its scope and features are changing and being revisited by the regulators, the users as well the power providers because now there are several actors playing in this space, who are responsible for bringing this up. The traditional network with upstream power flow to downstream power consumption has changed due to the distributed generation at all voltage levels due to renewable integration in the grid and further fueled by rooftop solar and wind power at domestic or community scale. This modified scenario not only increases the number of PCCs in the network, but it also offers no control of many nonstandardized equipment introduced in the grid, including a lot of power converters and inverters, which might not simply comply with the several PQ standards that exist in respective countries (EN 50160), IEEE 1159, IEEE 519, IEC 62052-11, IEC 62053-21/22/23, and IEC 62052-21 and IEC 62054-21, etc.

The voltage deviation as it is seen during the transition from medium voltage (MV) to low voltage (LV) is illustrated in Figure 18.13, where, in the modified scenario, there are new actors positioned in the form of eV

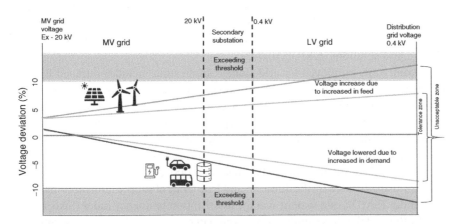

Figure 18.13 An illustration of voltage deviation passed from MV upstream to LV downstream.

Figure 18.14 Geospatial representation PQ data obtained from different meters connected in the network.

charging network, data centers, etc. as consumers with bidirectional power flow capabilities, as well as integration of distributed energy resources (DERs). The voltage fluctuation at the receiving end of the domestic consumers experiences either as an increase or decrease, depending on the in-feed or demand side fluctuation.

The challenge also comes from a lot of eV chargers that range from tens of kW to UFCs that operate in a MW range. In addition, a lot of fluctuating power is introduced in the electricity-supply system by the RES that result in high frequency emissions, otherwise called as supraharmonics [51] that has a range from 2 to 250 kHz. The source of these emissions is largely due to semiconductor switching from converters of different sizes. Supraharmonics have the potential of impacting sensitive instruments such as smart meter gateway, light dimmers, and programmable logic controller (PLC) communications.

Therefore, it becomes important to specify the tolerance zone and unacceptable zone of these deviations (as suggested in Figure 18.13) so that the physical equipment remains protected against unwanted fluctuation, or at least appropriate counter measures in terms of PQ filters are to be employed. For records, PQ problems cost Europe 150 billion Euros annually [52]. Therefore, it becomes paramount to employ PQ meters at various PCCs so that there is a better transparency observed for the cause and appropriate accountability to eliminate this can be established. A series of ground steps to address this problem and set ground rules were suggested in [53].

A big data framework can be set up by comparing the PQ data obtained from a series of PQ meters installed across a network, which could be presented on a geospatial scale and compared together with the single-line diagram (SLD) models in order to track a specific cause and effect. Different events can be classified on a historical basis to even plan for a periodical maintenance strategy, very similar to the condition-based maintenance adopted for the assets. A small depiction of the said concept is presented in a simulated environment, where Figure 18.14 shows the physical location of the PQ meters in the utility network and Figure 18.15 depicts various events registered in a time axis, which can be combined, correlated and analyzed further to even relate to the system faults those are captured separately in the supervisory control and data acquisition (SCADA) logs. This kind of concept opens up many new possibilities of combining the analysis together with the phasor measurement unit (PMU) data and continuous events can be nullified by revisiting some of the protection schemes and the SLD, which might not have been modified following the DER integration or FC additions.

18.3.4 A Path Forward Towards Holistic Condition Monitoring

The electric substation has been in existence from the inception of different voltage levels in generation, transmission, and distribution sectors. Traditionally most of the primary assets, namely the transformers, switchgear components, surge arresters, and reactors have been designed very robust to last beyond the suggested lifetime of 30–40 years, even during adverse climatic conditions. However, due to incorporated modernization of the grid in an effort to become smarter, more flexible and practical, their operability is now being put to test and are expected to deliver more than what they were used to. This is of course subjected to evaluation of the assets if they are in a condition to do so. As a result, a lot of smart interfaces are added to the substation equipment to understand

Figure 18.15 Classification of different events presented in a time scale.

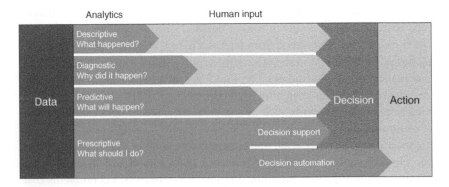

CATCON 2019 workshop power apparatus

Figure 18.16 Evolution of decision-making ability on an asset or a substation in a power system.

specific condition of one particular asset. This is done, mostly by collecting data from various sensors that is either inherently installed or retrofitted, out of need. This is a ploy to switch from a periodic maintenance regime to reactive and subsequently predictive and prescriptive maintenance and that counts a lot on the condition of the asset to suggest various recommendations. Figure 18.16 captures this evolution in the best way, which basically moves from top to bottom in historical perspective of handling maintenance, where the human inputs are considerably integrated into the smart interfaces in terms of machine intelligence and rather automated, so that the decisions are automatically taken based on the available intelligence offered to the system and thus the period of action is shortened. All these are possible with more information about the asset made available, that was historically not gathered, stored, and analyzed.

However, this is still the primary step, where the assets are understood at an individual level, without their gross interaction with the family of other devices being understood. This way, the entire substation fails to respond collectively and even communicate individual asset health for a smart demand management, even when such information is warranted. This is where the concept of holistic condition monitoring (HCM) pitches in [39]. In a modern substation environment, the asset operation and management responsibility are split between the operational technology (OT) and information technology (IT), so that from ICT perspective, there is a clear distinction of responsibility. This is represented in Figure 18.17. The IT part does bulk of the data handling, evaluation, and portfolio management (represented as #4, 5, 6) and has an interface to the external world and deals with the data and information received from the OT region. The OT side, on the other hand, is responsible for day-to-day business that deals with the assets, collection of data from the sensors attached to these assets and aggregating them over a network mechanism (represented as #1, 2, 3) and would require no direct intervention from the IT side.

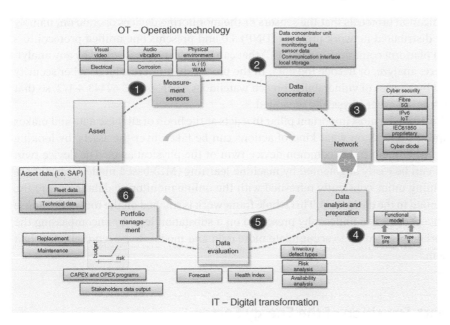

Figure 18.17 Division of responsibility in OT and IT domain in a utility environment. CAPEX = capital expenditure, OPEX = operational expenditure, SAP = system applications and products.

This is ensured by incorporating a one-way communication, depicted as #3 and is otherwise called a "Cyberdiode" that blocks information from the IT side from being received on the OT side.

Figure 18.18 explains the house of HCM, which is built on the foundation of a host of sensors that are collected from different assets in a substation in an incongruous manner. Those sensors could be connected on asynchronous platform, communication protocol, collecting data at different periodicity and thus have different X and Y-label. The scope of the house, however, is to utilize the data from the sensor base by building three independent, yet supportive pillars which have distinct features of using a unified data model, a modular gateway architecture and a versatile analytics platform.

The unified data model tries to connect the sensor data and their relevance to the physical assets in a common information model (CIM) environment, which can be dictated by, e.g. IEC 61970 standard. Such a unification makes the data models vendor independent, thus evaluating the connection between the sensor and the assets, based on their individual property mapping. The modular gateway architecture develops a common architecture

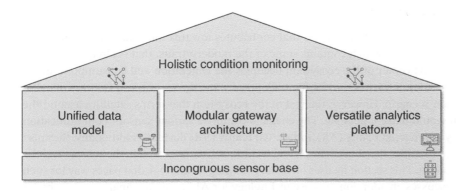

Figure 18.18 A concept of holistic condition monitoring: an essential step towards a reliable grid solution.

that assembles different communication protocols that the sensors or the monitoring devices operate on, namely Modbus TCP/IP, IEC 61580, and distributed network protocol (DNP) 3.0 and presents one unified protocol to a consuming platform, e.g. an open platform communication (OPC) that can be universally accepted by any analytics platform that wants to visualize, analyze, or decode the data. In the process it enforces a lot of cyber security classification, essentially dictated by a series of vulnerability breach watchdogs, such as IEC 62443-4-1/2, so that the data between the OT and IT side are cyber-physically separated.

The analytics platform is the third and the most important pillar that acts as the brain of all these data and makes them useful by offering prescriptive remarks on what kind of actions can be taken over the assets, by looking into the mutual interaction and fitting them into a common device twin of the physical asset. The device twin is built on reliability models and can be easily augmented by machine learning (ML)-based models, once there is access to the data of all performing units, constantly refreshed with the online monitoring data to ensure the latest recommendations are presented in the dashboard. This whole framework is stitched on the roof of the house which is termed as the HCM concept. This is a philosophy presented on a substation level by encompassing the associating assets, can be appropriately scaled up to accumulate more data and models to be representative of a ring network or a larger subsystem.

18.4 Ways towards Smart Transition of the Energy Sector

The RES in recent times gives a positive headache to the rest of the integration system due to the variability it brings in its production process towards clean energy transition but is at the same time compounded with predictability and operational controllability issues. These zero-marginal cost resources generate the energy flexibility options which did not exist in the past, as the reserve for the fossil-based sources had to be made available in the supply chain, before the production could happen. Notwithstanding the uninterrupted availability of the RES (of course subject to geographical and climate conditions, translated to 365 days of the year), the limited predictability feature returns a surplus or deficit situation for the amount of energy added to the grid at a given point. This creates a gap on the already promised evacuation of power and expects means to deal with the surplus or deficiency. That's where enter a lot of transactional players, who tend to take advantage of this situation and take certain amount of risks for influencing the spot pricings. Due to these locational marginal prices, the price of electricity can even go negative, to maintain the orderly addition of the promised capacity. The transmission bottlenecks and ample access to distributed generation in a given grid, might even result in exacerbating the situation, benefiting the transactional players in one way or the other.

18.4.1 Creating an All-Inclusive Ecosystem

An all-inclusive ecosystem will encompass a lot of verticals that contributes towards the definition of smart systems. Figure 18.19 offers a broad ecosystem which contains some of the modern traits, that is already discussed in different stages in this chapter. The center of the ecosystem occupies the generation and consumption units, which of course requires a transmission infrastructure for the bulk power evacuation, unless it is a case of collocated generation and consumption scenario. However, the rest of the ecosystem that forms satellites around the center, are various enablers that would make it possible. The enablers comprise various sensors, communication technology, further supported by augmented reality (AR) and virtual reality (VR) devices. Additionally, they also include all the intelligence needed to utilize the above sensors and techniques, without which, the sensors and the enabling technology bear no meaning. This section focuses on some emerging technology that paves way for such an ecosystem to thrive. It also discusses an adoption of those technology for AI-based data analytics for a better understanding of the incoming data and hence the state of the health (SoH) of the substation.

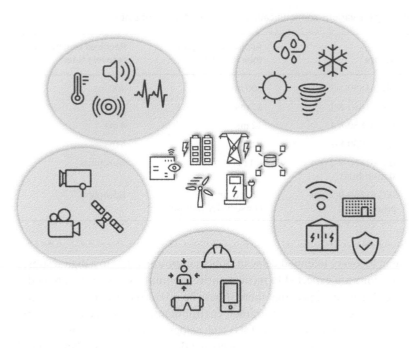

Figure 18.19 Modern power ecosystem.

18.4.1.1 Example of Sensor-Based Ecosystem

There is a plenty of IoT-based sensors that find their way through, into the utility space, due to the sheer virtue of being useful when used in combination with the current, voltage, gas or liquid profile of the insulating medium. The legacy sensors and monitoring devices mostly run-on proprietary communication protocols, e.g. modular advanced control for HVDC and SVC (MACH) for HVDC control and protection within Hitachi Energy [54], S7comm that runs between PLCs of Siemens, MOD5-to-MOD5 for peer-to-peer communication in MOD5 process control systems in Asea Brown Boveri (ABB), etc. There are also a series of standardized protocols which, of late, finds wider application in substation automation systems and utility environment such as IEC 61850, DNP3, MODBUS, profibus, controller area network (CAN) bus, EtherCAT, open platform communication Unified architecture (OPC-UA), MTConnect, IEC 62351. There are even more versatile ones that mostly connect the adjoining IoT networks over ANSI C12.18/21/22, IEC 61107, M-Bus, Zigbee, Modbus, device language message specification (DLMS)/IEC 62056.

The recent advancements in wireless technology have however brought in large-scale disruption even in the utility environment. There are perfect use cases which explains the retrofit sensor integration in a brown field installation, for instance a digital infrastructure requires digging the substation for laying such cables. This causes huge cost and availability issues and is better facilitated, if a wireless route is chosen, provided it offers the same level of efficacy and reliability in terms of payload transfer, packet loss, and other key performance indicators (KPIs).

The ubiquitous fusion of weather data into historical data analysis demands a lot of temperature, pressure, humidity, and similar measurements that need to be corroborated with meteorology data, that is freely available to download and compare across geographical region in order to give a direct analogy of the behavior of the asset against their corresponding ambient profile and thus can be compared against the other performing peer assets in the neighborhood.

368 | *18 Power and Energy Management in Smart Power Systems*

Table 18.1 Application recovery tolerated delay for various communicating devices in digital substation.

Communicating devices	Service	Application required tolerated delay	Required communication recovery time
SCADA to IED (MMS)	IEC 61850-8-1	800 ms	400 ms
IED to IED (Interlocking, IF8)	IEC 61850-8-1	12 ms	4 ms
IED to IED reverse blocking	IEC 61850-8-1	12 ms	4 ms
Protection trip (excluding Busbar)	IEC 61850-8-1	8 ms	4 ms
Busbar protection	IEC 61850-9-2 (on station bus)	<1 ms	Bumpless
Sampled values	IEC 61850-9-2 (on process bus)	Less than two consecutive Samples	Bumpless

Table 18.1 below details about application recovery tolerated delay in utility communication in a digital substation environment. Some of the communications are expected to recover entirely bumpless, when the protection command has to be sent and executed. The SV signals expect that there should be no consecutive packet loss, failing which the reconstruction becomes impossible. In such a constrained environment, it makes sense to test the latest communication technology such as wireless fifth generation (5G) technology network, if it is able to offer a viable alternative to route this delicate traffic in a wireless route, at least for the sake of a redundant fallback solution.

5G is based on a lot of accomplished frameworks such as small cell, millimeter (mm) wave, ultra-reliable low latency communication (URLLC) of around 1 ms radio latency, backhaul, radio area network (RAN) virtualization, massive mission critical communication (mMTC), such as one million devices/km^2 operating in an enhanced mobile broadband (eMBB) at up to 20 Gbps peak data rate [55]. This is currently further enhanced with digital representation of the sensors in cyber physical domains in 6G [56]. This is in the direction of increasing the observability with sensor data based on tag identity, time stamping, geospatial positioning giving further enhancements. 6G is expected to break into a sub-THz frequency range between 100 and 300 GHz with an expected radio latency as low as 100 μs, whereas 5G had an expected operating frequency range between 3.5 and 7 GHz (<10 ms latency) and further expanded up to 24 GHz for the mm Wave signals (~1 ms latency). The promises of driving extremely high data rate/capacity, extreme coverage, obviously at low energy and cost and high reliability is becoming ideal, for the industrial revolution; thus, finding their use for the utility industry, across the entire horizontal application areas presented in Figure 18.9.

5G was tested and validated for the most critical SV substation communication, together by Ericsson and ABB [57], where a 10 000 m^2 digital substation environment was simulated, that was surrounded by a prohibitive area of 50 m wide. The randomly distributed blocks acted as potential blocker for the line of sight (LoS) for obstructing the wireless communication. Eight SVs worth of 124 bytes of data at 4 kHz sampling rate was sent with the RAN latency requirement of 3 ms and reliability of data transmission of 99.99%. Seven eMBB sites were chosen for this work with 21 cells, with URLLC chosen as 3.5 GHz and the system bandwidth of 20 MHz. The simulation results show successful network performance of the bit rate and acceptable signal to noise ratio (SNR) (a gain of 25–40 dB is considered very good) and path gain both in the isolated network (Figure 18.20) and in presence of the eMBB network (Figure 18.21).

18.4.1.2 Utilizing the Sensor Data for Effective Analytics

Various supervised or unsupervised learning can be employed on a vast dataset to have a better understanding of the collected data from a big data perspective. This is important for a systemic aggregation of all the information,

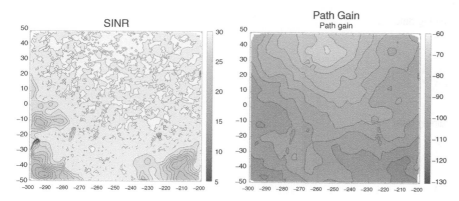

Figure 18.20 SNR and path gain for in isolated 5G network scenario.

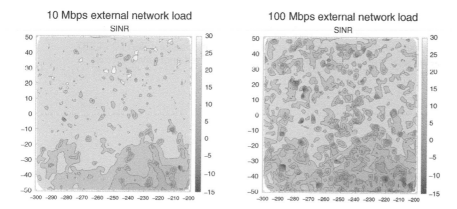

Figure 18.21 SNR in presence of eMBB at different network interference level. SNR = signal to noise ratio.

thereby drawing several regression analogies between data coming out of different assets. The ML techniques adopted for handling this data are classified in two categories. The supervised learning uses regression and classification algorithms, whereas the unsupervised learning doesn't have that luxury of assessing the data beforehand to be put in suitable buckets. The unsupervised learning therefore employs techniques such as clustering – based on splitting data following similar characteristics and dimensionality reduction (DR), to reduce the overlapping behavior- an illustration depicted in Figure 18.22 [58].

DR is achieved using techniques such as principal component analysis (PCA), multidimensional scaling (MDS), and T-distributed stochastic neighbor embedding (T-SNE), and it depends on the data, on which method to apply and the objective behind this DR. For example, T-SNE is widely used for visualization purpose, while preserving global and local structure. A representative outcome of classifying a series of dataset obtained from an HVDC station is presented in Figure 18.23 [59].

Using the learning data, it is possible to predict the future time series using traditional models such as autoregressive moving average (ARMA), or ensemble ML-based models, namely Random Forest (RF) or multivariate time series predictions. The results of prediction versus actual ARMA model are presented in [59]. Figure 18.24 shows the results obtained via RF ML model. The performance evaluation is obtained by the mean absolute error (MAE) and Root Mean square error (RMSE). The RMSE and MAE are both lower for the RF method in this particular example, standing at 0.065 and 0.040, compared to those numbers from the ARMA model (0.188 and 0.143, respectively).

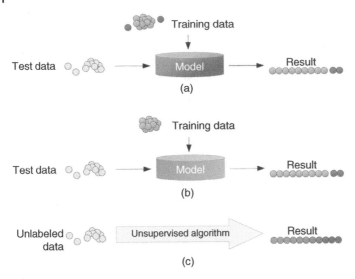

Figure 18.22 Illustration of training data and result for different supervised algorithms. Source: [58]/PLOS/CCBY. (a) Supervised anomaly detection. (b) Semi-supervised anomaly detection. (c) Unsupervised anomaly detection.

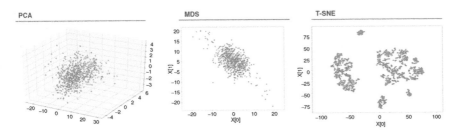

Figure 18.23 Representation of DR-based classification of HVDC data.

Usage of ML-based predictive models for time series data in a utility environment is still a niche topic, because of the associated complexities, such as sympathetic inrush event [38] still contributing to the overall development. Till all those associated complexities are suitably modeled and better understood, the challenge still remains with an accurate performance estimation of these models. The techniques seemingly work fine though in other application areas, namely healthcare, weather forecast even in aviation industries. Notwithstanding the above limitations, AI finds a nice inroad into the demand prediction and system stability analysis due to the observable prediction and the interactions coming from metrological and other forms of public domain data to improvise such models.

18.4.2 Modular Energy System Architecture

The energy storage system architecture in the Microgrid environment is getting a massive overhaul due to its real time communication to the control and network infrastructure and it is an encompassing technology that handles the RES and FCs in the same communication infrastructure. A visual depiction of an e-mesh™ is given in Figure 18.25 [60] that offers a host of connectivity and controllability options in a DER environment. Such concepts adhere to the comprehensive solutions energy storage systems together with the other grid constituents (either Microgrid or normal grid) bring on to the table. The PowerStore offers grid forming (GFM) architecture with seamless islanded transition possibilities, alongside being able to operate on the virtual generator mode (VGM). It thus is capable of supporting a grid-connected and off-grid applications. e-mesh utilizes MicroSCADA and remote terminal unit (RTU) platforms with connectivity integrations into other substation equipment in IEC 61850, DNP3,

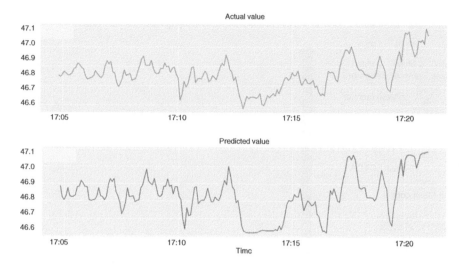

Figure 18.24 Comparison of predicted versus actual in random forest model.

OPC, Modbus. In such a scenario it offers a single aggregator function with inputs from RES, BESS, substation and Microgrid, while still working in unison with the existing SCADA infrastructure. The Energy management system (EMS) runs efficient optimization application by taking all those inputs from SCADA and control platforms for effective presentation to the e-mesh monitor. This includes detailed level of visualization coming from different sources and their respective segregation in a connected environment. The e-mesh monitor offers a cloud-based digital platform that exhibits various strategic as well as historic analytics insights and pushes that information to relevant stakeholders, such as maintenance, energy portfolio or service departments. This is a classic example of offering an end-to-end architecture in a software-as-a-service (SaaS) application framework.

Such modular energy system concepts are becoming more common these days with both OEMs as well as software companies like Google, Microsoft, Cisco, and Oracle, entering this space to offer a comprehensive SaaS package or playing roles in some of the verticals, mentioned in Figure 18.25, facilitating the data and information flow both south and northbound. There are also possible diversifications of these services expanding the connectivity to process control systems, and smart or PQ meters so that there is even a wider view of the energy scenario presented in a combined platform.

18.5 Conclusion

The future transformation of the electric power system is undeniable and yet partially unimaginable. The share of renewable by 2050 might reach 100% in some countries, with surplus power to play with. But the scenario can entirely change depending on some unforeseen situation, similar to some presented in Section 18.1. The energy storage systems will probably see a new definition, almost operating in a hybrid mode with generation and storage going together all the time. Most of the domestic consumption will be self-sustaining, even without the need of being part of a connecting grid. The transaction and participation in a local cum global energy market will almost be compulsory. There will be greater transparency in the entire grid, with power being available-on-demand in different forms and shapes. The green energy might prevail at least in 60% of the world.

Some of the emerging trends that we are currently observing and expected in future include-

1. The demand flexibility will be humongous, largely driven by technology and incentives for the voluntary participation. The concern of personal data utilized by a lot of different aggregators who are or will be part of smart home automation will be driven by standardization, towards anonymity and data normalization to avoid intruding into the user privacy.

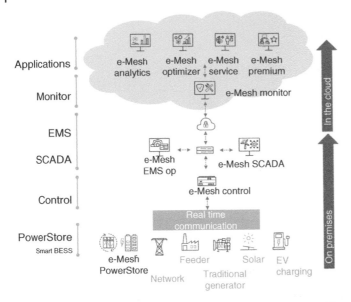

Figure 18.25 e-Mesh architecture for grid edge solutions from Hitachi energy.

2. At the same time, the cyber concerns of the grid will be more at loom and so will be the need for various standardizations, in order to meet the compliance standards. The world might be a better place to live in.

3. Energy forecast is never foolproof. It is subjected to change and hence capacity planning can never be accurate. This will give headroom for an effective market capitalization, depending on the seriousness of the actor and the demand for the energy.

4. HVDC as a technology is getting employed at more interconnection points, for bidirectional power flow and multiterminal setpoint to at least be partially available in the event of a failure, loss of redundancy.

5. Data is becoming a central part of many interplays between several actors, participating in energy transactions, asset optimization, health assessment, system wide analytics and holistic CM.

6. Digitalization of the grid will continue to drive the next three decades, when the grids across the world make efforts to follow the decarbonization route. Nevertheless, coal will still continue to be a dominant player as an energy source to fuel the industrialization and address the short- and medium-term deficit.

7. Energy storage will become a standard service similar to generation and might capture as high as 50% of the installed capacity space for at least an hour of capacity. Different new forms of storage will evolve beyond BESS, compressed gas storage, liquid hydrogen (LH_2), etc. and will become the single most commodity to be traded in the power world.

8. Advanced sector coupling (such as HVAC with electricity sources) architectures and control strategies facilitating an all-inclusive 100% RES (e.g. on-site) targeting energy intensive consumers and applications, viz Data Centers and Power-to-X will emerge.

In summary, a more decentralized grid is expected to be seen, ubiquitous access to clean energy and a fair contribution to the green transition of the planet from the electric power systems. The long-distance transmission for bulk power transition will transform to ultra-HVDC, due to its technical and commercial advantage. Offshore platforms for wind will be adopted in different parts of the world as they will dictate the bulk energy transition. We will see a lot of enablers for dynamic balancing, periodic overloading, and more voluntary decongestion of the grid due to the associated incentive it would bring along for the prosumers. All in all, technology, legislations, and political engagement will drive a smart, efficient, and green transition of the future electric grid.

References

1 Wikipedia Article (2022). *Russian Invasion of Ukraine.* https://en.wikipedia.org/wiki/2022_Russian_invasion_of_ Ukraine.

2 Anderson, R.J. (2008). *Europe's Dependence on Russian Natural Gas: Perspectives and Recommendation for a Long-Term Strategy.* European Center for Security Studies.

3 Banila, N. (2022). ENTSO-E syncs Moldova, Ukraine power systems to continental Europe grid. *SeeNews*, 16 March 2022.

4 Prokip, A. (2021). Integration of Ukrainian power system to ENTSO-E: challenges and opportunities. *CEEnergy News*, 25 March 2021.

5 Alvik, S. and Irvine, M. (2020). *The Impact of Covid-19 on the Energy Transition.* DNV Consulting.

6 NY Times Editorial Coronavirus world map: tracking the global outbreak. *The New York Times*, updated 24 April 2022.

7 Yasmina Abdelilah, Heymi Bahar, Trevor Criswell et al. (2021). Covid-19 and the resilience of renewables. IEA report 2021.

8 Waldholz, R. (2020). Coronavirus slump will lead to more hours of negative power prices in Germany – study. *Clean Energy Wire*, 06 April 2020.

9 Dickie, G. (2020). The Arctic is a death spiral. How much longer will it exist? *The Guardian*, 13 October 2020.

10 Morris, C. (2021). COP 26: How much are poor countries getting to fight climate change?. *BBC News*, 14 November 2021.

11 Pierre-Olivier Gourinchas (2022). War sets back the global recovery. World Economic outlook, International monetary fund, April 2022.

12 Appunn, K. and Russell, R. (2021). Set-up and challenges of Germany's power grid. *Clean Energy Wire*, 10 June 2021.

13 Cassie Fountain, Timothée Degrace (2018). UK- France HVDC interconnector. AQUIND, PINS Project Reference No. EN020022, 17 September 2018.

14 Hausler, M. Nordlink: green electricity from Norway. https://www.kfw.de/stories/environment/renewable-energy/nordlink/.

15 Project 110 - Norway-Great Britain, North Sea Link, NATIONAL DEVELOPMENT PLAN PCI 1.10.1, 23 NOVEMBER 2017, https://tyndp.entsoe.eu/tyndp2018/projects/projects/110.

16 Hitachi Energy Communication Center. Statcom - SVC Light brochure. https://www.hitachienergy.com/offering/product-and-system/facts/statcom/svc-light.

17 Goasduff, L. (2022) Gartner forecasts 6 million electric cars will be shipped in 2022. *Stamford, Conn*, 26 January 2022.

18 Sahoo, S., Zhao, X., and Kyprianidis, K. (2020). A review of concepts, benefits, and challenges for future electrical propulsion-based aircraft. *Aerospace* 7: 44. https://doi.org/10.3390/aerospace7040044.

19 Anwar, S., Zia, M., and Rashid, M. et al. (2020) Towards ferry electrification in maritime sector. https://doi.org/10.3390/en13246506.

20 Schwab, A., Thomas, A., Bennett, J., and et al. (2021). Electrification of aircraft: challenges, barriers and potential impacts. *NREL*, October 2021.

21 ABB Communication Center. Terra HP – charger. https://new.abb.com/ev-charging/produkte/car-charging/high-power-charging.

22 Hubner, I. (2021). Electric trucks: ultrafast charging with 1 MW. *elektroniknet.de*, 20 October 2021.

23 Adamson, J. (2017). Carbon and the cloud, hard facts about data storage. *Stanford Magazine*, 15 May 2017.

24 Eric Masanet and Nuoa Lei How much energy do data centers really use? *Energy Innovation Policy & Technology LLC*, 17 March 2020.

25 George Kamiya (2021). Data centres and data transmission networks. *Tracking report - IEA*, November 2021. https://www.iea.org/reports/data-centres-and-data-transmission-networks.

26 Malev, M. (2021). The climate is overheating – time for green data centers! *Digital Marketing Exposition and Conference* 22: 24.

27 See, A.V. (2022). Number of data centers in selected European countries. https://www.statista.com/statistics/878621/european-data-centers-by-country/, 16 February 2022.

28 Fiona Harvey (2017). "Tsunami of data" could consume one fifth of global electricity by 2025. *The Guardian*, 11 December 2017.

29 EIA Staff (2021). Use of energy explained. *US Energy Information Administration*, 02 August 2021.

30 Frede Blaabjerg, Simon Round (2021). Power electronics: revolutionizing the world's future energy systems. *Hitachi Energy Perspective*, 26 August 2021.

31 Kabalci, E. Solid state transformers with multilevel inverters. *Science Direct* 249–266. https://doi.org/10.1016/B978-0-323-90217-5.00005-8.

32 Hashiesh, F. (2020). Power Quality – Enabling a stronger smarter and greener grid, *Hitachi Energy*.

33 Ning Lin (2021). The timeline and events of February 2021 Texas electric grid blackouts. *The University of Texas at Austin Energy Institute Report*, July 2021.

34 Joshua W. Busby, Kyri Baker, Morgan D. Bazilian et al. (2021). Cascading risks: understanding the 2021 winter blackout in Texas. *Energy Research and Social Science*, vol. 77, July 2021. https://doi.org/10.1016/j.erss.2021.102106

35 Industrial Strategy Staff (2019). GB power system disruption – 9 August 2019. *Energy Emergencies Executive Committee: Interim Report, Department for Business, Energy and Industrial Strategy*, September 2019.

36 Watch How Hackers Took Over a Ukrainian Power Station, wired.com, Released on 06/20/2017, https://www.wired.com/video/watch/watch-hackers-take-over-a-ukranian-power-station.

37 Sahoo, S., Bengtsson, T., and Abeywickrama, N. et al. (2017). Monitoring power transformer performance, usage and system event impacts – a case study, *IEEE CatCON Conference*, 16–18 November 2017.

38 Sahoo, S., Abeywickrama, N., Bengtsson, T., and Saers, R. (2019). Understanding the sympathetic inrush phenomenon in the power network using transformer explorer, *IEEE CatCON Conference*, 21–23 November 2019.

39 Sahoo, S., Weitz, P., and Schnittker, H. et al. (2021). Improved health assessment of Substation using holistic condition monitoring. *ETG Congress 2021*, pp. 1–6.

40 Meier, S. (2014). Enabling digital substation. *ABB Review* (4). https://library.e.abb.com/public/d2bb717efd6f82cc83257dc50034841d/06-10%204m472_EN_72dpi.pdf.

41 Bengtsson, T., Abeywickrama, N., Saers, R., and Sahoo, S. (2019). Power transformer performance monitoring presented in SCADA, *ABB Review*, 19 November 2019.

42 Heimisson, B. (2021). Wide area monitoring and control come to Iceland, *T&D World*, 01 June 2021.

43 Techsheet of Battery Technology from entso-E, 2022, https://www.entsoe.eu/Technopedia/techsheets/battery-technology.

44 Luthy, M. (2021). A new type of battery made from concrete. *Onezero* 28 October 2021.

45 Hydrogen storage, office of Energy efficiency & renewable energy, Department of Energy, USA, https://www.energy.gov/eere/fuelcells/hydrogen-storage.

46 Patonia, A. and Poudineh, R. *Amonia as a Storage Solution for Future Decarbonized Energy Systems*. The Oxford Institute for Energy Studies.

47 Hughes, P. (2019). Porous polymer offers methane storage solution. *Chemistry World, Royal Society of Chemistry* 12 July 2019.

48 Räuchle, K., Plass, L., Wernicke, H.-J. et al. (2016). Methanol for renewable energy storage and utilization. *Energy Technology* 4 (1): 193–200. https://doi.org/10.1002/ente.201500322.

49 Rullaud, L. and Gruber, C. (2020). Distribution grids in Europe, facts and figures 2020, *Eurelectric powering people*, December 2020.

50 Balazs Jozsa, "Study on the quality of electricity market data of transmission system operators, electricity supply disruptions, and their impact on the European electricity markets," European commission final report, March 2018.

51 Zavoda, F., Bollen, M.H.J., Langella, R. et al. (2018). CIGRE/CIRED C4.24, Power quality and EMC issues with future electricity networks, *CIGRE Technical Brochure* 719.

52 EPRI Product ID 1006274 (2013) The cost of power disturbances to industrial and digital economy companies, *Palo Alto: Primen/EPRI*.

53 Sahoo, S. (2019) Power quality scenarios in Sweden, *Energiforsk report*. https://energiforsk.se/media/27406/elkvalitetsarbete-i-sverige-2019-energiforskrapport-2019-633.pdf.

54 Hitachi Energy Communication Center (2021). It's time to connect, *HVDC light brochure, Hitachi Energy*.

55 Wikipedia article on 5G communication, https://en.wikipedia.org/wiki/5G, (UTC).

56 Erik Ekudden (2021) Technology trends for the future network platform, Ericsson Technology Review.

57 Kumari, N., Chernogorov, F., Ashraf, I. (2019). Enabling process bus communication for digital substations using 5G wireless system. In: *2019 IEEE 30th Annual International Symposium on Personal, Indoor and Mobile Radio Communications (PIMRC)*, Istanbul, Turkey, pp. 1–7. https://doi.org/10.1109/PIMRC.2019.8904287.

58 Goldstein, M. and Uchida, S. (2016). A comparative evaluation of unsupervised anomaly detection algorithms for multivariate data. *PLoS One* 11 (4): e0152173. https://doi.org/10.1371/journal.pone.0152173.

59 Borhani, M. (2020). Anomaly detection using machine learning approaches in HVDC power system. Masters thesis, Malardalen University, Sweden, 20 May 2020.

60 e-mesh™: Infinite insight, Hitachi energy article, https://www.hitachienergy.com/offering/solutions/grid-edge-solutions/our-offering/e-mesh.

Index

a

Adaptive Fuzzy Logic Controller   129
Adroit Neural Network   179, 279
ANFIS   184, 193
Application Programming Interface (API)   328
Augmentation of PV-Wind Hybrid Technology   179
Automatic Metering Infrastructure   4

b

Bidirectional Buck–Boost Converter   118
Black Start   112
Blockchain Technologies   328
Blockchain Technologies for Smart Power Systems   331

c

common information model (CIM)   365
Communication Network   5
condition monitoring (CM) techniques   357
Contingency Analysis   32
Conventional Power Systems   1
Current Stress   52
Cyber Security   11

d

DC-Link Topologies   68
Decoupled Load Flow Method   18
Deep Reinforcement Learning   207, 210, 212, 280
Develop Fuel Cell Control Systems (FCCS)   303
Diesel generators   1, 100
Distributed generation (DG)   1
distributed network protocol (DNP)   366
DVR   203, 236

e

e-STATCOMs   355
Eigenvalues   166

Electric Vehicle

Electric Vehicle Charging Infrastructure   10
Electric Vehicles   274, 281
Electricity Market   216
Energy Storage Systems   8, 99, 109, 111
Energy Storage Systems for Smart Power Systems   99
ERCOT   355

f

Firefly Algorithm (FA)   248
FPGA Controller   119
Fuzzy Logic Controller   192

g

Gauss Seidel Method   18
Generic object-oriented system-wide events (GOOSEs)   359
grid solutions (GRIDSOL)   361

h

H-Bridge Inverter   54
Hardware in the Loop (HIL)   291
Harmonic Analysis   20
Harmonic Filters   236
HIL for Electric Vehicles   301
HIL for HVDC and FACTS   299
HIL for Renewable Energy Systems (RES)   296
holistic condition monitoring (HCM)   364
Home Area Network (HAN)   5
HVDC Transmission Lines   141

i

Intelligent Transportation and Transportation   278
Intermittency   1, 2, 99, 101, 157
Internet of Things   260
IoT in Smart Cities   275

Artificial Intelligence-based Smart Power Systems, First Edition.
Edited by Sanjeevikumar Padmanaban, Sivaraman Palanisamy, Sharmeela Chenniappan, and Jens Bo Holm-Nielsen.
© 2023 The Institute of Electrical and Electronics Engineers, Inc. Published 2023 by John Wiley & Sons, Inc.

k

Karush–Kuhn–Tucker Conditions   220

l

LCCs   353
Load Flow Analysis   17
low frequency demand disconnection (LFDD)   356
LSTM Time Series   285

m

Machine Learning Approaches   157, 279
Manufacturing Message Specification (MMS)   359
massive mission critical communication (mMTC)   368
Meter Data Acquisition System   5
Minimum Damping Ratio   166
Model Batteries and Develop BMS   302
Model Fuel Cell Systems (FCS)   303
Model-Based System Engineering (MBSE)   302
Modes of Operation of MCBC   42
MPPT Control   129, 130
multidimensional scaling (MDS)   369
Multilevel Cascaded Boost Converter   37, 40
MVDC   353

n

NARMA-L2 Controller   190
network control centers (NCCs)   359
Newton Raphson Method   18

o

open platform communication (OPC)   366
Optimization Techniques   247
original equipment manufacturers (OEMs)   359

p

Phasor Measurement Unit   6, 311, 313
PHIL   297
Pitch Angle Control Techniques   189
PMSG Wind Turbine   129
Power Electronic Converters   66
Power Quality Conditioners   233, 235
Power System Models   159
Practical Byzantine Fault Tolerance (PBFT)   330
precision time protocol (PTP)   359
Predictive Controller Technique   183
principal component analysis (PCA)   369
programmable logic controller (PLC)   361

Proof of Authority (PoA)   330
Proof of Stake (PoS)   329
Proof of Work (PoW)   329
PSS-UPFC   160
PV-Wind Hybrid Power Generation   180

s

SCADA   312
Short Circuit Analysis   19
Smaller ledger   332
Smart Contracts   336
Smart Distribution Systems   9
Smart Grid   245
Smart House   261
Smart Meters   4
Smart Power System   233
Solar Photovoltaic   65, 99, 218, 311
Spider Monkey Optimization (SMO)   250
Standards of Power Quality   244
Steady-State Analysis   16
Supercapacitor   115
Sustainable Energy Solutions   245
SVC   355
system average interruption duration index (SAIDI)   361
system average interruption frequency index (SAIFI)   361

t

T-distributed stochastic neighbor embedding (T-SNE)   369
TCSC   354
Time-Domain   170
Transactive Energy   335

u

ultra-reliable low latency communication (URLLC)   368

v

V2G   282
Voltage Stability   26
Voltage Stress   49

w

Wide Area Monitoring and Control   2
Wind Energy Conversion System   131
Wind Power Generation Plant   187

Printed and bound by CPI Group (UK) Ltd, Croydon, CR0 4YY

27/10/2024

14580137-0005